Research for Development

The series Research for Development serves as a vehicle for the presentation and dissemination of complex research and multidisciplinary projects. The published work is dedicated to fostering a high degree of innovation and to the sophisticated demonstration of new techniques or methods.

The aim of the Research for Development series is to promote well-balanced sustainable growth. This might take the form of measurable social and economic outcomes, in addition to environmental benefits, or improved efficiency in the use of resources; it might also involve an original mix of intervention schemes.

Research for Development focuses on the following topics and disciplines:

Urban regeneration and infrastructure, Info-mobility, transport, and logistics, Environment and the land, Cultural heritage and landscape, Energy, Innovation in processes and technologies, Applications of chemistry, materials, and nanotechnologies, Material science and biotechnology solutions, Physics results and related applications and aerospace, Ongoing training and continuing education.

Fondazione Politecnico di Milano collaborates as a special co-partner in this series by suggesting themes and evaluating proposals for new volumes. Research for Development addresses researchers, advanced graduate students, and policy and decision-makers around the world in government, industry, and civil society.

THE SERIES IS INDEXED IN SCOPUS

More information about this series at http://www.springer.com/series/13084

Giuliano Dall'O'
Editor

Green Planning for Cities and Communities

Novel Incisive Approaches to Sustainability

Springer

Editor
Giuliano Dall'O'
Dipartimento di Architettura
Ingegneria delle Costruzioni e Ambiente
Costruito (DABC)
Politecnico di Milano
Milan, Italy

ISSN 2198-7300 ISSN 2198-7319 (electronic)
Research for Development
ISBN 978-3-030-41071-1 ISBN 978-3-030-41072-8 (eBook)
https://doi.org/10.1007/978-3-030-41072-8

This Springer imprint is published by the registered company Springer Nature Switzerland AG
The registered company address is: Gewerbestrasse 11, 6330 Cham, Switzerland

Contents

Chapter 1
Introduction

Giuliano Dall'O'

Abstract This introductory chapter describes the objectives and the purposes of this book, giving the reader a general framework for an initial understanding of the proposed model of green planning of cities and communities. In the first part of the chapter, the critical elements that determine the need to radically change the approach to the planning of cities and communities are highlighted. Subsequently, the methodological criteria with which the book handles the theme of green planning are defined. Finally, this chapter examines the structure of the book, defining various parts that constitute it. The book is conceived to form a useful instrument both for learning about and acting on the subject matter.

1.1 The Environmental, Social and Economic Criticalities in the Sustainable Development of Cities

The UN Secretary-General Ban-Ki-moon in a message to the Governing Council on April 23, 2012 stated that "Our struggle for global sustainability will be won or lost in cities" (U.N. 2012).

Living in a city is attractive from many points of view: cities offer great job opportunities, make all kinds of services available and are also places of cultural growth and social cohesion.

The world is becoming increasingly urbanized. Since 2007, more than half the world's population have been living in cities and that share is projected to rise to 60% by 2030. It is estimated that by 2050, 68% of the world population will live in urban areas (U.N. 2018).

Cities and metropolitan areas are powerhouses of economic growth, contributing about 60% of global gross domestic product (GDP).

Cities, however, are places of great conflicts and contradictions: the high population density not only offers opportunities but generates critical issues of social,

G. Dall'O' (✉)
Architecture, Built Environment and Construction Engineering, ABC Department, Politecnico di Milano, Via G. Ponzio, 31, 20133 Milan, Italy
e-mail: giuliano.dallo@polimi.it

G. Dall'O' (ed.), *Green Planning for Cities and Communities*,
Research for Development, https://doi.org/10.1007/978-3-030-41072-8_1

economic and environmental nature (cities account for about 70% of global carbon emissions and over 60% of resource use) (U.N. 2019).

Urban cities occupy only a small part of the world's surface; in fact, cities occupy only 2% of the land surface area. However, each city has its own history, marked by the way it extended and grew; thus it is not surprising to find cities that expanded and occupied new land even in periods in which their population was decreasing (Terradas 2001).

Although the area occupied by the city is relatively small, the areas where the cities are located are the best on our planet in terms of land morphology and environmental resources, generally being in the most strategic areas.

The cities also represent a greater threat to the rest of the land: previous urban planning which was egocentric paid little attention to peri-urban spaces, i.e. non-urbanizable land (Gomez and Salvador 2006).

In 2015, countries adopted the 2030 Agenda for Sustainable Development and its 17 Sustainable Development Goals (SDGs). In 2016, the Paris Agreement on climate change entered into force, addressing the need to limit the rise of global temperatures. From Table 1.1, which shows the list of the 17 SDGs, it is possible to observe that many of the SDGs contain objectives that directly or indirectly concern the green planning of cities and communities: SDG 11, in particular, defines the goal of "Make cities and human settlements inclusive, safe, resilient and sustainable".

Other SDSs concern issues are directly related to the sustainable development of cities: among these, the SDG 3 defines the goal of ensuring healthy lives and promoting well-being for all ages, the SDG 6 defines the objective of ensuring availability and sustainable management of water and sanitation for all, and the SDG 7 is concerned with ensuring access to affordable, reliable and modern energy for all. Then, there is the SDG 13 which indicates the need to implement urgent action to combat climate change and its impacts, the SDG 14 which deals with preserves and sustainability use of oceans, seas and marine resources for sustainable development and finally the SDG 15 which aims to protect, restore and promote sustainable use of terrestrial ecosystems, sustainability manager forests, combat desertification, and halt and reserve land degradation and biodiversity loss.

An important goal for the SDGs is to promote an overall and shared strategy at the global level to combat climate change.

The irreversible effects of a further increase in the average temperature are of great concern. The Agenda for Sustainable Development 2030 of 2015 negotiated with the Paris Agreement aims to contain the increase in global average temperature below the 2 °C threshold above pre-industrial levels and to limit this increase to 1.5 °C, since this would substantially reduce the risks and effects of climate change.

Decarbonization is an objective that can be achieved through energy efficiency and the use of renewable energy sources: the changes taking place are tangible; however, the path is long and the times we have available to reach the goal are not compatible with the current trend.

The circular economy, understood as an economic system designed to be able to regenerate itself by itself thus also guaranteeing its eco-sustainability, is the most coherent answer; however, the transition must be accelerated.

Table 1.1 List of the Sustainable Development Goals (SDG) with the objectives concerning green planning (GP) highlighted

#	SDG title	Objectives	GP
1	No poverty	End of poverty in all its forms everywhere	
2	Zero hunger	End of hunger, achieve food security and improved nutrition and promote sustainable agriculture	
3	Good health and well-being	Ensure healthy lives and promote well-being for all at all ages	✓
4	Quality education	Ensure inclusive and equitable quality education and promote lifelong learning opportunities for all	
5	Gender equality	Achieve gender equality and empower all women and girls	
6	Clean water and sanitation	Ensure availability and sustainable management of water and sanitation for all	✓
7	Affordable and clean energy	Ensure access to affordable, reliable sustainable and modern energy for all	✓
8	Decent work and economic growth	Promote sustained, inclusive and sustainable economic growth, full and productive employment and decent work for all	✓
9	Industry, innovation and infrastructure	Built resilient infrastructure, promote inclusive and sustainable industrialization and foster innovation	✓
10	Reduced inequalities	Reduce inequality within and among countries	
11	Sustainable cities and communities	Make cities and human settlements inclusive, safe, resilient and sustainable	✓
12	Responsible consumption and production	Ensure sustainable consumption and production patterns	
13	Climate action	Take urgent action to combat climate change and its impacts	✓
14	Life below water	Conserve and sustainably use the oceans, seas and marine resources for sustainable development	✓
15	Life and land	Protect, restore and promote sustainable use of terrestrial ecosystems, sustainably manage forests, combat desertification and halt and reserve land degradation and biodiversity loss	✓

(continued)

Table 1.1 (continued)

#	SDG title	Objectives	GP
16	Peace, justice and strong institution	Promote peaceful and inclusive societies for sustainable development, provide access to justice for all and build effective, accountable and inclusive institutions at all levels	
17	Partnership for the goals	Strengthen the means of implementation and revitalize the global partnership for sustainable development	

In recent years, it has become increasingly evident that due to growing demand, the scarcity of freshwater is becoming a danger for the sustainable development of human society. The water on the planet is 97% seawater and therefore unusable for humans, agriculture and industry. The entire demand for water for human use is therefore 3%, a value that is reduced to 2% if we remove that part which is unusable because it is in the form of ice at the poles (Mancuso 2017).

If we consider the increase in population and the continuous improvement of standards, the availability of water represents a critical element for which it is necessary to provide answers rapidly.

The SDGs defined in 2015 define very ambitious objectives which must be achieved to avoid an environmental situation so critical as to be irreversible. To achieve these objectives, strategies are needed that must be adopted and pursued at all levels. A radical change in city and community planning through a "green planning" approach can be an effective response since therein a single action is able to consider multiple aspects while ensuring concrete sustainable development.

1.2 Green Planning of Cities and Communities: Towards an Holistic Approach

The above-mentioned critical environmental factors of our planet, fully recognized and shared by the scientific community, are probably the result of the convergence of many factors. Humanity, however, has the greatest responsibility, obviously with an enormous inequality between industrialized countries and developing countries. The International Resource Panel Report of the United Nations (2017) says that extraction of material resources—biomass, fossil fuels and non-metallic minerals—from the Earth could reach 88.6 billion tonnes in 2017, or three times that used in 1970.

International Resource Panel co-chairs Janez Potocnik and Izabella Teixeira in a joint statement, said "the amount of natural resources used is closely linked to the amount of final waste and emissions generated through their use" and "Effective pollution control must also look to minimise raw material use, thereby decreasing final waste and emissions".

A positive fact is that we have now all the elements, all based on scientific evidence, on the seriousness of the problem, not only in qualitative terms but also in quantitative terms. A clear and objective reference framework, constantly updated by the scientific community, and in particular by the United Nations Intergovernmental Panel on Climate Change (IPCC), is an indispensable instrument for knowing the baseline to define strategies, to implement actions and above all for measuring objectively the results obtained.

The environmental emergency of the Earth, which has unequivocally demonstrated that the economic development model adopted so far is wrong, can and must be considered not only a great emergency to be overcome but also, and above all, a great opportunity for humanity to radically change development strategies: decarbonization, circular economy and enhancement of environmental resources are the new paradigms of sustainable development to which there are no alternatives. The big challenge is that this revolution, based essentially on a new alliance between humanity and nature, should not lead humanity to renounce its well-being, health and quality of life, overcoming at the same time the current economic and social inequalities.

Given that in future years, the world population will be concentrated in urban and metropolitan areas (68% of the population by 2050), the great "green" revolution must start from there.

The topics addressed in this book fit into this precise scenario: the goal is to provide a compendium for those (public administrators, architects, designers, urban planners) who will have to take charge of planning the new neighbourhoods but also the re-planning of the existing cities and communities.

To the term "green planning" that we use speaking about planning strategies of cities and communities, we give a broad and inclusive meaning that considers both the needs of human beings and the needs of nature as a biosphere: in our consideration, "green planning", therefore, is simultaneously "green", "sustainable" and "smart".

The green planning of cities and communities is based on green development defined as an approach that carefully considers the social and environmental impacts of development. The paradigms of the green development are as follows:

- *re-naturalization of cities* intended to be understood as respect for the intrinsic value of nature with the minimization of damage to the ecosystem (e.g. integration of buildings and urban infrastructure with the natural environment);
- *efficient use of resources* intended as the ability to guarantee human needs by containing waste of resources and reducing waste (implementation of the circular economy principles);
- *decarbonization* intended as the implementation of all actions that can reduce greenhouse gas emissions with the objective of zero emissions (e.g. energy efficiency and renewable energy sources);
- *sensitivity of the community and culture understood* as the ability to plan, recognizing the uniqueness of the cultural values that each community hosts;

- *risk assessment and disaster recovery* considered as the ability to plan the urban spaces able to react to unexpected natural events, such as earthquakes or floods;
- *universal design* intended as the ability to plan a city considering access to spaces and service to all the citizens.

The management of all these aspects that contribute to the green planning of the cities requires a holistic approach which is indispensable in order to enhance all possible synergies, through the synergistic contribution of all technical, economic and even sociological skills.

The new green and smart cities, to work in an efficient and sustainable way, will have to be lived by citizens capable of changing their lifestyles. This change is already underway, stimulated by an increasingly widespread awareness of the importance of sharing the principles of environmental sustainability.

The spread of renewable energy sources, which for new buildings is becoming the "new normal", and the tendency, at least for those who live in the city, to renounce the use of private means of transport powered by endothermic engines in favour of public transport or from electric vehicles are some of the examples that testify to the beginning of the green revolution.

Effective green city planning aims to accelerate this revolution by redefining new urban assets, often a few decades ahead. Considering the long time required to energy retrofit existing buildings and redefine urban green infrastructures compatible with the objectives of sustainable development (one can think, e.g. of the radical change in infrastructures for sustainable mobility), it is necessary to plan today the green strategies for the cities of 2050.

Playing ahead of time considering the technological evolution of the coming years is the winning strategy for those involved in urban planning.

The good news is that in recent years many things have changed. In the Sustainable Development Goals Report 2019 (U.N. 2019), there have been improvements, although not sufficient, compared to 2015, the reference year.

The technologies for renewable energy sources (solar thermal, photovoltaic (PV) solar, wind and biomass) have long been mature technologies and often competitive with traditional sources that use fossil fuels.

The building market, stimulated by the European directives on energy efficiency, starting from the Energy Performance of Building Directive (EPBD) of 2002, has radically changed: almost zero-energy buildings represent the current standard in all EU countries. New building technologies (e.g. the Energiesprong project) allow existing buildings to be upgraded energetically at competitive costs and with minimal inconvenience for the occupants (https://energiesprong.org.).

The market for sustainable mobility technologies is growing rapidly: at least in the cities and metropolises, electric mobility will become the new standard in just a few years. The associated infrastructures will evolve rapidly, and 5G will accelerate the evolution towards autonomous driving vehicles (in cities such as Oslo-DK, this is already happening).

Many cities have already successfully implemented green planning strategies that can be taken as examples and replicated in other contexts.

The economic element is another positive aspect to consider. The traditional economy, based on models that aim at ever-increasing consumption of energy and resources, is evolving into the "green economy" which finally recognizes a market value for everything that is more compatible with sustainable development. The exponential spread, on a voluntary basis, of international environmental certification protocols for buildings, neighbourhoods and cities, such as LEED® or BREEAM®, highlights a change in the real estate market.

If on the one hand one must necessarily confront environmental emergencies, the SDGs constitute an important point of reference, on the other one can count on a context, the current one, which offers all the tools, all the technologies and all the good practices for implementing a green planning compatible with the needs not only of today but also of the next objectives at 2030 and 2050. The purpose of this book is to collect in a single work, a complete reference framework for the green planning of cities and communities.

1.3 Synopsis of the Chapters

This book is structured into four sections: many of the contributions presented show the results of practical experience (research projects, studies, consultancy work) which provide operational content to the theoretical and methodological elements presented in the text that are enriched in each section by case studies.

1.3.1 Strategies for Green Planning of Cities and Communities

This section provides a current framework on the topic of sustainable planning for cities and communities. The topics covered, defined in detail in the chapters below, will illustrate the different strategies, international projects and methodologies adopted.

The building sector is responsible for the energy consumption of over a third of the world's energy resources. Energy consumption, which generates a significant environmental impact, is greater in cities where the population density is high.

- Chapter 2 provides a general overview of the different strategies to be adopted for the energy and environmental planning of cities and communities, highlighting the complexity of the theme and the need to adopt a holistic approach while maintaining the fundamental principles.
- Chapter 3 deals with the issue of the re-naturalization of cities and defines criteria to enhance the relationship between urban environment and natural environment outside and inside the city.

- Chapter 4 deals with the topic of risk assessment and disaster recovery considered as a needful contribution to green planning.
- Chapter 5 deals with a fundamental theme of the environmental question: the circular economy. Green planning can give the circular economy a significant boost, thus speeding up the process.
- Chapter 6 is devoted to the universal design, intended as the capacity to plan an inclusive city in which access is guaranteed to all citizens (a city with barriers cannot be considered sustainable, green and smart).
- Chapter 7 provides a framework for the methodologies and tools available for design and managing sustainable energy planning in green city and community.

1.3.2 Energy Efficiency and Sustainability in Urban Planning

Decisive action on improving the energy and environmental efficiency of buildings becomes a priority element in urban planning strategies. For this reason, this section is dedicated to the theme of sustainability in the building sector, which focuses on all the aspects which underlie a concrete policy that can lead to an ever-increasing number of zero-energy buildings or, in any case, high building efficiency. In this section, many research experiences applied to real case studies are discussed.

- Chapter 8 focuses on the methodologies and tools that can be used to assess energy efficiency on an urban scale. The chapter in particular deals with the issue of estimating energy needs in various sectors aimed at drafting an energy balance on the basis of which to construct the improvement strategies for global energy and environmental efficiency.
- Chapter 9 deals with the potentials of the energy retrofit strategies in the construction sector considering not only the technical aspects but also the economic and social ones.
- Chapter 10 is dedicated to the theme of integrating renewable energy sources (solar thermal and photovoltaic, biomass, wind) in urban planning. The chapter provides the state of the art of this topic, reports some international case studies and highlights potential and criticality.

1.3.3 Sustainable Mobility for Cities and Communities

Mobility within cities plays an essential role in green planning. In recent years, the paradigms have changed: zero impact transport has been enhanced (e.g. the use of bicycles and the related infrastructures), public transport has been enhanced, and even "private" mobility is constantly evolving both in its usage (one example is car sharing) and in terms of the technologies employed (transition to electric mobility).

The diffusion of renewable energy sources does not only concern the building sector but also mobility which can benefit from the synergies that emerge (e.g. the use of photovoltaic energy).

All these elements are considered in this section which illustrates the changes underway, highlighting the new relationship between public mobility and private mobility. This part is structured into two chapters which are summarized below.

- Chapter 11 deals with the topic of the infrastructures for sustainable mobility within cities, with particular reference to electric mobility. The authors discuss both collective mobility and individual mobility, therefore ranging from underground railways to electric scooters. In the final part, the general criteria for the planning of sustainable mobility are provided.
- Chapter 12 is dedicated to the changes in the behaviour of citizens and digital technologies that can facilitate their behaviours on sustainable mobility. The chapter focuses on the concept of Mobility as a Service (MaaS), which is the final evolution of the concept of ownership towards that of the availability of a service.

1.3.4 Rating Systems for Green Planning of Cities and Communities

If urban environmental sustainability is increasingly becoming an attractive element of the green economy, environmental certification protocols are confirmed as the most effective tools for promoting a real estate market which finally recognizes the value of ecological and environmental choices, both by public administrations and private players. The environmental certification protocols, originally created to certify green buildings, in recent years have developed towards the measurement and certification of the sustainability of neighbourhoods and cities.

This section examines the application of rating systems in urban or neighbourhood settings. Starting from real case studies, synergies are highlighted, as well as critical elements, among these protocols and public planning tools.

- Chapter 13 deals with the most important environmental certification protocols and rating systems for green districts and cities.
- Chapter 14 deals with the most important environmental certification protocols and rating systems for green infrastructures.

1.3.5 The "Smart City" Approach to Improve Urban-Scale Sustainability

This section represents the complete evolution of the energy and environmental planning of cities and communities. The "smart city" approach to planning, in fact, introduces all the elements that characterize an intelligent evolution in city management: not only energy efficiency and environmental quality but also quality of lifestyle, a new economy, new services, the smart economy, etc.

In this section, the topic of smart cities, in which sustainability remains an element of strong attraction, is treated by referring to the state of the art and by proposing methodologies which have been successfully applied and validated in real cases.

- Chapter 15 deals with state-of-the-art technologies and services for smart cities; the chapter also deals with the economic, social and cultural issues related to the smart revolution, highlighting the close link that exists between smart cities and sustainable cities.
- Chapter 16 is concerned with the topics of smart and sustainable indicators used to define the smartness ratings of cities and communities.
- Chapter 17 is concerned with the application of Building Information Modelling (BIM) methodologies in the green planning of buildings and communities.

Acknowledgements The realization of this volume, whose purpose is to provide tools and methodologies to support the implementation of Sustainable Development Goals in city and community planning, would not have been possible without the contribution of many people whom I want to thank.

Firstly, I want to thank my colleagues who, with their research and subsequent publications, have been a source of inspiration for me in the theoretical systematization of the topics covered: among them, the professors Tridib Banerjee of the University of Southern California, Scott D. Campbell of the University of Michigan and Stefano Mancuso of the University of Florence.

I thank the colleagues who have provided written contributions, the professors Daniele Bignami, Elisa Bruni, Paola Caputo, Simone Ferrari, Monica Lavagna, Michela Longo, Roberto Mazzoncini, Claudio Somaschi, Marco Torri and Alessandro Zichi of the Politecnico di Milano and professor Alberto Arenghi of the University of Brescia.

I thank the colleagues with whom I met during the writing of the book, among them professor Marco Caffi, Vice-President of the European Regional Network of World Green Building Council.

I am grateful to Mark Izard who helped me in finding the correct way to express the concepts, but at the same time his help was for me a useful comparison in viewpoints on the technical aspects.

I thank Cristina, the lady of my life, not only for having encouraged and supported me in difficult moments but above all for having contributed, with her intelligence and her pragmatism to a continuous comparison on the topics covered in the book that have citizens as users.

Finally, I express my thanks to the Fondazione Politecnico and Springer Nature for agreeing to include this book in the prestigious series "Research for Development".

References

Gómez F, Salvador P (2006) A proposal for green planning in cities. Int J Sustain Develop Plann 1:91–109

Mancuso S (2017) Plant revolution, le piante hanno già inventato il nostro futuro. Giunti Editore, Firenze

Terradas J (ed) (2001) Urban ecology. Rubes, Barcelona

United Nations (2012) Our struggle for global sustainability will be won or lost in cities, says secretary-general, at New York event [Press release]. Available on: un.org/press/en/2012/sgsm14249.doc.htm. Accessed 20 Aug 2019

United Nations IRP (2017) Assessing global resource use: a systems approach to resources efficiency and pollution reduction. ISBN: 978-92-807-3677-9 DTI/2141/PA UNEP 162

United Nations (2018) Inform on the population. Department of Social and Economic Matters, Secretariat of the United Nations

United Nations (2019) The sustainable development goals report. New York

Part I
Strategies for Green Planning of Cities and Communities

Chapter 2
Green Planning of Cities and Communities: Theories, Strategies and Tools of a Complex Framework

Giuliano Dall'O'

Abstract If urban planning plays a fundamental role in the future of humanity, which is concentrating an increasing share of the population in cities, then green planning is the most effective response to the climate change emergency. This chapter analyses theories, strategies and tools that characterize the green planning of cities and communities useful for understanding and managing its complexity. The first section analyses the continuously evolving relationship between urban planning and sustainable development. The topic of sustainable development in green planning is seen as a comparison between two approaches, one more utopian and idealistic and one more pragmatic, both in the two fundamental declinations: the more anthropocentric and the more ecological. Subsequently, the more applicative aspects are dealt with: from the analysis of the sectoral planning tools, relations and synergies are highlighted with the aim of providing an integrated, inclusive model. The final part of the chapter proposes a planning strategy aimed at supporting a structured, integrated and effective green planning model for cities and communities.

2.1 Green Planning: Evolution of Ancient Paradigms in Urban Planning

Although the concept of "sustainable development" linked to climate change is relatively recent in urban planning, but more generally in architectural design, the research by architects, landscape planners and urbanists of such a relationship, sometimes codified, is not new. This has been between the shape of cities, or architecture, and the natural environment understood in all its meanings, from the choice of vegetation to environmental control through the use of natural resources (in particular solar energy and wind energy).

The definition of tools and methods for the green planning of the territory coherent with the current needs cannot ignore the analysis of what has been done, or simply

G. Dall'O' (✉)
Architecture, Built Environment and Construction Engineering, ABC Department, Politecnico di Milano, Via G. Ponzio31, 20133 Milan, Italy
e-mail: giuliano.dallo@polimi.it

G. Dall'O' (ed.), *Green Planning for Cities and Communities*,
Research for Development, https://doi.org/10.1007/978-3-030-41072-8_2

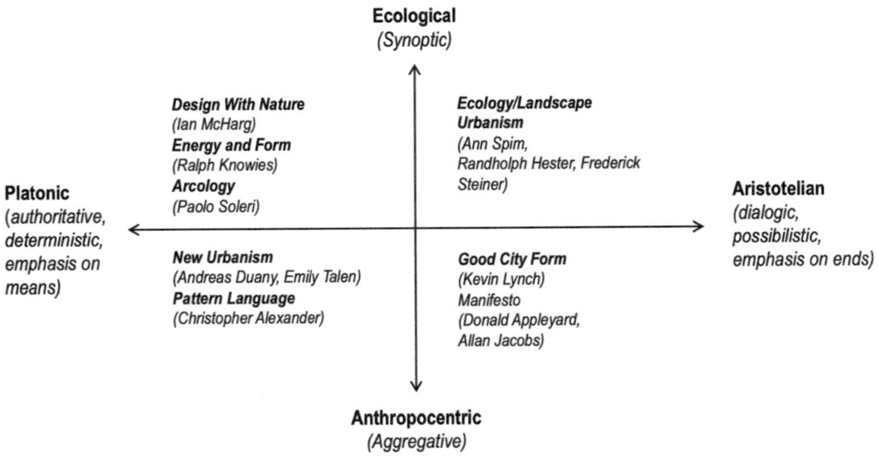

Fig. 2.1 Normative models of sustainability through design (Banerjee 2014)

theorized, before the theme of sustainable development linked to climate change, effectively supported by the scientific community, became the dominant theme of all environmental, economic and social choices on a universal scale.[1]

The definition of tools and methods for the green planning of the land coherent with the current instances cannot ignore the analysis of what has been done, or simply theorized.

Interesting in this regard is the work of Banerjee (2014), Professor in Urban and Regional Planning at the University of Southern California whose title is eloquent: "Urban design and sustainability: looking backward to move forward". Objective of that paper is a critically review of the urban design theories and movements to reveal how their theoretical roots relate to sustainability.

Banerjee argues that "as is common practice, in the tradition of design, much of this legacy in defining, practicing and achieving sustainability goals is normative and speculative, and relatively few of this tenets have been applied in practice and are thus yet to be formally vetted through empirical studies through empirical studies or formal evaluation of the outcomes".

The author frames the various contributions and theories within a four-quadrant matrix structured along two axes.

The first (horizontal) axis compares two approaches, the Platonic and the Aristotelian, while the second (vertical) axis compares two ways of conceiving planning: one anthropocentric and the other ecological (Fig. 2.1).

Regarding the first axis, it is useful to recall the differences between the Platonic approach and the Aristotelian approach. Although Aristotle was a disciple of Plato,

[1]The definition now widely shared of "sustainable development" is that contained in the Brundtland Report drawn up in 1987 by the World Commission on the Environment and Development.

the two philosophers have two different ideas about optimal dimensions and the governmental order of cities.

As far as Plato is concerned, the theoretical intent to plan an ideal polis takes on a philosophical dimension expressed in two dialogues, on the Republic and the Laws. He does not go as far as to define the completeness of architectural forms: the only aesthetic notation concerns the urban scheme for which Plato deems any solution of absolute regularity to be deplorable.

Aristotle dedicates an entire work to the administration of the polis, the "Politics" in which he analyses the political realities starting from the organization of the family, understood as the basic nucleus of society, to move on to the different types of constitution.

Central for Aristotle is the reference to nature: man is a "political animal" and as such is naturally led to join his own kind to form communities. The fact that man is born of "logos" fits well with his innate sociability. Unlike Plato, for Aristotle politics has a certain autonomy with respect to philosophy: the politician and the legislator can perform their task well thanks to their practical wisdom.

Aristoteles bases his reflections on the experience of real cities (city states), and emphasized the process by which the ideal form and size may be achieved (Banerjee 2014).

As for the second axis (from anthropocentric towards ecological), the first approach involves the synoptic thinking that involves the wider ecosystem, while the second focuses on human goals and on the consequences of the results of design actions: in other words, it puts the man at the centre of the interests of design choices.

The quadrants of the matrix define four major conceptual approaches possible to sustainable design:

- Ecological and Platonic;
- Anthropocentric and Platonic;
- Ecological and Aristotelian;
- Anthropocentric and Aristotelian.

The first quadrant (**Ecological and Platonic**) includes design approaches strongly related to nature. In this quadrant, we find the theories of Ian McHang (Design With Nature), of Ralph Lewis Knowles (Energy and Form) and of Paolo Soleri (Arcology).

Ian L. McHarg (1920–2001), a Scottish landscape architect and writer on regional planning using natural systems, was the founder of the department of landscape architecture at the University of Pennsylvania in the USA. His book "Design with Nature" (McHarg Ian 1969) that continues to be one of the most widely celebrated books on landscape architecture and land-use planning pioneered the concept of ecological planning and set forth the basic concepts that were to develop later in geographic information systems (GIS).

Going against the Judeo-Christian traditions of the Bible which says that man must have dominion over the earth,[2] McHarg affirms that for man to survive, this

[2]The position of the Catholic Church towards the environment has changed for many years. The last document concerning the environment is the encyclical of Pope Francesco "Laudato si" published

idea must be taken only as an allegory and not as literally true and states that the man is a "planetary disease", who lived without due regard for nature.

Ralph Lewis Knowles (1928), American professor emeritus of Architecture and fellow of the American Solar Energy Society is a leading theorist of solar access design. He created the concept of the "solar envelope" and championed solar access planning. The concepts expressed by Knowles concerning access to sunlight, contained in his work "Energy and Form: An Ecological Approach to Urban Growth" (Knowles 1978) are still today contained in national and regional legislation and in municipal building codes.

A theorist who partly puts into practice the concepts of "design with nature" is Paolo Soleri (1919–2013) architect, writer, sculptor, urban planner and Italian artist. Just graduated, and in 1947, he moved to the USA where he met and attended Frank Lloyd Wright.

In 1956, he moved to Arizona where he founded Arcosanti (1970), a prototype city for 5000 people, based on the concepts of "arcology" (architecture and ecology) (Soleri 1969). His model of city is based on the restraint of energy and environmental resources aimed at protecting the environment, thus setting out an ethical path for the future of man.

The second quadrant (**Anthropocentric and Platonic**) includes design approaches related to the "New Urbanism" (Andrés Duany and Emily Talen) and to the "Pattern Language" (Christopher Alexander).

The New Urbanism is an urban movement which promotes pedestrian areas that contain the mix of urban uses.

Developed in the USA since 1980, it continues to reform many aspects of real estate development and urban design.

The proposed model, described in the Charter of New Urbanism, holds that the traditional city, with its mix of functions, density and integration of different transport systems, constitutes a much more efficient way of developing a lively community rich in cultural interactions.

New Urbanism strategies are aimed at reducing road congestion, overbuilding and the conversion of urban areas: it also includes strategies aimed at historical preservation, road safety, green building and the development of brownfield sites.

LEED® (Leadership in Energy and Environmental Design) environmental certification protocols draw many of their concepts into the Charter of the New Urbanism (CNU) (see Table 2.1).

Duany worked as a guest professor in many institutions and received two honorary doctorates. His conception of the new urban model is contained in the book "Suburban Nation: The Rise of Sprawl and the Decline of the American Dream" (Duany et al. 2000).

Another important exponent of the New Urbanism Movement is Emily Talen, professor of urbanism at the University of Chicago. Her research is devoted to urban

---

in 2015. The encyclical, in addition to expressing great concern for the environmental emergency (considers the environment the common home), states the need to stimulate an ecological education and spirituality for the development of new beliefs, new attitudes and lifestyles.

Table 2.1 Principles enunciated in the chart of new urbanism (The region: metropolis, city and town) (https://www.cnu.org/who-we-are/charter-new-urbanism)

#	Description
1	Metropolitan regions are finite places with geographic boundaries derived from topography, watersheds, coastlines, farmlands, regional parks and river basins. The metropolis is made of multiple centres that are cities, towns and villages, each with its own identifiable centre and edges
2	The metropolitan region is a fundamental economic unit of the contemporary world. Governmental cooperation, public policy, physical planning and economic strategies must reflect this new reality
3	The metropolis has a necessary and fragile relationship to its agrarian hinterland and natural landscapes. The relationship is environmental, economic and cultural. Farmland and nature are as important to the metropolis as the garden is to the house
4	Development patterns should not blur or eradicate the edges of the metropolis. Infill development within existing urban areas conserves environmental resources, economic investment and social fabric, while reclaiming marginal and abandoned areas. Metropolitan regions should develop strategies to encourage such infill development over peripheral expansion
5	Where appropriate, new development contiguous to urban boundaries should be organized as neighbourhoods and districts, and be integrated with the existing urban pattern. Non-contiguous development should be organized as towns and villages with their own urban edges, and planned for a jobs/housing balance, not as bedroom suburbs
6	The development and redevelopment of towns and cities should respect historical patterns, precedents and boundaries
7	Cities and towns should bring into proximity a broad spectrum of public and private uses to support a regional economy that benefits people of all incomes. Affordable housing should be distributed throughout the region to match job opportunities and to avoid concentrations of poverty
8	The physical organization of the region should be supported by a framework of transportation alternatives. Transit, pedestrian and bicycle systems should maximize access and mobility throughout the region while reducing dependence upon the automobile
9	Revenues and resources can be shared more cooperatively among the municipalities and centres within regions to avoid destructive competition for tax base and to promote rational coordination of transportation, recreation, public services, housing and community institutions

design and the relationship between the built environment and social equity. She is also the editor of several volumes, among these "New Urbanism and American Planning: The Conflict of Cultures" (Talen 2005).

This quadrant also includes the important contribution of Christopher Alexander (1936), British–American architect and design theorist, currently professor emeritus at the University of California, Berkeley. Alexander is perhaps best known for his book "A Pattern Language" (Alexander et al. 1977). Arguing that users are more sensitive to their needs than any architect could be, he produced and validated a "pattern language" to empower anyone to design and build at any scale. Alexander's work has also influenced the development of "agile software development" intended as approach to software development under which requirements and solutions evolve through the collaborative effort of self-organizing and cross-functional teams and their customer(s)/end user(s).

The third quadrant (**Ecological and Aristotelian**) includes more fundamentalist design approaches related to the Ecology or Landscape Urbanism inspired by traditional landscape architecture (Anne Spirn, Randholph Hester and Frederick Steiner).

Anne Whiston Spirn is an American landscape architect, photographer and author: her work promotes community-oriented spaces that are functional, sustainable, meaningful and artful but also resilient (Spirn 2011). She makes reference to the writings of Leon Battista Alberti in the fifteenth century and more recently to such pioneers as George Perkins Marsh, Frederick law Olmsted and Lewis Mumford (Banerjee 2014).

Randolph T. Hester is a professor in the Department of Landscape Architecture and Environmental Planning at the University of California at Berkeley. Hester is also a sociologist, practicing landscape architect and co-director of Community Development by Design, a neighbourhood planning organization focused on community participation and input (Hester 2009).

Frederick R. Steiner is an American ecologist who currently serves as the Dean and Paley Professor for the University of Pennsylvania School of Design. Fellow of the American Society of Landscape Architects and the American Academy in Rome, Steiner is an expert in ecological planning, historic preservation, environmental design, green building and regional planning, all of which are discussed in "The Living Landscape". His important contribution to landscape architecture is the book "An Ecological Approach to Landscape Planning" (Steiner 2000).

The last quadrant (**Anthropocentric and Aristotelian**) mainly includes a design approach related to the "Good City Form" (Kevin Lynch), a design strictly related to the anthropocentric prospective. The contributions of Donald Appleyard and Alan Jacobs can reasonably be included in this quadrant (Appleyard and Jacobs 1987).

Kevin Andrew Lynch (1918–1984) was an American urban planner and architect. He graduated in urban planning at the Massachusetts Institute of Technology in 1947, having also done an internship in Frank Lloyd Wright's Taliesin studio, and began his research and teaching work at MIT where he became an assistant in 1949, associate professor from 1955 and full professor since 1963.

Lynch concentrates his research activity on the study of people's perception of the urban landscape, therefore from an absolutely anthropocentric position. His scientific contributions range, in a vast conceptual field, from environmental psychology to the geography of perception (Lynch 1981).

Lynch, with his studies, has shown that people perceive the urban space they frequent or live in through common elements and mental patterns, creating their mind maps through the use of five categories: paths, margins, neighbourhoods, nodes and references.

The contribution of Banerjee, discussed in this paragraph, has allowed us to construct a theoretical path on the contributions that many authors have expressed, regarding the relationship between urban planning and the environment, even in

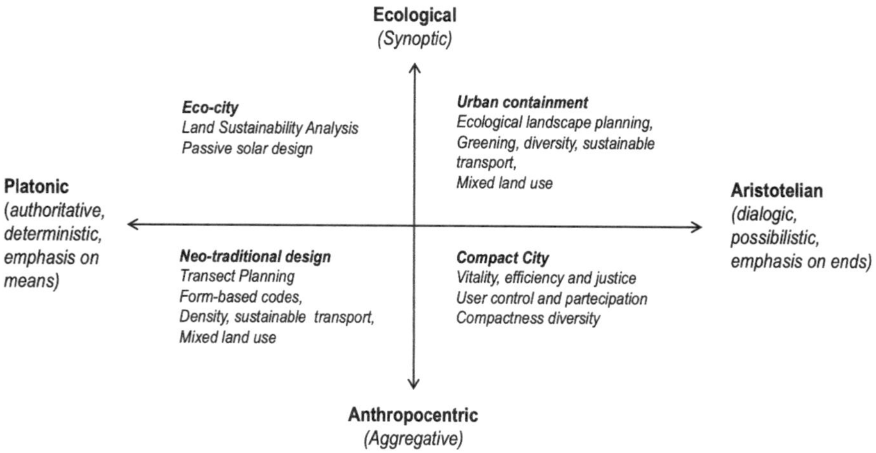

Fig. 2.2 Form models of cities inserted in the matrix of Fig. 2.1: towards an integrative approach (graphic elaboration from Banerjees 2014)

periods in which the emergence of sustainability as we know today, it was not emerging.

Of particular interest is the redefinition, by Banerjee, of the two-dimensional matrix and of the four quadrants of Fig. 2.1 useful for their macro-classification according to the different approaches to urban planning. In his document, the author hopes for a supplementary framework by providing an overview that is shown for convenience in graphic form in Fig. 2.2.

2.2 Sustainability in Green Planning Between Utopias and Pragmatism

2.2.1 Conferences of the Parties: Utopian but Essential Strategies for Sustainable Development

The concept of “sustainable development”, the main attractor of all the environmental policies of recent decades and, therefore, also of green planning of cities, is quite ambiguous: Pankaja and Nagendra (2015) argue that “it is now enshrined on the masthead of Environment magazine, featured on 8,720,000 Web pages, and enmeshed in the aspirations of countless programs, places, and institutions”.

The most widely accepted definition is that proposed by the World Commission for the Environment and Development was established by the United Nations General Assembly in 1982, better known as the “Brundtland Commission”: “Humanity has the capacity to make development sustainable—to ensure that it meets the needs of the present without compromising the capacity of future generations".

The inspiring principles of the contribution, "Our common future", published in 1987 date back to a few years before: in the Stockholm Conference on the human environment of 1972, where the conflicts between environment and development and in the "Strategy World Conservation" programme of the 1980 International Union for Nature Conservation, which advocated conservation as a means to help development and specifically for sustainable development and the use of species, ecosystems and resources.

As Brundtland argued that "*The environment does not exist as a sphere separate from human actions, ambitions, and needs and attempts to defend it in isolation from human concerns have given the very word "environment" a connotation of naivety in some political circles. The word "development" has also been narrowed by some into a very limited focus, along the lines of "what poor nations should do to become richer", and thus again is automatically dismissed by many in the international arena as being a concern of specialists, of those involved in questions of "development assistance." But the "environment" is where we live; and "development" is what we all do in attempting to improve our lot within that abode. The two are inseparable*" (Brundtland 2004).

The publication of the report marked the beginning of an important path: important international meetings then followed, starting with the United Nations Conference on Environment and Development (UNCED) in Rio de Janeiro in 1992 (the so-called Earth Summit), which published a declaration of principles, a detailed agenda 21 of the desired actions, international agreements on climate change and biodiversity, and a declaration of principles on forests.

Ten years later, in 2002, at the world summit on sustainable development in Johannesburg, South Africa, the commitment to sustainable development was reaffirmed. Since then, the leaders of the United Nations Framework Convention on Climate Change (UNFCCC) have written the history of the fight against climate change, between successes and failures, milestones and agreements of convenience.

Among the successive conferences of the parties (COP), it is useful to recall the most important ones.

The first real breakthrough takes place at COP 3 (1997), with the approval of the Kyoto Protocol, the first treaty in the world to reduce greenhouse gas (GHG) emissions. The commitments envisage a first phase of emission reduction for the 2008–2012 period compared to 1990 levels.

The COP 18 in Doha (Qatar) in 2012 managed to secure a second season for the Kyoto Protocol (expiring the same year), extending it until 2020; season from which, however, most of the industrialized nations have been lost by the wayside.

The twenty-first COP in Paris (1015) brought home the first major achievement, namely a shared global climate pact. The overriding objective is to keep the temperature rise "well below 2 °C", with the recommendation to do more (for a scenario below 1.5 °C). One of the key provisions of the agreement is the creation of a review mechanism for the various countries' commitments: the reviews will take place every five years, with a view to progressively increasing its ambition.

The 2016 Marrakech COP 22 closed with the approval of the Marrakech Alliance for global climate action. The assembly drafted the draft of a common plan for the implementation of the Paris Agreement, the first set of rules by which national reduction commitments will have to be relaunched: the goal is to create a shared system to judge the effectiveness of the climate policies and measure cuts in emissions.

Sustainable development as a concept, as an objective and as a movement and is now fundamental to the mission of countless international organizations, national institutions, businesses, cities and communities (Pankaja and Nagendra 2015).

2.2.2 *Sustainability in Urban Planning: A Continuous Commitment for Solving Contradictions*

The environmental topics covered by the Conferences of the Parties (COP) discussed in the previous paragraph are fundamental as they allow compatibly with the commitments of the Nations involved, the definition of strategies at a global level and the monitoring of the health of our planet over the years.

For the evaluation of the state of the environment at global level (e.g. the emissions of GHGs and the values of the average temperature of the planet), for the definition of the necessary objectives to be achieved through environmental indicators and for the monitoring of the results over the years, predictive complexes models are necessarily used.

The achievement of the objectives of improving environmental conditions, however, takes place through concrete actions, promoted by national, regional environmental policies but also environmentally conscious local, such as sustainable or green planning of cities and communities.

The definition of environmental strategies at the global level and the promotion of bottom-up planning actions are theoretically two sides of the same coin, in that they contribute to achieving a common goal which is sustainable development as a contribution to solving climate changes.

Realistically, the gap between top-down planning and bottom-up planning is large and difficult to fill and the conviction that in order to achieve a climate improvement objective it is sufficient to define the new improvement objectives in the COPs highlights an approach that is not constructive.

The real war on climate change is fought from the bottom up through bottom-up planning actions that must be implemented in concrete measures (e.g. the energy retrofit of existing buildings, the construction of new housing models or the use of renewable energy sources, the passage from the generation of centralized energy to distributed energy or the re-greening of urban areas).

The complexity in the implementation and management of bottom-up actions is in any case considerable and the technologies and the skills to act are not enough: it is necessary to face economic, organizational and social aspects, but it is above all necessary to affront and manage the inevitable clashes between different interests.

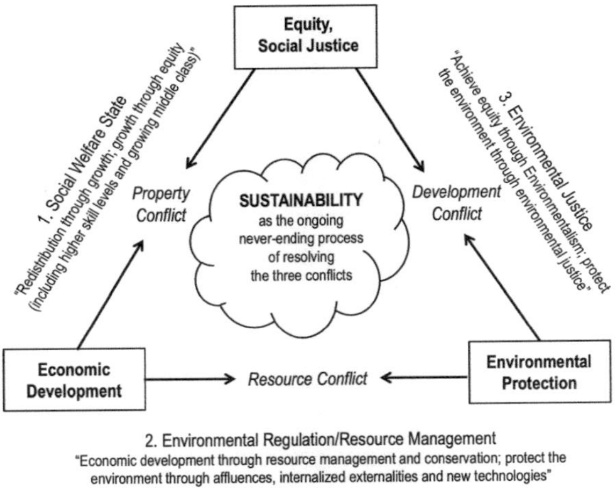

Fig. 2.3 "Planner's triangle" representing the three fundamental priorities of planning (green cities, growing cities, just cities), the three social conflicts and the three broad social and political institution to manage these conflicts (Campbell 2016)

A pragmatic and realistic theory, which also involves a new definition of "sustainable development" is that proposed by Scott D. Campbell Professor of Urban Planning at the University of Michigan (USA) (Campbell 1996).

According to this theory, urban planners work within the tension generated between three fundamental objectives: environmental protection, economic development and social equity, located at the three vertices of a triangle at the centre of which sustainable development is placed (Fig. 2.3).

The socially constructed view of nature proposed by Campbell puts into question the vision of these conflicts as a classic battle between "man against nature" or its current variation, "job against environment". The triangular model proposed is used to ask whether sustainable development, the current object of interest for planning, is a useful model to drive planning strategies.

The three types of priorities are present in three perspectives:

- The *planner of economic development* sees the city as a place where production, consumption, distribution and innovation take place. The city competes with other cities for markets and new industries. Space is the economic space of motorways, market areas and commuter areas.
- The *environmental planner* sees the city as a resource consumer and a waste producer. The city is in competition with nature for scarce resources and land being thereby always a threat to nature. Space is the ecological space of greenways, river basins and ecological niches.
- The *equity planner* sees the city as a place of conflict over the distribution of resources, services and opportunities. The competition is within the same city, between different social groups. Space is the social space of community, neighbourhood organization, trade unions: the space of access and segregation.

Campbell argues that there are other important ways of viewing the city, including the architectural, psychological and circulatory (transport); and one could conceivably construct a rectangle, a pentagon or polygons more complex than that of a planner. The triangular shape itself is not proposed as the geometric structure underlying the planner's world. Rather, it is useful for its conceptual simplicity.

Finally Campbell considers the implications of this point of view for planning. An important and constructive aspect of this theory is that the triangle shows not only conflicts, but also the potential complementarity of interests.

The former is inevitable and require planners to act as mediators, but the latter is one in which planners can be particularly creative in building coalitions between previously separated interest groups, such as workers and environmentalists, or community groups and companies. To this end, planners must combine their procedural and substantive skills and thus become central actors in the battle for growth, environment and social justice.

Campbell's first work published in 1996 was updated by his most recent paper (Campbell 2016) in which the Author, after about twenty years, substantially confirms the concepts of the previous publication.

At the end of the paper, Campbell states that "*The tension between growth and conservation persists: Our profession works to both aggressively expand the boundaries of the metropolitan region and erect bulwarks to conserve the natural and historic landscapes. …. Finally, the sustainability narrative remains vibrant and vital within planning because it has evolved in the past two decades through its very engagement with social justice, grounded in the ongoing practice of planning and designing both a greener and more equitable built environment. It is this productive collision of the environmental and community activist movements and the ongoing efforts to recombine and reconcile these two traditions with their divergent histories, values, and communities that has fueled the thoughtful advancement of sustainability planning. The sustainability movement will continue to be powerful as long as it creates a commons where planners, their allies, and their adversaries can debate the hard questions, negotiate compromises in the distribution of natural and human wealth, and creatively explore alternative urban futures*".

2.3 Design Concepts for Sustainable and Green Cities

The topic of the green planning of the city and of the communities should interface with those concepts of design which are the sectorial design criteria that have a direct or indirect effect on the sustainability of the territory, and the form of the built environment. All these aspects are analysed by Yosef Rafiq Jabareen who in his paper (Jabareen 2006) identifies sustainable urban forms and their design concepts by proposing a sustainable urban form matrix to help planners assess how different urban forms take account of sustainability.

2.3.1 Compactness

The compactness of the built environment, also referable to urban continuity and connectivity (and in these cases should be occurring adjacent to existing urban structures), is a design strategy that allows for more sustainable urban forms. The compactness of urban space can reduce the distances necessary for the transportation of energy, water, materials, products and people.

The intensification of the built form also includes not only the occupation of previously undeveloped urban land but also the redevelopment of existing buildings or the recovery of existing abandoned and degraded areas: in these cases the goal is to limit the occupation of new areas not yet urbanized.

2.3.2 Density

The density expresses the relationship between people or housing units and the occupied area. The relationship between density and urban character is also based on the concept of vital thresholds: with certain densities (thresholds), the number of people within a given area becomes sufficient to generate the interactions necessary to make urban functions or activities viable.

The density and type of housing influence sustainability through differences in energy consumption; materials; and land for housing, transport and urban infrastructure. The high density and use of integrated territory not only preserves resources, but also provides compaction that encourages social interaction. From the energetic point of view, the urban intensity allows one to have buildings that, for the same volume, have a lower dispersing surface. In a high-density city, it is also possible to use network energy infrastructures (e.g. district heating or district cooling), producing and employing the necessary energy more efficiently.

Urban density, on the other hand, limits the use of renewable energy sources, in particular solar thermal and PV solar, and could be in conflict with passive solar design.

2.3.3 Mixed Land Uses

Zoning, a tool used in urban planning consisting of dividing the territory into homogeneous areas from a functional point of view, has generated negative effects both at the environmental and at the social level. For this reason, urban planners have turned to a different approach: the mixed use of land.

A functional diversification takes place within the city, the neighbourhood or in any case the urban areas. With this approach within the same area, multiple functions coexist: residential, commercial, industrial (small industry), artisan craft and institutional. This choice generates a more stimulating urban environment from the social point of view but also a more sustainable environment from an environmental point of view. The coexistence of more functions, necessary for the vital functioning of the city, avoids large displacements for citizens: many of the journeys can be made on foot or by bicycle, thus limiting the environmental impact due to mobility.

A point to be considered in the green design of cities is that of diversity both in the forms and in the aesthetic aspect of the buildings and in the size of the spaces. Aesthetic, functional and dimensional diversity are the characteristics that make old cities which have developed over time through an overlap of attractive styles: the same concept should be applied to current cities that often, unfortunately, are characterized by a strong formal standardization which makes them unattractive and unsustainable.

2.3.4 Passive Solar Design

Passive solar design defines a bioclimatic design approach through which the design of buildings is carried out via two strategies: reducing energy consumption (e.g. through the thermal insulation of the building envelope) and using renewable energy sources (e.g. through the design of doors and windows that exploit the direct gains due to solar radiation). Through passive solar design, the building behaves like a solar collector with a direct exploitation of energy. In design strategies, it must be considered that solar radiation in summer can generate an increase in energy consumption (increase in summer thermal loads): for this reason, it is necessary to provide all the measures which can reduce the effect of solar radiation (e.g. through the use of shielding systems).

Urban passive solar design implies particular morphological choices: if solar energy is the reference energy source, the layout of the city must be organized so that the buildings have access to direct sunlight (right to the sun). The orientation of buildings and urban density are elements that must be considered.

The passive design of cities must also favour green areas and natural ventilation: the aim is to guarantee a comfortable urban ecosystem both in winter and in summer, limiting the phenomenon of "heat islands". Passive solar design can be considered in contradiction with the compact concept design: a more compact urban fabric, in fact, limits the use of solar energy.

A design that stimulates a relationship with the climate can no longer neglect the fact that the climatic conditions can sometimes be the cause of disasters. A green design must carefully consider resilience, understood as the ability of the city to react effectively with respect to potentially devastating natural "stresses" (e.g. earthquakes or floods).

2.3.5 Sustainable Mobility

Mobility is probably the main problem for environmental debates related to urban form: the evolution of cities and urban forms will depend to a large extent upon the implementation of new technologies that are rapidly evolving (e.g. electric mobility and in near future self-driving vehicles linked to the development of 5G technology) and to the changes in the paradigms that are already underway in some cities. For those who live in the city, the behavioural model, which includes for the possession of a vehicle, will have to change to the benefit of a model that envisages intensive use of public transport, cycle paths and use of supplementary on-demand private services (car-sharing, bike-sharing). Electric mobility will make the use of renewable energy sources (in particular solar PV) possible, thus eliminating emissions.

The policies for sustainable urban development, also considering the evolution taking place in the world of work which favours working from home, should, therefore, include measures to reduce the need for movement and provide favourable conditions for forms of transportation that are energy efficient and respectful of the environment. Land use planning plays a key role in achieving these goals.

2.3.6 Greening

The greening of the city, or green urbanism, is an important design concept for the attainment of a sustainable urban form. The advantages that can be obtained are considerable: it contributes to the maintenance of biodiversity, to the improvement of the urban physical environment by reducing pollution, moderating the extremes of the urban climate and contributing to sustainable urban drainage systems. The greening of the city also has positive psychological effects on citizens, making the environment more pleasant. The improvement of the urban image generates economic advantages since the city becomes more attractive. Greening also has health benefits and an educational function as a symbol or representation of nature.

According to Beatley (2000), a city exemplifies green urbanism if it strives to live within its ecological limits, is designed to function in a similar way to nature, strives to achieve a circular rather than linear metabolism, aims at local and regional self-sufficiency, facilitates more sustainable lifestyles and emphasizes a high quality of community life and neighbourliness.

2.4 Green Planning: Synergies Between Planning Tools Towards a Common Goal

To combat climate change, it is necessary to define, plan and implement strategies which are consistent with the international objectives that are often incorporated by the states and made operational, sometimes mandatory, in national and possibly regional legislation.

The green planning of cities and communities must be developed in a coherent way through sectoral legislative tools.

A complete frame of reference is shown in the diagram in Fig. 2.4 which illustrates the relationships between green planning tools, urban planning tools and regional, national and international guidelines.

From the diagram, it is possible to observe the complexity of the matter. In fact, there are many planning tools and since they are promulgated by various administrative entities or by different public and private bodies, there are often overlaps or inconsistencies. This is what emerges from a research work done by Dall'O' et al. (2013) in which some planning tools approved in a small town near Milan (Italy) are analysed and compared.

From the diagram, one can also observe how there may be synergies between the various planning tools which could and should be exploited.

The proposed diagram has the advantage of "systemising" for the first time a complex and articulated framework: the proposed green planning model is the result of choices made at a local level (city or neighbourhood) but consistent with a strategic vision of sustainable planning developed at different levels (i.e. think globally but act locally).

At the highest level of environmental strategies are the international agreements on climate change defined at the Conferences of the Parties (COP) by the participating nations. On the same level are the Sustainable Development Goals (SDGs) developed by the United Nations. Internationally, there may also be actions at the continental level, for example, the European Union Directives.

At a lower level, one observes legislation or guidelines adopted at national and/or regional level on energy and environmental issues.

These legislative frameworks are important and must be considered in the green planning, for two reasons: they declare the commitments taken at a higher level than nationally and/or regionally and, more significantly, they are often mandatory.

In the diagram, the operational scope of sustainable green planning on an urban scale or on a community scale is delimited by the central ellipse. In practice, there may be two areas: one urban and one suburban (neighbourhood scale or part of a neighbourhood scale).

Suburban planning is very important because it can affect interventions at a lower scale than those of the associated city which can be implemented, thanks to their lesser size and lower economic commitment, with greater ease.

Urban planning tools on an urban or suburban scale can be classified into three categories:

- mandatory planning tools as provided by national and/or regional legislation (e.g. town planning scheme, building code, mobility plan, etc.);
- those voluntary planning tools which are managed within national or international projects (e.g. Sustainable Energy (and Climate) Action Plan);

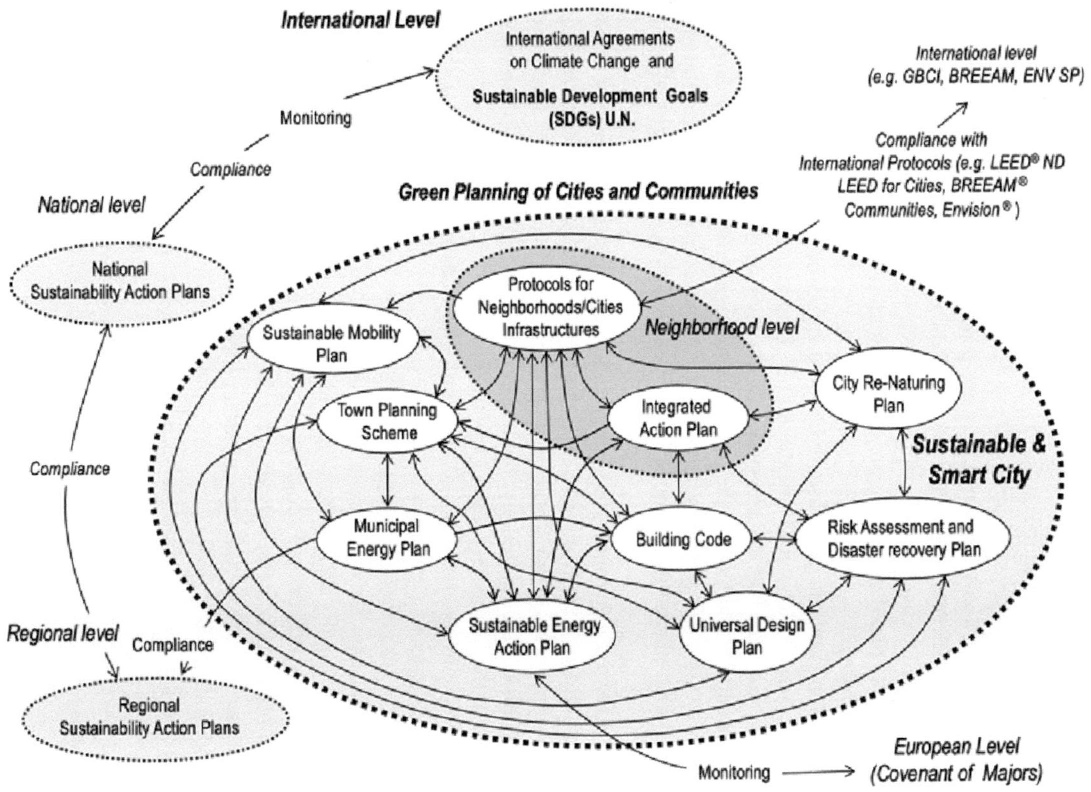

Fig. 2.4 Relationships between green planning tools, urban planning tools and regional, national and international guidelines (updated and expanded from Dall'O' et al. 2013)

- those voluntary planning tools which are managed by private institutions (e.g. environmental certification protocols for buildings, neighbourhoods and cities).

A short description of the objectives of the green planning tools considered in Fig. 2.4 is reported in Table 2.2 that also shows the compatibility of the Sustainable Development Goals for each tool.

In the diagram of Fig. 2.4, the green city is also called the smart city. At first sight, this may appear to be an error but that is not so. Since the Smart Cities seem more fashionable, one naturally asks oneself: is there a difference between the two names? Everything depends on the meaning which is attributed to the term "smart". A widely accepted meaning, proposed by Dall'O' et al. (2017) is that the Smart City is a "smart" planned city, managed in an "intelligent", environmentally sustainable and inclusive way. All these features are contained in the Green City, so it is possible to state, according to this vision, that the "green" concept is inclusive of the "smart" concept. Therefore, a green city is both sustainable and smart. In today's urban configurations, smart technologies are becoming widespread, supported by ICT (Information and Communications Technology) systems: thanks to these infrastructures, the platforms offered by the world of Internet and related services can be used, from the IoT (Internet of Things) to the coming 5G essential to promote autonomous driving of vehicles.

Technological innovation does not conflict with green choices but, on the contrary, provides an indispensable support to manage the city, its functions and its services in a sustainable way. With reference to the diagram in Fig. 2.2, it can be stated that the green planning of cities and communities cannot be considered a single and independent action plan but an inclusive cluster plan that encompasses all the sector plans reported in Table 2.2 which also highlights the synergies with the SDGs.

This strategy is a good way to promote green planning that takes into account all the aspects related to improving the sustainability of the urban environment in its different components. These include not only the environmental and ecological but also the social, economic and safety aspects.

Actions to combat climate change, fundamental attractors of green policy policies, are transformed into an indispensable opportunity to promote new planning that can resolve the numerous conflicts discussed in Sect. 2.2.2.

Also interesting is the harmonized coexistence and the synergy stimulated between the three categories of planning tools:

- mandatory planning tools help to ensure consistency with international agreements;
- voluntary planning tools stimulate participation and involvement by citizens and all stakeholders;
- the environmental certification protocols such as LEED® or BREEAM®, as well as spreading a culture of sustainability among the stakeholders, also stimulate the real estate market since the certified buildings increase in value and the certified neighbourhoods become more attractive for investors.

Table 2.2 Synergies between green planning tools and sustainable development goals

Green planning tool	Description	SDGs#
National Sustainability Action Plan	It defines the set of actions (laws, guidelines, technical reports, etc.) that Countries implement to combat climate change in line with international objectives and commitments	6, 7, 11, 13, 14, 15
Regional Sustainability Action Plan	It defines the set of actions (laws, guidelines, technical reports, etc.) that Regions implement to combat climate change in line with international objectives and commitments	6, 7, 11, 13, 14, 15
Town Planning Scheme	It is the fundamental tool for urban planning. Defines the rules for creating or modifying the urban layout (roads, public and private spaces, buildings, urban infrastructures, etc.)	6, 7, 11, 13, 15
Building Code	It is a tool that defines the functional, structural, energy and environmental requirements and performances for new buildings and buildings to be redeveloped. It also contains the rules for the provision of public and private outdoor spaces	6, 7, 11, 13, 15
Sustainable Mobility Plan	It is a green evolution of the classic transport plan. Defines the rules for sustainable urban mobility with less environmental impact	7, 10, 11, 13
Urban Energy Plan	It defines the set of actions and strategies that a city adopts for the optimal use of energy accelerating the transition to a model that, through energy efficiency, distributed generation and the use of renewable energy sources drastically reduces emissions of climate-altering and polluting gases	7, 11, 13

(continued)

Table 2.2 (continued)

Green planning tool	Description	SDGs#
Integrated Action Plan	It defines strategies and rules to be applied for the redevelopment of neighbourhoods and small urban areas. It takes into account the peculiarities of the place where the interventions will be carried out, enhancing the environmental and socially sustainable choices	7, 11, 13
City Re-Naturing Plan	It defines strategies and rules to apply for a progressive greening of the city	3, 11, 13, 14, 15
Risk Assessment and Disaster Recovery Plan	Action plan aimed at making a city more resilient to external events that are dangerous for the city and its inhabitants (such as earthquakes or floods)	3, 11
Universal Design Plan	The tool defines rules and strategies to guarantee spaces and services of the city accessible to all the citizens	4, 5, 8, 10
Sustainable Energy (and Climate) Action Plan	It is the key document in which the Covenant signatory (of the Covenant of Mayors for Climate & Energy) outlines how it intends to reach its CO_2 reduction target by 2020. It defines the activities and measures set up to achieve the targets, together with timeframes and assigned responsibilities	3, 6, 7, 11, 13, 14, 15
Environmental Certification Protocol	Evaluation protocols for the environmental sustainability of buildings, neighbourhoods or entire cities, drawn up by non-profit associations or organizations, which assign a global score based on a score. The certification of the level obtained is made by an independent certification body	3, 6, 7, 11, 13, 14, 15

2.5 Defining a Green Planning Strategy

Progressing to a more operational step, it is important to organize a systematic path which permits the design, implementation and management of green planning: this path, schematized in Fig. 2.6, can be designated as the Green City Strategic Plan (CGSP). Green city planning is the primary responsibility of local governments at various scales with the participation of all relevant local stakeholders and professional expertize as required.

The Green City Strategic Plan will be the result of a process of consultation that has to be defined at the outset and carried out through well-defined steps.

The key element of the CGSP is green growth, defined by (Global Green Growth Institute 2019): "*Green growth provides strategies to sustain economic expansion while protecting the environment and ensuring socially inclusive development. In the urban setting, green growth is defined through negotiating amongst stakeholders a shared vision and mission for green city development, as well as specific green growth goals related to resilience to climate change and natural hazard impacts, energy and resource efficiency and related savings, greater and more equitable access to urban services and welfare, poverty reduction, and urban competitiveness Such goals have to be identified at the outset of the Green City Strategic Planning process as the most relevant ones for the local context*".

Consistently with what was discussed in Sect. 2.4, the GCSP can be conceived as a set of actions that must simultaneously satisfy sectoral needs: in Fig. 2.5, the key urban sectors relevant to Green City Strategic Plan are reported.

A typical flow chart of a Green City Strategic Plan is illustrated in Fig. 2.6. This defines four macho phases: Audit Phase, Design Phase, Implementation Phase and Monitoring Phase.

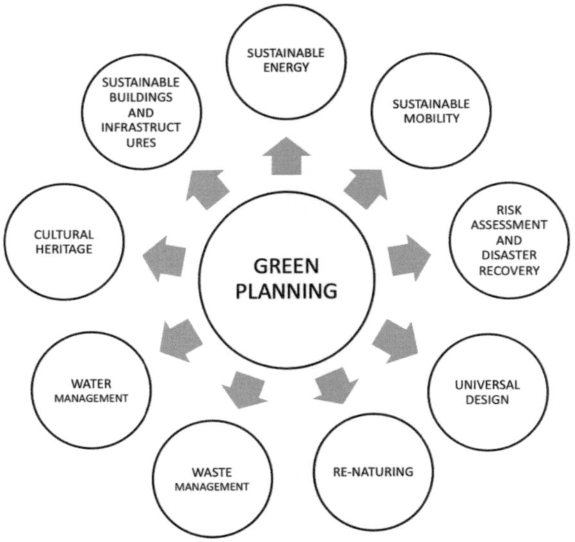

Fig. 2.5 Key urban sectors relevant to a Green City Strategic Plan

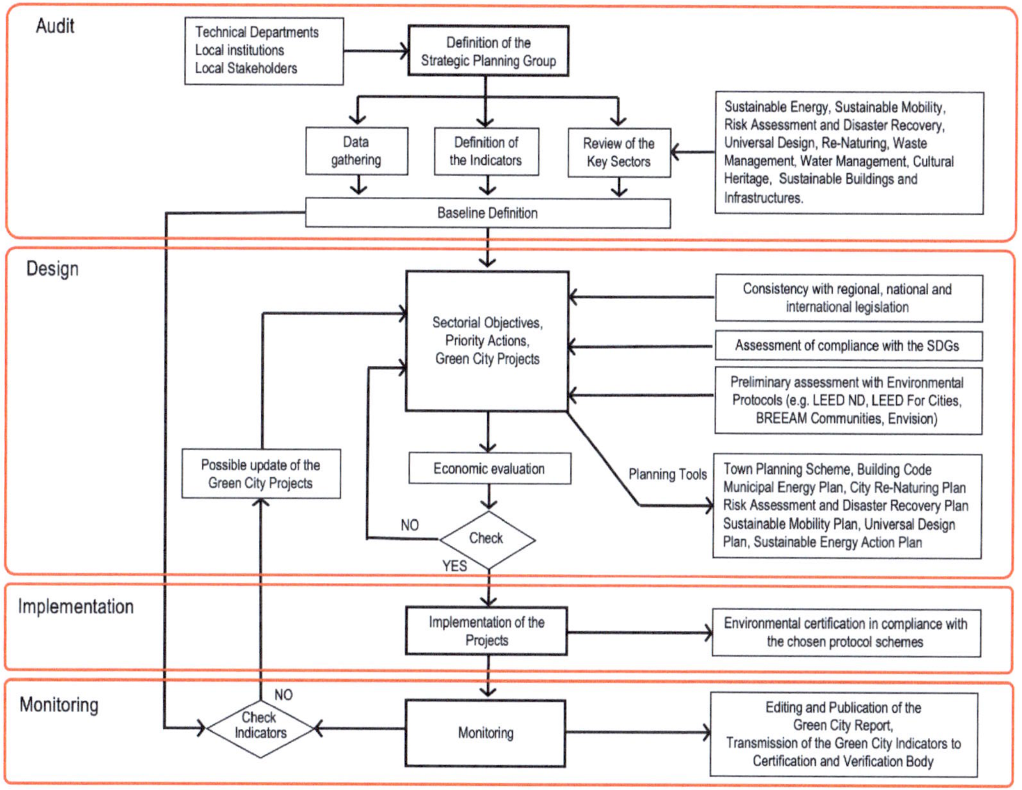

Fig. 2.6 Flow chart for Green City Strategic Plan for cities and communities: the actions are divided into three macrophases: audit, design, implementation and monitoring

2.5.1 Audit Phase

The audit phase deals with two aspects: the definition of the work group and, above all, the analysis of the current situation that constitutes the baseline, starting point and reference point on the basis of which to implement the actions (projects). For the development of a Green City Strategic Plan, it is important to create an operative team involving different individuals, characterized by clearly defined positions of leadership, thus able to take decisions. The top level of leadership of a GCSP requires a steering committee constituted by representatives from:

- Technical Departments (e.g. City Planning Department, Building Department, Environmental Department, Mobility Department, City Manager, Mobility Manager).
- Local Institutions (e.g. Provincial Government, Regional Government, National Energy Agencies, Environment Agencies, Public Universities, Public Research Institutes, Ministry of the Environment, Ministry of Energy, Minister of Infrastructure, Minister of Economy).
- Local Stakeholders (e.g. Non-government organizations, Civil society organizations, Private Universities/Academia, Research Institutions, Trade Associations, No-Profit Associations, Green Building Councils).

A Planning Secretariat should also be appointed, whose responsibility is to manage and coordinate day-to-day operations. The chair of the steering committee should be entrusted to a high profile person (e.g. the Mayor) preferably supported by a representative.

Once the steering committee is established, the audit phase should consider three aspects:

- Data gathering;
- Review of the key sectors;
- Definition of the urban sustainability indicators.

The collection of data within the Administration is also a useful opportunity to put order in the information substantially available, to define the state of the art of those actions and projects already started and those already completed and above all to start a dialogue and a permanent comparison between the different departments. This represents an important added value from a management point of view.

Particular attention must be paid to the definition of the urban sustainability indicators since they should be consistent with all the green planning tools: the goal is to define indicators that can be used for additional planning tools, avoiding duplication. An interesting contribution on this delicate topic is contained in the "Indicators for Sustainable Cities" report published by the European Commission (2018) which in its introduction describes a useful explanation of what the indicators are: "*Urban sustainability indicators are tools that allow city planners, city managers and policymakers to gauge the socio-economic and environmental impact of, for example, current urban designs, infrastructures, policies, waste disposal systems, pollution and access to services by citizens. They allow for the diagnosis of problems and pressures, and thus the identification of areas that would profit from being addressed through good governance and science-based responses. They also allow cities to monitor the success and impact of sustainability interventions*".

2.5.2 Design Phase

The design phase defines the objectives and strategies, intended as projects, which must be adopted to achieve them. At this stage, it is important to:

- define realistically achievable goals;
- define objectives that are not in conflict with legislative instruments and with guidelines approved by higher-level administrations (e.g. regional or national);
- define strategies consistent with the sustainable development goals (SDGs) defined by the United Nations.

In choosing the actions to implement, the steering committee should ask itself the following questions:

- Is the proposed action strategic with respect to the needs identified in the audit phase?
- Is the implementation time compatible with the needs identified during the audit phase?
- Are there technical skills available within the Technical Departments capable of managing the action both from a technical and an economic point of view?
- Is it feasible from an economic point of view? In other words, are the financial resources available to cover the costs of implementation but also of management and maintenance?

If the objectives to be achieved can be clear, the paths to follow, and, therefore, the actions to be implemented in order to attain those objectives, are various. The resultant choice of what is best done, must be in the best interests of the community.

An evaluation of the design choices in green planning can be made using different approaches in accordance with the scheme proposed by Mondini (2018) (Fig. 2.7): Environmental Impact Assessment, Cost–Benefit Analysis and Multi-criteria Decision Analysis.

As regard the third approach, interesting is the contribution of Dall'O' et al. (2013b) in which the authors apply a Multi-Criteria Methodology to Support Public Administration Decision Making Concerning Sustainable Energy Action Plans.

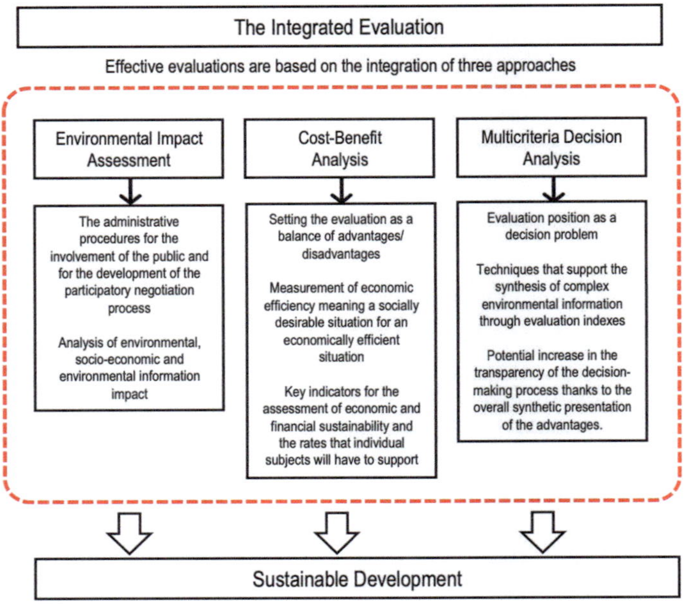

Fig. 2.7 Main classic evaluation approaches employed in the pursuit of sustainable development (Mondini 2018)

2.5.3 Implementation Phase

In this phase, the projects are implemented: considering their complexity, the implementation may take a few years. The implementation of an environmental certification system compliant with a sustainability protocol (e.g. LEED® ND, LEED® for Cities, BREEAM® Community, Envision®) makes it possible to follow the implementation of the project with great care, ensuring consistency between the objectives defined in the design phase and their actual accomplishment.

2.5.4 Monitoring Phase

By having a monitoring phase, it is possible to verify whether the actions implemented were truly effective in achieving the objectives: this is assisted by a verification over time of the partially implemented actions.

The green planning strategy, although it may include the implementation of several sectoral projects, cannot be considered a macroproject, with a beginning and an end, but rather defines a continuous path of improvement that must be constantly verified through the monitoring of the achieved results.

This monitoring is performed by means of a comparison between the indicators periodically detected (e.g. every year) and the indicators obtained from the baseline that becomes the reference base. It is, thus, the presence of a monitoring phase which makes the difference between a project and a process.

Through a precise and detailed control, it is possible to identify the partial or global ineffectiveness of some actions, identify their causes and make corrections. All is done according to the typical Quality Systems approach which, using the Deming cycle (PDCA),[3] is a recursive processes.

The indicators built on the basis of the information collected during the monitoring phase can be used to update shared databases, for example, the international Application for Resilient Communities (ARC) platform developed and managed by the GBCI.[4]

To be successful, a Green Planning Strategy must be inclusive and shared. The involvement of citizens and stakeholders is fundamental to promote sustainable development, a process that offers endless opportunities for growth but at the same time generates economic and social conflicts. Campbell, it should be recalled, defines sustainability as "*the ongoing never ended process on resolving the three conflicts: development conflict, resource conflict and property conflict*" (Campbell 2016).

Conflict management can only take place on the basis of objective information: the indicators obtained during the assembly phase demonstrate whether certain choices have been effective or not and constitute an objective basis for comparisons and for the management of any conflicts.

It is, therefore, advisable that periodically the data obtained from the monitoring be made available to the community through a "Green Planning Report" to be disseminated through traditional means of communication and via the Internet. The steering committee, amongst its activities, should promote a public Web platform containing all information regarding the Green Planning Strategy.

2.6 Conclusions

An Italian plant physiologist, Stefano Mancuso, has published a book with a captivating title: "The Plant Revolution: How plants have already invented our future", in which he analyses in depth how the plant world works and how the plant world is organized always to ensure the sustainability of our planet (Mancuso 2017).

Plants are energy self-sufficient using only solar energy, they do not impact the environment, rather they help us to remove the climate-changing gases which mankind produces, with waste by implementing circular economy strategies, they

[3]PDCA, an acronym for Plan-Do-Check-Act, is a four-step iterative management method used for the control and continuous improvement of processes and products.

[4]Green Business Certification Inc. (GBCI) is the only certification and credentialing body within the green business and sustainability industry to exclusively administer project certifications and professional credentials of LEED®, EDGE®, GRESB®, Parksmart, PEER®, SITES®, TRUE and WELL®.

are also resilient, adapting, within certain limits, even to the climate changes. The plants also have a distributed intelligence and are organized "on the net".

By imagining a strategy for sustainable development, one could start with this model or at least derive ideas from it. Mancuso's reflections are interesting and should stimulate us to learn from nature ideas that could be useful for green planning in our cities.

Climate change asks us for important and decisive actions and the International Panel on Climate Change (IPCC), through objective monitoring of the environmental situation at a global level, conveys to us the concern that we are not far from the point of no return (or "tipping point"), i.e. from an environmental situation that no longer permits recovery. If the international community shares these concerns, the strategies which can be implemented are various.

If the problem is determined by growth, degrowth could be the solution. Some minority movements choose this path and call this strategy "happy degrowth".

Humanity cannot share this strategy which would mean relinquishing many things. An aspect to consider, then, is the legitimate expectation of social improvement, and, therefore, of growth, of that part of the world's population which currently lives below the poverty line.

A more realistic approach is to implement relatively quickly all possible strategies to reduce the environmental damage caused by anthropization without rules and regulations.

In the last few years, many things have been done: almost zero energy buildings are being built, cars are being manufactured which impact far less from an environmental standpoint, strategies are being implemented to promote the economic economy and protocols are spreading on a voluntary basis. International environmental certification for buildings, neighbourhoods and cities has allowed the real estate sector to reward economically those who invest in sustainability.

This is the road: to trace a path in which strategies take into account the enhancement of nature, a common good, but at the same time, use technological innovation and scientific knowledge to re-establish a sustainable relationship between mankind and the environment.

The purpose of this chapter, which shares a conscious growth strategy, is that of being a contribution to the topic of green planning for cities and communities.

References

Alexander C, Ishikawa S, Silverstein M, Jacobson M, Fiksdahl-King I, Shlomo A (1977) A pattern language: towns, buildings, construction. Oxford University Press, New York

Beatley T (2000) Green urbanism: learning from European cities. Island Press, Washington, DC

Banerjee T (2014) Urban design and sustainability: looking back-wards to move forward. In: Mazmanian DA, Blanco H (eds) Elgar companion to sustainable cities: strategies, methods and outlook. Edward Elgar, Cheltenham, UK, pp 381–396

Brundtland GH (2004) Sustainable development—a global perspective on ecology, economy & equity. In: 4th annual Peter M. Wege lecture. University of Michigan, Ann Arbor, MI

Campbell D S (1996) Green cities, growing cities, just cities? Urban planning and the contradictions of sustainable development. J Am Plann Assoc 62(3):296–312
Campbell SD (2016) The planner's triangle revisited: sustainability and the evolution of a planning ideal that can't stand still. J Am Plann Assoc 82(4):388–397D
Dall'O' G, Galante A, Sanna N, Miller K (2013a) On the integration of leadership in energy and environmental design (LEED)® ND protocol with the energy planning and management tools in Italy: strengths and weaknesses. Energies. MDPI
Dall'O' G, Norese MF, Galante A, Novello C, (2013b). A multi-criteria methodology to support public administration decision making concerning sustainable energy action plans. Energies. MDPI
Dall'O' G, Sarto L, Panza A, Bruni E, Khayatian F (2017) Evaluation of cities' smartness by means of indicators for small and medium cities and communities: a methodology for Northern Italy. Sustain Cities Soc. Elsevier
Duany A, Plater-Zyberk E, Speck J (2000) Suburban nation: the rise of sprawl and the decline of the American dream. North Point Press, New York. ISBN 0-86547-606-3
European Commission (2018) Indicators for sustainable cities. Available on: https://ec.europa.eu/environment/integration/research/newsalert/pdf/indicators_for_sustainable_cities_IR12_en.pdf. Accessed 20 Aug 2019
Global Green Growth Institute (2019) Green city strategic planning methodology: a guide for the development of a strategic green city strategic plan. Available on https://gggi.org/report/cambodian-green-city-strategic-planning-methodology/. Accessed 20 Aug 2019
Hester RT (2009) Design for ecological democracy. Landsc J 28(2):235–236. https://doi.org/10.3368/lJ.28.2.235
Jabareen Yosef R (2006) Sustainable urban forms: their typologies, models and concepts. J Plann Educ Res 26(1):38–52
Jacobs A, Appleyard D (1987) Toward an urban design manifesto, planner's notebook. J Am Plann Assoc. Available on: https://pdfs.semanticscholar.org/293a/15cd8ad1e63d3676e577dca120872a80e771.pdf. Accessed 20 Aug 2019
Knowles RL (1978) Energy and form: an ecological approach to urban growth. The MIT Press
Lynch K (1981) A theory of good cities form. The MIT Press, Boston, MA
McHarg Ian L (1969) Design with nature. (Wiley Series in Sustainable Design Book 6). English edition, 25th edn
Mancuso S (2017) Plant revolution, le piante hanno già inventato il nostro futuro. Giunti Editore, Firenze
Mondini G (2018) An integrated approach for assessing environmental damage and (inter) generational Debt in the definition of territorial transformation policies, in integrated evaluation for the management of contemporary cities. Green Energy and Technology, Springer Nature
Pankaja MS, Nagendra HN (2015) Green city concept—as new paradigm in urban planning. Int J Eng Sci (IJES) 4(10):55–60
Soleri P (1969) Arcology: the city in the image of man. Available on https://www.organism.earth/library/document/76. Accessed 20 Aug 22 2019
Steiner FR (2000) The living landscape: an ecological approach to landscape planning, 2nd edn. McGraw-Hill
Talen E (2005) New urbanism and American planning: the conflict of cultures. Taylor & Francis Ltd
Whinston AW (2011) Ecological urbanism: a framework for design of resilient cities. In: Steward TA, Pickett Mary L. Cadenasso, Brian P. McGrath (eds) Draft of a chapter for resilience in ecology and urban design. Springer, Berlin. Available on https://annewhistonspirn.com/pdf/spirn_ecological_urbanism-2011.pdf. Accessed 20 Aug 2019

Chapter 3
Renaturing Cities: Green and Blue Urban Spaces as Paradigms of Urban Planning

Giuliano Dall'O'

Abstract Green in cities does not only play an aesthetic role but also contributes greatly to making the urban environment more sustainable in all senses of the term: environmental, energy-wise, economic and social. The purpose of this chapter is to tackle the issue of green in cities by highlighting the connections that exist between having a nature base or renaturing of cities and sustainable planning, stimulating the necessary in-depth analysis of a topic that must constitute an information resource for those who deal with green planning. The chapter, after an introduction that highlights the need to seek a new alliance between urban spaces and natural spaces, deals with the issue of urban infrastructure, highlighting its potential. The final part of the chapter deals with the issue of urban heat islands providing some improvement strategies.

3.1 Urban Spaces and Natural Spaces Toward a "New Alliance" for Sustainability

Cities and metropolises represent a conquest for humanity which still today in urban spaces seeks the great opportunities that they offer: work, social cohesion, services, culture. Today, cities and metropolises are home to around half of the world's population, and in 2050, 68% of the world's population will reside in metropolitan areas (U.N. 2018).

When comparing the urban space with natural space, considering anthropization as a process in conflict with the natural ecosystem, from the point of view of environmental sustainability, a paradox becomes apparent: If the urban environment is the opposite of the natural one, could a "return to nature" and therefore relinquishing urban living not be the solution to promote sustainable development?

Those who live in the city often dream of "escaping" and choosing a rural life in a house in the boundless countryside, in direct contact with nature.

G. Dall'O' (✉)
Architecture, Built Environment and Construction Engineering, ABC Department, Politecnico di Milano, Via G. Ponzio, 31, 20133 Milan, Italy
e-mail: giuliano.dallo@polimi.it

G. Dall'O' (ed.), *Green Planning for Cities and Communities*,
Research for Development, https://doi.org/10.1007/978-3-030-41072-8_3

However, considering the current levels of population and the projected increase over the coming years, this model of living is unsustainable from an environmental point of view.

An isolated house in the countryside can be energetically self-sufficient and therefore sustainable, but it cannot be the model of sustainable living. The concentration of the population in urban spaces offers significant advantages, for example, starting from mobility. The strategies defined by the "New Urbanism" movement identify its own urban model as the real chance to combat climate change (CNU 2012).

If the urban model is a winning model to contrast climate changes, how is it possible to overcome the paradox that sees it as contraposed to the natural ecosystem? The answer is simple in form and complex in content, through a "new alliance" between man and the environment, thus between urban and natural spaces.

Mancuso (2017) claims that plants have already invented our future. The green revolution can start right here: from learning from plants, the most obvious expression of the natural ecosystem. Stimulated by reading Mancuso's work, we tried, through a simple exercise, to compare the needs and strategies of some principles of green planning with what plants can teach us. The result is a Table 3.1 that provides interesting insights for our work. When we are seeking and largely implementing in order to promote and accelerate sustainable development, plants have always implemented.

Learning from plants, therefore, can be an interesting starting point to consolidate our theories and to promote the aforementioned "alliance" with the natural ecosystem of which we are a part.

Natural spaces within cities are not only aesthetically beautiful spaces but must be considered urban infrastructures to be designed, engineered and maintained. The renaturing of cities can be a slow process and requires considerable investments: It is a process that will necessarily change the appearance of cities, making urban spaces more pleasant and environmentally sustainable.

Renaturing of cities is a long-term urban planning process which guarantees benefits, although, on the downside, it may require managing and overcoming conflicts between public interest and private interest.

The new spaces needed to host the new green urban infrastructures can be obtained through the enhancement of degraded urban areas or abandoned industrial areas: Fig. 3.1 illustrates an example of renaturalization: The abandoned area, where the Alfa Romeo plant was located, in a central area of Milan, has been transformed into a green urban space of considerable environmental value.

Figure 3.2 shows another interesting case: the case study of ecological district b01 "City of tomorrow" in Malmö (Sweden). "*Hit by the economic decline of the industrial sectors in the 1960s and 1970s, both areas, Augustenborg and Bo01 (a small site in Western Harbor district), struggled with ecological and socioeconomic difficulties in subsequent decades. In order to maintain the housing areas, a total transformation was necessary—Bo01 required complete renovations and Ekostaden (eco-neighborhood) Augustenborg had the need for revitalization.*"

"*Ecological improvements of the local rainwater system for instance and of the housing stock were carried out in combination with approaching social issues, too.*

Table 3.1 Plants have already invented our future: a comparison between the environmental strategies adopted by plants and possible strategies for green planning (inspired by Mancuso 2017)

Issue	Plant strategies	Green planning strategies
Energy	Chlorophyll photosynthesis is the biochemical process underlying plant survival: Through this phenomenon, sunlight is captured through chlorophyll and transformed into chemical energy, which is essential to synthesize glucose molecules and release oxygen	The target of decarbonization can be pursued through energy efficiency and the progressive replacement of fossil energy sources with renewable energy sources starting from solar energy. Plants provide us directly with an important renewable energy source: biomass
Waste management	In the natural ecosystem, plants do not generate waste and always provide for the implementation of circularity strategies	The waste generated by human activities must be drastically reduced through the lower production of packaging, the recycling of waste and through the implementation of the circular economy, understood as an economic system designed to be able to regenerate on its own, thus also guaranteeing its eco-sustainability
Urban air quality	Plants do not pollute the air but, on the contrary, contribute greatly to improving air quality in urban spaces. The plants act as carbon dioxide accumulators, fixing the carbon during the photosynthetic process and storing the excess in the form of leaf and wood biomass. The leaves of the trees intercept and remove many of the pollutants present in the air, such as carbon monoxide and dioxide, ozone, monoxide and nitrogen dioxide, sulfur dioxide and particulate matter PM10	Activate a process of decarbonization and drastic reduction of toxic emissions for man and nature through strategies that reduce the energy consumption of fossil fuels: from green buildings to sustainable mobility. Forestation interventions in cities must be widely supported for overall improvement of environmental quality

(continued)

Table 3.1 (continued)

Issue	Plant strategies	Green planning strategies
Environmental adaptation, resilience	Plants have extraordinary adaptability thanks to which they can live in extreme environments	Implement risk assessment and disaster recovery strategies to counter the environmental effects sometimes generated by the effects of climate change
Relationship with the subsoil	The relationship with the ground is fundamental through the root system, the most important portion of the plant. It is a physical network whose apexes form a front in continuous advancement	In green planning strategies, the ground is a resource: In addition to hosting infrastructure networks, the subsoil can function as a source of energy and as a thermal flywheel: typical application is the ground source heat pump
Modularity and networking	Plants are modular and flexible systems that internalize the logic of the network. This strategy allowed them to resist and develop over time	Green planning, thanks to new technologies, favors network-based information and energy systems: just think of the Internet or distributed energy generation that is contracted to the classic model of centralized generation

One basic principle was the participation of the local population in planning and implementation" (https://use.metropolis.org/case-studies/city-of-tomorrow).

The good news is that this change in urban areas is already underway as shown in the report "Spatial analysis of green infrastructure in Europe," prepared by the European Environmental Agency (EEA 2014) which provides a useful framework to inspire new projects.

The new models of sustainable mobility will assist this change: The decrease in individual mobility in favor of public means of mobility or alternative mobility (e.g., car-sharing, bike-sharing) will make it possible to transform traditional urban streets into green streets: Many cities have drawn up design manuals which are an excellent source of inspiration (City of Philadelphia 2012; City of Dallas 2016).

The US Environmental Protection Agency makes available a conceptual guide to effective green street design solutions: An example of a commercial green street is shown in Fig. 3.3 (US EPA 2009) from which it is possible to note that the greening of the street involves a reduction in the section of the road width compared to the original size.

These green roads and green corridors will be the new arteries for mobility, not only for vehicles, but also for pedestrians and cyclists, and will become the infrastructures that will connect the points of interest inside and outside the urban area, facilitating the relationship between the city and the extra-urban natural areas.

Fig. 3.1 In the former Alfa Romeo industrial area in the center of Milan, a park with a pond was built, and with the rubble from demolition of the disused premises hills were created. *Credits* The Author 2019

Fig. 3.2 Green and blue urban spaces for the ecological district b01 "City of tomorrow" in Malmö (Sweden). *Credits* The Author 2013

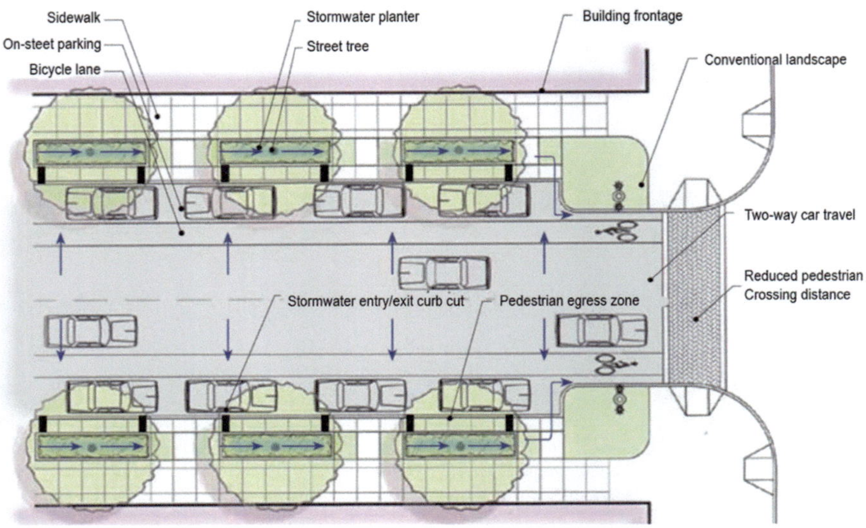

Fig. 3.3 Example of a commercial green street with permeable paving (US EPA 2009)

3.2 Green Urban Infrastructures

The new alliance between urban spaces and natural spaces is based on a concept of ecological, functional and social integration of urban infrastructures and natural or **green infrastructures**. Pötz and Bleuze (2011) defines Urban Green Infrastructure (UGI) as "*a network providing the ingredients for solving urban and climatic challenges by building with nature*".

On May 2013, the European Commission published a strategy to promote green infrastructure—essential to the functioning of cities and regions—and make it mainstream in EU policy areas. The strategy notes the potential for green spaces to make a major contribution to sustainable development, by enhancing social cohesion, supporting the economy and adapting to a changing climate, and highlights the importance of green infrastructure solutions in cities, where more than 60% of the EU population resides (EC 2013).

UGI planning is a strategic planning approach that aims to develop networks of green and blue spaces in urban areas, designed and managed to deliver a wide range of ecosystem services and other benefits at all spatial scales. Due to its integrative, multifunctional approach, UGI planning is capable of addressing a set of four urban challenges (Hansen et al. 2017):

- Adapting to climate change
- Protecting biodiversity
- Promoting a green economy
- Increasing social cohesion.

The GREEN SURGE Project[1] states that UGI planning is based on four core principles:

(1) *Green–gray integration*—intended as combining green and gray infrastructure UGI planning and integration of urban spaces with other infrastructures, such as transport systems and utilities.
(2) *Connectivity*—intending as the creation of green space networks for connectivity, involves creating and restoring connections to support and protect processes, functions and benefits that individual green spaces cannot provide alone.
(3) *Multifunctionality*—intended as delivering and enhancing multiple functions and services which combine different functions to enhance the capacity of urban green space to deliver multiple benefits.
(4) *Social inclusion*—intended as collaborative and participatory planning aimed at obtaining cooperative, socially inclusive processes.

A nature base can be used to provide important services for communities by protecting them against flooding or excessive heat, or to help improve air, soil and water quality. When is nature is harnessed by people and used as an infrastructural system, it is called "green infrastructure" (Benedict and McMahon 2006).

Blue infrastructure refers to urban infrastructure relating to water. Blue infrastructure is commonly associated with green infrastructure in the urban setting and may be referred to as "**blue–green**" infrastructure when in combination. Rivers, streams, ponds and lakes may exist as natural features within cities, or be added to an urban environment as an aspect of its design (Grellier et al. 2017) (Fig. 3.4).

In the urban context, green (and blue) urban spaces are different, ranging from urban forests and woodlands in parks to rooftop gardens.

Some of these spaces are, in general terms, already considered in the practice of urban planning, for example, city parks, but the others, in particular the private green areas like gardens, or indeed urban agricultural lands, have only attracted interest, although their diffusion constitutes good practice in the actions to renature cities. The GREEN SURGE Project has contributed to this knowledge gap by developing a green space typology made up of 44 elements, in eight groups, and linking them to scientific evidence on their corresponding ecosystem services (Table 3.2). This provides an important basis for understanding the functional connections between green spaces and the surrounding built environment.

Many of the green spaces shown in Table 3.3 are present in all cities: Their presence, however, is not sufficient for one to consider them as green urban infrastructures. The UGIs can be defined as such if there is a functional aggregation project behind the green spaces.

Using the principles of connectivity and multifunctionality, it is possible to determine which of these spaces are part of the UGI network of the city and where it

[1]The GREEN SURGE project is a collaborative project between 24 partners in 11 countries. It is funded by the European Commission Seventh Framework Program (FP7). GREEN SURGE will identify, develop and test ways of linking green spaces, biodiversity, people and the green economy in order to meet the major urban challenges related to land use conflicts, climate change adaptation, demographic changes and human health and wellbeing (https://greensurge.eu).

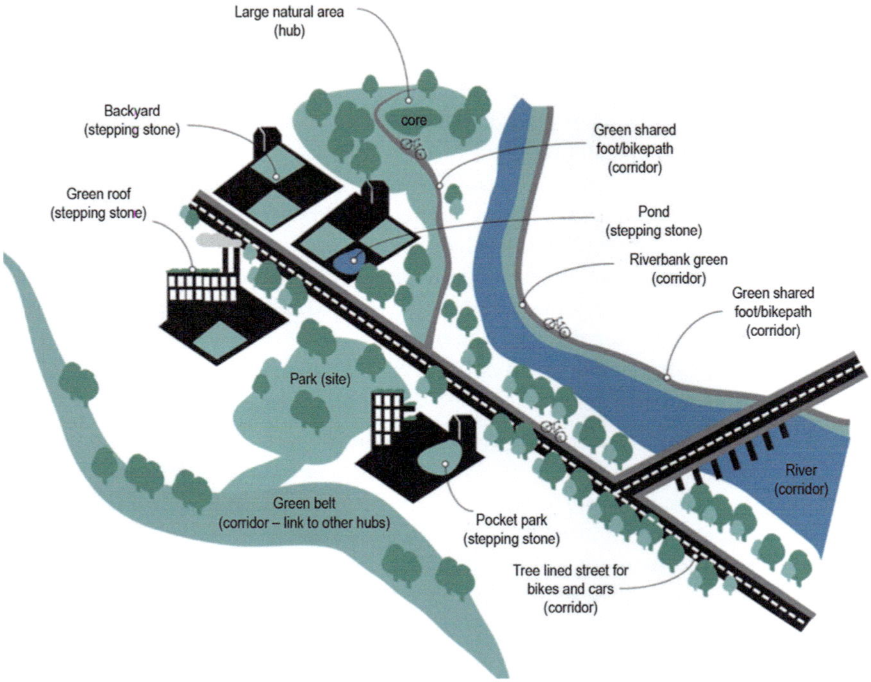

Fig. 3.4 Elements of a green infrastructure network (Hansen et al. 2017), Design: Eleanor Chapman

is necessary to improve the quality of the existing elements or invest in new and strengthened connections.

The goal is to create an infrastructural network that integrates with the urban infrastructures without interruptions or barriers: a green network that makes the city functional, which connects it internally with gray infrastructures and outside with other infrastructures, green and gray, of the suburban territory.

Although the social, environmental and regulatory context of urban areas may change, the design of green urban infrastructures has a significant impact on the city's green planning. The matrix shown in Table 3.3 defines the relations between the four core principles of UGI planning and the urban challenges (Hansen et al. 2017). In the following paragraphs, some examples of urban infrastructures are discussed.

3.2.1 Urban Green Corridors and Greenways

Urban green corridors or **greenways** are an essential component of the cities' ecological networks, being a powerful response to environmental problems. They

Table 3.2 Green space typology, made up of 44 green spaces types clustered in eight groups (Hansen et al. 2017)

Group	Green space typology
Building greens	Balcony green; Ground-based green wall; Façade-bound green wall; Extensive green roof; Atrium
Allotments and community gardens	–
Natural, semi-natural and feral areas	Forest (e.g., remnant woodland, managed forests, mixed forms), Shrublands; Abandoned areas; Rocks; Sand dunes; Sand pit, quarry, open cast mine; Wetland, bog, fen, marsh
Parks and recreation	Large urban park; Historical park/garden; Pocket park; Botanic garden/arboretum; Zoological garden; Neighborhood green space; Institutional green space; Cemetery and churchyard; Green sport facilities; Camping areas
Agricultural land	Arable land; Grassland; Tree meadow/orchard; Biofuel production/agroforestry; Horticulture
Private, commercial, industrial and institutional green/green space connected to gray infrastructure	Bioswale; Tree alley and street tree, hedge; Street green and green verge; Private garden; Railroad embankment; Green playground; school ground
Riverbank green	–
Blue spaces	Lake, pond; River, stream; Dry riverbed; Canal; Estuary; Delta; Coast

have a dual purpose: In addition to their ecological role in creating a natural pollution-free image and helping people to live in better communities, they provide access routes and improve the quality of life, particularly social life (Aly and Amer 2010).

The greenways can be of various types and can be characterized by various widths and combinations. Their purpose is to create a network linking green infrastructures with significant ecological, recreational, cultural and historical features: They can connect parks, nature reserves, urban farms or protected areas.

The greenways have the main purpose of offering opportunities for walking, cycling: In a sustainable project, the network must be developed in order to guarantee continuity and must be structured so as not to be interrupted by obstacles. For example, wherever possible, systems to cross major roads or highways must be constructed. A typical example is the one shown in Fig. 3.5 which illustrates a pedestrian and cycle connection bridge between two green areas.

The main function of these green urban infrastructures is to facilitate urban mobility: A dynamic greenway network can constitute an important part of a sustainable transport system.

Table 3.3 Four core principles of UGI planning can each help to address a range of challenges, including those examined in GREEN SURGE (Hansen et al. 2017)

	Integration	Connectivity	Multifunctionality	Social inclusion
Climate change	Green–gray measures for flood retention or urban cooling	Connected green structures that enhance natural ventilation and cooling	Regulating services that contribute to climate change adaptation as an integral part of planning for multifunctionality	Inclusion of groups vulnerable to climate change impacts in UGI planning
Biodiversity	Habitat provision, supporting native plants as one of the co-benefits of green–gray solutions	Networks for ecological connectivity	Protecting ecological functions and habitat as an integral part of planning for multifunctionality	Fostering awareness among all groups of the value of biodiversity
Green economy	Reduced management costs through integrated green–gray systems; avoided costs through risk mitigation	Promotion of sustainable transport systems, e.g., walking and biking to lessen environmental impacts	Cost-effective UGI solutions through providing multiple benefits in the same space	Promotion of a green economy, through cocreation, comanagement and co-governance of urban green spaces
Social cohesion	Consideration of the usability and amenity values of integrated UGI measures to promote social cohesion	Provision of equitable access to urban green spaces	Provision of UGI to meet identified demands and needs of all groups	Consideration of vulnerable and less-vocal groups' needs and their empowerment through collaborative planning

Green corridors bring people closer to nature by providing immediate benefits. From an environmental point of view, they prevent soil erosion and absorb water, thus improving drainage and protecting from the urban heat island effect. Considering green corridors only as connecting elements is an understating their importance: From the social point of view, they can offer people recreational places where they can play, meditate or simply rest.

Greenways can also recover and redevelop urban infrastructure that is no longer used. The most interesting case is that of the High Line in Manhattan, New York City, a 2.33 km elevated linear park, greenway and rail trail created on a former New York Central Railroad spur on the west side of Manhattan.

Fig. 3.5 Example of a pedestrian and bicycle bridge that connects two greened areas bypassing a busy road. *Credit* The Author 2019

The abandoned spur has been redesigned as a "living system" drawing from multiple disciplines which include landscape architecture, urban design and ecology. Since opening in 2009, the High Line has become an icon of contemporary landscape architecture (Keller 2011).

The success of the High Line has inspired other cities around the world to redevelop obsolete infrastructure as public spaces.

The success of the High Line has inspired other cities in the world to redevelop obsolete infrastructure as a public spaces. The project stimulated real estate development in adjacent neighborhoods, increasing property values and prices along the way: A demonstration that green redevelopment measures can also be economically viable.

3.2.2 Urban Forests

Trees in the cities represent the most visible green infrastructures: Green planning cannot do without the inclusion of trees in cities: They can be planted along avenues, in squares, or in parks where urban *forests* can be created.[2]

Trees in the city could be integrated into the architecture: This solution is called "**vertical forest**." The first example of the vertical forest of two residential towers of 110 and 76 m height, designed by the Architect Stefano Boeri, was built in the center of Milan in 2014, on the edge of the related island neighborhood, hosting 800 trees (each measuring 3, 6 or 9 m), 4500 shrubs and 15,000 plants from a wide range of shrubs and floral plants distributed according to the sun exposure of the facade. On flat land, each vertical forest equals, in amount of trees, an area of 20,000 m^2 of forest. 75,000 m^2 of urban densification is the equivalent of an area of a single family of nearly 75,000 m^2 [3] (Fig. 3.6).

From an environmental point of view, trees contribute greatly to improving the urban microclimate: improving air quality, reducing rainwater runoff and accommodating wildlife.

The absorbed solar energy is diverted from the leaves of deciduous trees in summer, while in winter it is only filtered by the branches of deciduous trees. With trees present, it is possible to moderate the "heat island" effect.

Trees also influence wind speed and direction. The more compact the foliage on the tree or on the group of trees, the more effective is the windbreak function. Trees slow down or absorb precipitation, sleet and hail, offering protection to people, cars and buildings.

The leaves filter the air, eliminating dust and other particles. The rain then washes away the pollutants into the ground. Through chlorophyll photosynthesis, which is the biochemical process underlying plant survival, sunlight is captured through chlorophyll and transformed into chemical energy, which is essential to synthesize glucose molecules and release oxygen.

The leaves also absorb other atmospheric pollutants—such as ozone, carbon monoxide and sulfur dioxide—and release oxygen.

Studies conducted both on models and on real cases in urban canyons (the set of roads and buildings that make up the urban fabric) have shown that the air quality in the metropolitan area is strongly conditioned by the presence of vegetation and its structure (Gromke and Ruck 2007; Gromke 2011). The results show, for example, how certain parameters, such as the type, height and diameter of the canopy of trees, represent key factors capable of conditioning air quality. For example, one measures higher levels of particulate matter in streets characterized by dense rows of trees, compared to roads with trees placed in random and random order.

The presence of trees in an urban land decreases the amount of stormwater runoff and pollutants that reach local waters (EPA 2013):

---

[2]The term "forest" rather than "wood" or "woodland" in the context of the urban spaces might seem exaggerated to the uninitiated, but this term is now fully established in design vocabulary.

[3]Source on Vertical Forest: (https://www.stefanoboeriarchitetti.net/en/project/vertical-forest/).

Fig. 3.6 Vertical forest, designed by Stefano Boeri, is a model for a sustainable residential building, a project for metropolitan reforestation contributing to the regeneration of the environment and urban biodiversity without the implication of expanding the city upon the territory. *Credits* The Author 2019

- trees reduce stormwater runoff by capturing and storing rainfall in their canopy and releasing water into the atmosphere;
- tree roots and leaf litter create soil conditions that promote the infiltration of rainwater into the soil;
- trees help slow down and temporarily store runoff and reduce pollutants by taking up nutrients and other pollutants from soils and water through their roots;
- trees transform pollutants into less harmful substances.

Particular attention must be paid to the choice of species, favoring indigenous species and using genetic strains of local origin. These in fact normally show the best adaptation to the climatic conditions.

The best interventions from an ecological point of view are those aimed at the creation of neo-ecosystems capable of maintaining themselves, through the spontaneous reproduction of individuals, in the absence of external inputs. These interventions, which can lead to the establishment of real urban forests, are those which also ensure considerable economic savings, given their ability to perpetuate themselves without human intervention.

3.2.3 Urban Agriculture

For those who live in the city, the possibility of having a vegetable patch or kitchen garden and growing vegetables is a dream that has become reality in many cases. Among urban green infrastructures, urban agriculture spaces are spreading, gaining much interest.

This interest is due mainly to an increase in awareness of the benefits of a healthy life in contact with nature. There is a growing commitment of associations and public administrations/local authorities toward a solution that more than others promotes sustainable development, creates economic networks of solidarity and, more specifically, has proved to be widely capable of favoring the rediscovery of social ties between people (Fig. 3.7).

Urban agriculture is defined as the production of crop and livestock goods within the city and towns, generally integrated into the local urban economic and ecological system. The activities can include the cultivation of vegetables, medicinal plants, spices, mushrooms, fruit trees and other productive plants (Lin et al. 2017).

Urban gardens can be classified according to the ownership of the area in private and public. They can also be individual, that is, managed independently by a single user, or collectively: This solution is preferable since it allows everyone involved to socialize. The areas can be found within the urban fabric or in suburban areas: Urban gardens can therefore also be an advantageous manner of recovering spaces from degraded or no longer used areas of the city. The spaces, in some cases, can

Fig. 3.7 Example of a community garden in Milan. *Credits* The Author 2019

also be recovered in or on the flat roofs of buildings, in which case the presence of this infrastructure permits an increase in the thermal insulation of the roof.

The municipal administrations often equip the urban garden areas and then rent them out at reasonable tariffs to the citizens, as individuals or more frequently as groups in specific associations.

The beneficiaries, typically nonprofessional growers, receive these spaces in concession for one or more predefined purposes, the primary one being the production of flowers, fruit and vegetables which will serve to satisfy the needs of the assignees.

The advantages of having an urban garden in the city are various: from the rediscovery of the value of the land to the cooperation between citizens and farmers to produce fresh fruit and vegetables. These initiatives are of help to the new generations, as they make them sensitive to more sustainable and "green" city ideas. They are also valuable for adults and the elderly who, through urban gardens, have the opportunity to do outdoor physical activity and produce nutritious foods without the use of chemicals and pesticides.

3.3 Urban Greening to Reduce Heat Island Effect

3.3.1 The Urban Heat Islands

Urban areas are usually warmer than their rural surroundings; this phenomenon is known as the "urban heat island" (UHI)." UHI can affect communities by increasing summertime peak energy demand, air-conditioning costs, air pollution, greenhouse gas emissions, heat-related illness and mortality and reducing water quality.

Properties of urban materials, in particular solar reflectance, thermal emissivity and heat capacity, also influence the development of urban heat islands, as they determine how the sun's energy is reflected, emitted and absorbed (US EPA 2008).

Surface temperatures have an indirect but significant influence on air temperatures, especially at the boundary layer closest to the surface. In natural areas, for example, parks, surface temperatures are colder, and in those conditions, even the air temperatures are lower.

From Fig. 3.8, that illustrates differences in surface and atmospheric temperatures comparing daytime and nighttime, one can observe a progressive variation between downtown and the rural areas.

Atmospheric heat islands vary much less in intensity than surface heat islands. On an annual mean basis, air temperatures in large cities might be typically 1–3 °C warmer than those of their rural surroundings (Oke 1997).

Elevated summertime temperatures in cities increase the energy demand for cooling and add pressure to the electricity grid during peak periods of demand, which generally occur on hot, summer weekday afternoons, when offices and homes are

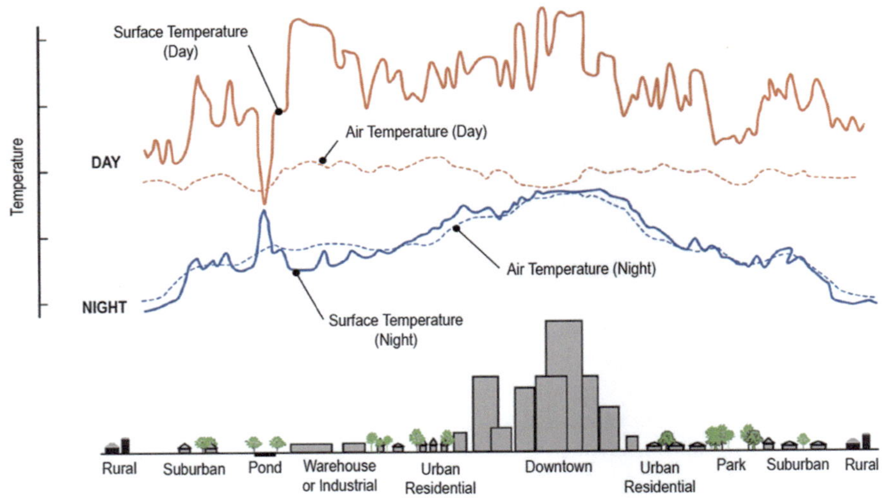

Fig. 3.8 Variations of surface and atmospheric temperatures (US EPA 2008, modified from Voogt 2002)

running cooling systems, lighting and appliances. This peak urban electric demand increases 1.5–2% for every 0.6 °C increase in summertime temperature (Akbari 2005).

From the energy point of view, the increase in outdoor air temperature in summer requires more power for summer air conditioning.

From a thermodynamic point of view, the use of air-conditioning equipment leads to a greater heat transfer to the outside. In the environmental energy balance, the greater heat transferred by the installations contributes to incrementally the external temperature of the air and therefore the UHI.

The increase in the value of the external air temperature requires a greater use of the air-conditioning systems and so on. The UHI, therefore, generates a negative environmental effect that is continuously fed. Higher temperatures can increase the energy demand, which causes higher levels of air pollution and greenhouse gas emissions.

Table 3.4 summarizes the factors that determine the heat island phenomenon. For further information on this topic, see (US EPA 2008).

Table 3.4 Factors that create urban heat island (US EPA 2008)

Factors communities are focusing on
• Reduced vegetation in urban regions: reduces the natural cooling effect from shade and evapotranspiration • Properties of urban materials: contribute to absorption of solar energy, causing surfaces, and the air above them, to be warmer in urban areas than those in rural surroundings
Future factors to consider
• Urban geometry: The height and spacing of buildings affects the amount of radiation received and emitted by urban infrastructure • Anthropogenic heat emissions: contribute additional warmth to the air[a]
Additional factors
• Weather: Certain conditions, such as clear skies and calm winds, can foster urban heat island formation • Geographic location: Proximity to large water bodies and mountainous terrain can influence local wind patterns and urban heat island formation

[a]Although communities currently can lower anthropogenic heat emissions through energy efficiency technologies in the building and vehicle sectors, this compendium focuses on modifying vegetative cover and surface properties of urban materials, as they have long been regarded as urban heat island reduction strategies. An emerging body of the literature on the role waste heat plays in urban heat island formation, though, may lead communities to focus on anthropogenic heat in the near future

3.3.2 The Strategies for Reducing the Heat Island Effect

The heat island effect can be substantially reduced by implementing urban green planning strategies. If the increase in temperature is determined by the concentration of gray infrastructure, the integration of urban green infrastructures, contributing to reducing the environmental gap between the urban ecosystem and the rural ecosystem, contributes systematically to a decrease in the outside air temperature.

Renaturing of cities is a slow process: Since some cities have been formed over centuries of history, it is impossible to pretend to change everything in a short time. Renaturing of cities is nonetheless the only strategy we have available.

Trees and vegetation help cool urban climates through shading and evapotranspiration. Leaves and branches reduce the amount of solar radiation which reaches the area below the canopy of a tree or plant.

The amount of sunlight transmitted through the canopy varies based on plant species. In the summertime, generally, 10–30% of the sun's energy reaches the area below a tree, with the remainder being absorbed by leaves and used for photosynthesis, and some being reflected back into the atmosphere.

In winter, the range of sunlight which is transmitted through a tree is much wider (10–80%) because evergreen and deciduous trees have different wintertime foliage, with deciduous trees losing their leaves and allowing more sunlight through. (Huang et al. 1990). An interesting compendium of strategies for reducing Urban Heat Island is given in (US EPA 2008b).

In urban environments, ground areas with traditional pavements usually cover a percentage of the undeveloped area depending on the typology of the city. This

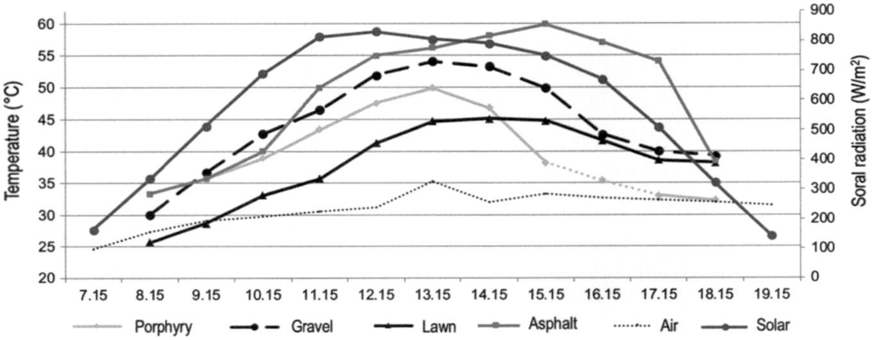

Fig. 3.9 Hourly temperatures of some outdoor pavement surfaces compared with air temperature and solar radiation: surfaces exposed to solar radiation. *Source* The Author 2014

choice is sometimes determined by the need to reduce maintenance costs: A green surface, in fact, often requires great care even if the current trend is to replace the classic lawns with natural systems that do not require maintenance.

Many studies have been done to evaluate the effect of solar radiation on the temperature of the ground surface covering normally used in cities: The purpose is to use materials that reduce the absorption effect of solar radiation.

Figures 3.7 and 3.8 show the surface temperatures of ground surfaces subject to solar radiation on a typical July day in Milan.[4]

The same graphs show the hourly values of air temperature (°C) and solar radiation (W/m^2) on the horizontal plane. The surfaces considered concern precisely: lawn, porphyry, gravel and asphalt. The graph of Fig. 3.9 concerns floors exposed directly to solar radiation, while the graph of Fig. 3.10 concerns shaded floors. The surveys were carried out simultaneously.

From the two graphs, it is possible to observe how there is a notable difference between the temperatures of the lawn and those of the other surfaces. Additionally across the same given surfaces, one observes the difference between those parts of the surfaces exposed to solar radiation and those in shadow.

To avoid the problem of overheating in summer, a solution for pedestrian or cycle areas is to replace traditional surfaces with **cool pavements**.

Understanding how cool pavements work requires understanding how the solar energy heats pavements and how the pavement influences the air above it. Properties such as solar energy, solar reflectance, material heat capacities, surface roughness, heat transfer rates, thermal emittance and permeability affect pavement temperatures.

Cool pavements and roadways are surfaces that use additives to reflect solar radiation unlike a conventional dark pavement. Cool pavements are made with different

[4]The surveys were carried out in the summer of 2014 as part of a research activity promoted by ABC Department—Politecnico di Milano, aimed at identifying a methodology to classify in a parametric way, considering the environmental aspects, the urban spaces.

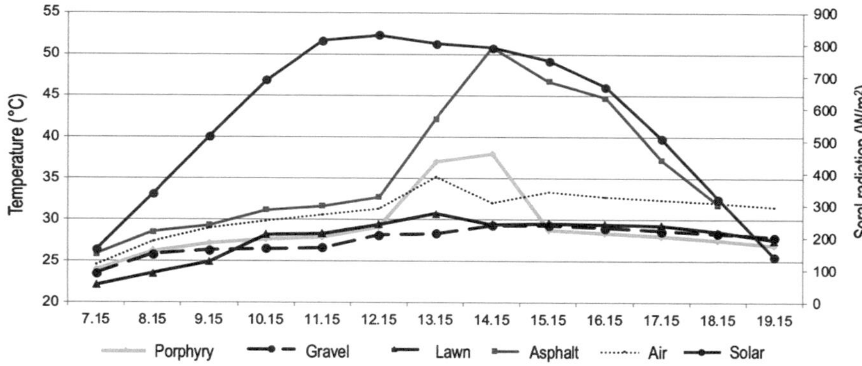

Fig. 3.10 Hourly temperatures of some outdoor flooring surfaces compared with the air temperature and with solar radiation: shaded surfaces. *Source* The Autor 2014

surfaces to increase albedo, thereby reflecting ultraviolet radiation out of the atmosphere. Increasing albedo reduces heat transfer to the surface and creates local cooling (US EPA 2012).

Similarly, it is possible to replace the traditional roof coverings using cool roofing technology. While traditional roofs, in contrast, can reach summer peak temperatures of 66–85 °C, thus creating a series of hot surfaces as well as warmer air temperatures nearby, **cool roofs**, made of highly reflective and emissive materials, can remain approximately 28–33 °C cooler than traditional materials during peak summer weather.

The use of cool roofs as a mitigation strategy brings many benefits, including lower energy use, reduced air pollution and greenhouse gas emissions and improved human health and comfort. A cool roof transfers less heat to the building below, so the building stays cooler and more comfortable and thus also uses less energy for cooling (US EPA 2008c).

Green roofs are a technology that can help communities mitigate urban heat islands effect.

A green roof or living roof is a roof (Fig. 3.11) of a building that is partially or completely covered with vegetation and a growing medium, planted over a waterproofing membrane. It may also include additional layers such as a root barrier and drainage and irrigation systems (Rodriguez 2011).

As with trees and vegetation elsewhere, vegetation on a green roof shades and insulates against heat from the air through evapotranspiration. These two mechanisms reduce the temperatures of the roof and the surrounding air. The surface of a vegetated rooftop can be cooler than the ambient air, whereas conventional rooftop surfaces can exceed ambient air temperatures by as much as 50 °C (US EPA 2008d).

The heat island effect is the cause of elements that contribute, directly or indirectly, to increase the air temperature. The regreening of the city contributes a lot to improving the environmental situation by reversing the cyclical evolution that, due to the necessary use of air-conditioning systems, continues to increase the temperature of

Fig. 3.11 Example of green roof installed in a tilted surface. *Source* The Autor 2014

outdoor environments. A city, however, consists not only of urban infrastructures but also of **buildings,** and their contribution to the summer energy balance is negative. Today's new buildings, very energy efficient and with little environmental impact (e.g., **zero-energy buildings** or ZEB), generate a very low environmental impact: An example we have seen is that of the vertical forest of Milan.

The critical situations concern many of the existing buildings, often conceived in periods in which there were no laws, regulations or indeed guidelines for green construction.

In order to obtain concrete results, in terms of improving the urban ecosystem, it is necessary to consider the possibility of requalifying the existing building stock from an energy and environmental point of view. The techniques available are many: In addition to the **green roofs**, already mentioned, it is possible to intervene on the facades through the application of a **green facade**. These renovation interventions, which also have an aesthetic role in addition to their functionality, are desirable as, in addition to making the buildings' image more modern, they contribute to making the city greener with the positive environmental, climatic and social effects already described.

Another aspect concerns urban mobility since vehicles with internal combustion engines, in addition to polluting, produce heat which contributes to the heat island effect.

The strategies to promote **sustainable mobility** (e.g., pedestrian, cycle and electricity) have a noticeable positive effect in combatting this phenomenon.

3.4 Conclusions

Tan (2017) states that "*increasing focus on urban green spaces (UGS) leads to them an important component of the physical makeup of cities. However, it is useful to be mindful that UGS implementation competes for precious land resources in cities, incur carbon and energy footprint, and can have long payback periods for net benefit to be achieved. The net benefit provided by UGS is thus assured by its mere presence; functional benefits need to be achieved through deliberate design.*" Tan Puay Yok suggests in particular that combining design with an understanding of the accumulated knowledge on urban ecology is an useful approach to increase the ecological function.

In the introduction to this chapter, we have called for a "new alliance" between urban spaces and green spaces whose function should no longer only be of an aesthetic nature.

The change of cities toward green models, which can be described as the renaturing of urban spaces, passes via this paradigm in which greening interventions are considered not as isolated episodes but as a network system. Structured routes are assured thanks to a greenways network, where they become not only an ecological but also a functional and social connection system of the overall urban structure.

Regreening the city must be conceived as a project that develops over time but follows an overall defined strategy.

The objective of this strategy is not only to combat global warming or climate change but also to make cities, within which the world's population will increasingly concentrate, the most beautiful places to live.

References

Akbari H (2005) Energy saving potentials and air quality benefits of urban heat island mitigation. Available on http://www.osti.gov/bridge/servlets/purl/860475-UlH-WIq/860475.PDF. Accessed 30 Aug 2019

Aly SSA, Amer MSE (2010) WIT, green corridors as a response for nature: greening Alexandria city by creating a green infrastructure network. Trans Ecol Environ 138, © 2010 WIT Press. www.witpress.com, ISSN 1743-3541 (online) https://doi.org/10.2495/dn100101

Benedict MA, McMahon ET (2006) Green infrastructure: linking landscapes and communities. Island Press, ISBN-13: 978-1559635585

City of Dallas (2016) Complete streets design manual. Available on: http://dallascityhall.com/departments/sustainabledevelopment/DCH%20documents/pdf/DCS-Design_Manual_TTRPC_092413_Final_UPDATE.pdf. Accessed 30 Aug 2019

City of Philadelphia (2012) City of Philadelphia, green streets design manual. Available on http://www.phillywatersheds.org/img/GSDM/GSDM_FINAL_20140211.pdf. Accessed 30 Aug 2019

Congress of the New Urbanism (CNU) (2012) Charter of the new urbanism available at https://www.cnu.org/who-we-are/charter-new-urbanism. Accessed 20 Aug 2019

Droguett BR (2011) Sustainability assessment of green infrastructure practices for stormwater management: a comparative energy analysis (M.S.). Available on https://search.proquest.com/docview/900864997/abstract/6E6468B55EF454FPQ/1. Accessed 30 Aug 2019

European Environmental Agency (2014) Spatial analysis of green infrastructure in Europe, EEA technical report, SSN 1725-2237. Available on: https://www.eea.europa.eu/publications/spatial-analysis-of-green-infrastructure. Accessed 30 Aug 2019

European Commission (2013) Building a green infrastructure for Europe. ISBN 978-92-79-33428-3. https://doi.org/10.2779/54125

Grellier J, White MP, Albin M, Bell S, Elliott LR, Gascón M, Gualdi S, Mancini L, Nieuwenhuijsen MJ, Sarigiannis DA, van den Bosch M, Wolf T, Wuijts S, Fleming LE (2017) BlueHealth: a study programme protocol for mapping and quantifying the potential benefits to public health and well-being from Europe's blue spaces. BMJ Open 7(6):e016188. https://doi.org/10.1136/bmjopen-2017-016188. PMC 5726080. PMID 28615276. Retrieved 16 June 2017

Gromke C (2011) Analysis of local scale tree-atmosphere interaction on pollutant concentration in idealized street canyons and application to a real urban junction. Atmos Environ 45:1702–1713

Gromke C, Ruck B (2007) Influence of trees on the dispersion of pollutants in an urban street canyon—experimental investigation of the flow and concentration field. Atmos Environ 41:3287–3302

Hansen R, Rall E, Chapman E, Rolf W, Pauleit S (2017) Urban green infrastructure planning: a guide for practitioners. GREEN SURGE. Retrieved from http://greensurge.eu/working-packages/wp5/. Accessed 30 Aug 2019

Huang J, Akbari H, Taha H (1990) The wind-shielding and shading effects of trees on residential heating and cooling requirements. In: ASHRAE winter meeting, American society of heating, refrigerating and air-conditioning engineers. Atlanta, Georgia

Keller J (2011) First drafts: James corner's high line park. The Atlantic. Archived from the original on July 13 2016. Retrieved June 5 2017

Lin BB, Philpott SM, Jha S, Liere H (2017) Urban agriculture as a productive green infrastructure for environmental and social well-being. In: Tan PY, Jim TY (eds) Greening cities: forms and function, advances in 21st century human settlements. Springer Nature Singapore Private Limited. https://doi.org/10.1007/978-981-10-4113-6_8

Mancuso S (2017) Plant revolution, le piante hanno già inventato il nostro futuro. Giunti Editore, Firenze

Oke TR (1997) Urban climates and global environmental change. In: Thompson RD, Perry A (eds) Applied climatology: principles & practices. Routledge, New York, NY, pp 273–287

Pötz H, Bleuze P (2011) Urban green-blue grids for sustainable and dynamic cities. Coop for life, Delft. ISBN 978-90-818804-0-4

Tan PY (2017) Perspectives on greening of cities through an ecological lens. In: Tan PY, Jim TY (eds) Greening cities: forms and function, advances in 21st century human settlements. Springer Nature Singapore Private Limited. https://doi.org/10.1007/978-981-10-4113-6_8

U.S. Environmental Protection Agency (2008) Urban heat island basics. In: Reducing urban heat Islands: compendium of strategies. Draft. https://www.epa.gov/heat-islands/heat-island-compendium. Accessed 30 Aug 2019

U.S. Environmental Protection Agency (2008b) Trees and vegetation. In: Reducing urban heat Islands: compendium of strategies. Draft. https://www.epa.gov/heat-islands/heat-island-compendium. Accessed 30 Aug 2019

U.S. Environmental Protection Agency (2008c) Cool roofs. In: Reducing urban heat Islands: compendium of strategies. Draft. https://www.epa.gov/heat-islands/heat-island-compendium. Accessed 30 Aug 2019

U.S. Environmental Protection Agency (2008d) Green roofs. In: Reducing urban heat Islands: compendium of strategies. Draft. https://www.epa.gov/heat-islands/heat-island-compendium. Accessed 30 Aug 2019

U.S. Environmental Protection Agency (2009) A conceptual guide to effective green streets. Available on: https://nacto.org/docs/usdg/2000_green_streets_epa.pdf. Accessed 30 Aug 2019

U.S. Environmental Protection Agency (2012) Cool pavements. In: Reducing urban heat Islands: compendium of strategies. Draft. https://www.epa.gov/heat-islands/heat-island-compendium

U.S. Environmental Protection Agency (2013) Stormwater to street trees, engineering urban forest for stormwater management. Available on. https://www.epa.gov/sites/production/files/2015-11/documents/stormwater2streettrees.pdf. Accessed 30 Aug 2019
United Nations (2018) Inform on the population. Department of Social and Economic Matters, Secretariat of the United Nations
Voogt J (2002) Urban heat Island. In: Munn T (ed) Encyclopedia of global environmental change, vol 3. Wiley, Chichester

Chapter 4
Disaster Risk Assessment, Reduction and Resilience: Their Reciprocal Contribution with Urban Planning to Advance Sustainability

Daniele F. Bignami

Abstract This chapter illustrates the need for the inclusion in sustainable urban planning of the problem of the impact of disasters on cities and communities: disasters that will continue to occur and may well even become more frequent because of urban expansion, increasing population densities and climate change effects. To achieve this result, first of all some basic tools about disaster terminology, disaster trends and disaster risk computation are provided. Subsequently, the latest international decisions at global level to face disasters and climate change and to promote sustainability are explained and elucidated, showing their connections and synergies. In the final part, some originally selected elements on how to put in place strategies and those actions which are no longer deferrable are proposed.

4.1 Introduction

The following pages describe one of the most important features of sustainable urban planning, focusing on issues related to different aspects, or pillars, of the disaster risk management (DRM) disciplines, such as disaster risk assessment (DRA), disaster risk reduction (DRR) and disaster resilience promotion (DrP), showing both their original place and their innovative role in the comprehensive framework of reference provided by this book.

Handling disaster events that strike cities and communities more or less recurrently requires an analytical and objective reading of the potential natural and anthropic calamitous phenomena in a manner similar to that related to energy and pollution issues. A political strategy, a multidisciplinary analysis and selection of the available tools of intervention, and an ability to monitor their evolution in an objective way (to verify the effectiveness of policies and solutions adopted) are therefore necessary also in this case; although, logically, this can be done only by following a detailed working path specifically and professionally conceived.

D. F. Bignami (✉)
Project Development Department, Fondazione Politecnico di Milano, Piazza Leonardo da Vinci, 32, 20133 Milan, Italy
e-mail: daniele.bignami@fondazione.polimi.it

G. Dall'O' (ed.), *Green Planning for Cities and Communities*,
Research for Development, https://doi.org/10.1007/978-3-030-41072-8_4

A green planning strategy which was to disregard disaster risk themes could trigger a failure or an underperformance in the competition within and between cities and communities, not only, but surely, for local and global sustainability. Indeed, to give some examples: disasters suspend development programmes and actions by destroying or damaging existing assets and lifelines, and interrupting the planning of the new ones; disasters worsen poverty, disturb small business, continuity of services and industrial production; disasters reduce human capital owing to casualties, injured and missing persons; disasters upset the balance of ecosystems and their connections with human health, food production and the economy.

Consistently, the goal of this chapter is to increase the use of disaster risk knowledge in decision-making processes related to sustainability of construction and the built environment sector, with the aim of improving the quality, at different levels, of designs, plans, policies, regulations and investments.

It is a fact that reducing the impact of natural and man-made hazards in urban areas helps in protecting the environmental matrices, assets, development gains and economic expansion of cities and communities, on the basis of a desirable long-term view, since the existing levels of risks are already "the results of decades or centuries of unplanned growth and an almost complete absence of proper risk considerations" (UNDP 2010). At the same time, this chapter will show how a green and sustainable approach to urban planning and design contributes, with many and complementary synergies, to reducing disaster risk, giving, from a holistic point of view, additional efficiency and proficiency to such a forward-looking strategy.

4.2 Basic Elements of a Disaster Risk Management Tool-Box for a Steady Dialogue with Sustainable Urban Planning

Disaster risk and sustainability are no theoretical issues or affairs; rather, in order to adopt countermeasures in an efficient and correct way, a well-displayed DRM "tool-box" is needed, to avoid unproductive or generic approaches or even counterproductive choices and effects.

Cities and communities, or, to be more accurate, the built environment and its peri-urban spaces, often to be considered from an extended point of view (Geneletti et al. 2017), are the domain where a series of events may become threats to human beings and their activities, goods and heritage. Moreover, the threats are increasing and this relates at least to the fact that the world is progressively becoming more and more urbanized and that fast-growing cities and urban areas around the world increase the risks of disasters owing to the impacting economic growth and rapid expansion of the population (Amaratunga et al. 2017).

Taking account of the objective of this book, three basic components to be included in a DRM tool-box have been selected, starting from those useful for a multi-disaster approach, the role of a green planner being that to handle sustainability with the ambition of a holistic approach. That ambition would not be possible without a proper vocabulary, able to guide in the selection of the theme and sub-theme of disaster to be tackled and fully treated; without a "compass" to give bearings towards the comprehension of data and trends on disasters, or a terrain on which to operate; without the computational rules for quantifying risks and related components to be compared with the available resources in the framework of the strategy in progress. In a period like the present, of growth in the concept of smart cities, based on the power of technology, it is useful to remember that technology, although valuable, is not the sole solution for knowledge-based decision-making for DRR (Norton et al. 2015).

4.2.1 Terminology

Initially, since disaster risk management disciplines are, relatively speaking, a field of study and action in its "youth", both for practitioners and researchers of urban planning and sustainability, it is useful to resume here the basic concepts to address this specific branch of knowledge correctly. Indeed, not rarely one can meet excellent professionals and researchers coming from different disciplines who unconsciously address these topics in what may seem an improper manner, due to the fact that they have their "origin" in other fields of study of which they maintain the major part of the related vocabulary and of the associated methodological schemes.

With the purpose of providing a necessary set of terms to be shared, it is suggested to put in the DRM tool-box the recommendations of the report of the intergovernmental expert working group on indicators and terminology relating to disaster risk reduction established by the General Assembly of UN—United Nations[1] (UN 2016), as a sort of "official reference" available to avoid any mismatch and discrepancy between schools of thought and technical approaches.

On the basis of the cited report, *disaster risk management*, the frame-concept of this chapter, "is the application of disaster risk reduction policies and strategies to prevent new disaster risk, reduce existing disaster risk and manage residual risk, contributing to the strengthening of resilience and reduction of disaster losses".

This very clear definition is helpful in anticipating the different main parts of the task of a planner during the elaboration of a strategy to manage the disaster risk of a community or of a city: it is not the same challenge to manage disaster risk in an existing urbanized context as that in a new area to be urbanized. Consistently, with the offered definition of disaster risk management, the report adds an annotation, a suitable kind of "corollary" to better specify the set of actions of disaster risk management, DRM that "can be distinguished between *prospective* disaster risk

[1] The group has been established as part of the path of implementation of the Sendai Framework for Disaster Risk Reduction 2015–2030, coherent with the work of the Inter-Agency and Expert Group on Sustainable Development Goal Indicators.

management, *corrective* disaster risk management and *compensatory* disaster risk management, also called residual risk management".

Prospective disaster risk management is defined as the set of activities which "address and seek to avoid the development of new or increased disaster risks. They focus on addressing disaster risks that may develop in future if disaster risk reduction policies are not put in place". The cited examples, well focused on the goals of this chapter, are the actions or the investments for "better land-use planning or disaster-resistant water supply systems".

Corrective disaster risk management activities are presented as tools to "address and seek to remove or reduce disaster risks which are already present and which need to be managed and reduced now". In this case, the ideal examples are the arrangements or the use of resources for "the retrofitting of critical infrastructure or the relocation of exposed populations or assets".

Compensatory disaster risk management activities, in their turn, are those that "strengthen the social and economic resilience of individuals and societies in the face of residual risk that cannot be effectively reduced". They include, for instance, preparedness and response activities, but also financing instruments, such as national contingency funds, insurance and reinsurance systems and contracts (such as processes of risk transfer, to shift abroad or to other financial entities the consequences of a disaster, paying the probabilistic allocation of a huge loss).

Reading from the same source, *disaster risk assessment*, the first pillar, chosen to methodologically describe main aspects of disaster risk management, is "a qualitative or quantitative approach to determine the nature and extent of disaster risk by analyzing potential hazards[2] and evaluating existing conditions of exposure and vulnerability that together could harm people, property, services, livelihoods and the environment on which they depend". In other words, the report rationally adds an annotation, underlining that the disaster risk assessment frame of action needs to include: the identification of hazards; a risk analysis, summarized as "a review of the technical characteristics of hazards such as their location, intensity, frequency and probability; the analysis of exposure and vulnerability, including the physical, social, health, environmental and economic dimensions"; and an evaluation of the effectiveness of the available means of coping in their capabilities to face likely risk scenarios.

The second (but not in order of importance) key aspect is *disaster risk reduction*, that, following the cited UN terminology, is "aimed at preventing new and reducing existing disaster risk and managing residual risk". In practice, disaster risk reduction is the "policy objective" of disaster risk management, and its goals and objectives should be defined in disaster risk reduction strategies and plans, defined on the basis of

[2]Hazards may be natural or anthropogenic in origin. Natural hazards are predominantly associated with natural processes and phenomena. Anthropogenic hazards, or human-induced hazards, are induced entirely or predominantly by human activities and choices. This term does not include the occurrence or risk of armed conflicts and other situations of social instability or tension which are subject to international humanitarian law and national legislation. Hazards include biological, environmental, geophysical, hydrometeorological and technological processes and phenomena.

a reliable disaster risk assessment to, at least, "contribute to strengthening resilience and therefore to the achievement of sustainable development".

The third basic definition is that of *resilience*, which, when one is referring to disasters, is described by the same UN working group as "the ability of a system, community or society exposed to hazards to resist, absorb, accommodate, adapt to, transform and recover from the effects of a hazard in a timely and efficient manner, including through the preservation and restoration of its essential basic structures and functions through risk management". Concerning resilience, rather more than in the case of other terms, it is relevant to add and underline that in literature there is no single consistently agreed definition, owing to its multifaceted nature (Tiernan et al. 2019). Nevertheless, we can be confident that for our domain this UN definition can be correctly applied.

Finally, it is important to emphasize that *residual risk* is not something negligible. On the contrary, it is "the disaster risk that remains even when effective disaster risk reduction measures are in place, and for which emergency response and recovery capacities must be maintained". Appropriately, and in parallel, residual risk should be acceptable or, sometimes, even tolerable, but the two terms are not always synonyms, where *acceptable risk* is not an easy, or banal, concept, summarized by the UN report specifying that "the extent to which a disaster risk is deemed acceptable or tolerable depends on existing social, economic, political, cultural, technical and environmental conditions". Generally, the level of acceptability of a risk is higher in a poor country than in a rich one (Hallegatte et al. 2017).

4.2.2 Data on Disasters

The maximum understanding of the potential consequences and losses deriving from threats affecting cities and communities is one of the bases for a modern and sustainable risk management, being a very significant part of the reconstruction of event scenarios. Risk analysis and assessment are therefore crucial to the implementation of risk reduction and resilience strategies, independently of the extent of the territory considered and studied. In this way, the analytical reading of the potential natural and anthropic calamitous phenomena will allow the monitoring of their evolution and impact in an objective way, to verify the effectiveness of policies and solutions adopted. A basic knowledge to handle data on disaster is therefore needed and must be put in the DRM tool-box.

Assuming as correct this point of view, one can observe that international agencies, institutions and companies, who dedicate their mission to disaster risk management, periodically examine information about the occurrence and severity of disasters, often starting from global data. This kind of statistics regularly underlines important evidence, showing significant human costs and economic losses (often in relation to gross domestic product—GDP) and fairly regularly highlights alarming tendencies or worries.

Before reviewing this kind of data, one must emphasize the need to handle global and local statistics attentively, in order to extrapolate correct and, whenever possible, reliable and undeniable information.

Primarily, it is important to know that data on disasters are usually reported in different manners, as data on damages, or as data on number or frequencies of event and type of disaster; case by case they are geographical (or georeferenced), more or less detailed or aggregated and/or illustrated in absolute values or percentage of other numbers (i.e. GDP).

When dealing with data on damages, it is a prerequisite to distinguish between types of damage which can be *direct* or *indirect* and *tangible* or *intangible*, as they are defined in literature (Messner et al. 2007; Thieken et al. 2008; Merz et al. 2010). This classification is essential since it is related to data which are collectable as referred to every single disastrous event, being concretely the elements of which databases are composed.

Direct damage occurs as a result of the physical effect of the calamitous phenomena on damageable goods; *indirect damage* is caused by the interruption of human activities and is related to the additional costs induced by the emergency.

Tangible damage is represented by all the economic losses that can be evaluated by applying monetary units. *Intangible damages* are those for which it is almost impossible to evaluate in monetary terms: logically, they can be both direct and indirect.

Combining these cases, one obtains four classes of damages. *Direct tangible damage*, as, for example, that which occurs for private buildings and contents; destruction of infrastructure such as roads, railroads; erosion of agricultural soil; destruction of harvest; damage to livestock; evacuation and rescue measures; business interruption inside the stricken area; clean-up costs (often this kind of damages is measured as replacement or repair values). *Direct intangible damage*: loss of life; injuries; loss of memorabilia; psychological distress, damage to cultural heritage and historical assets; negative effects on ecosystems and environmental degradation or pollution. *Indirect tangible damage*: disruption of public services outside the stricken area; production losses induced in the activity of companies outside the stricken area (e.g. suppliers of damaged companies); suspension of commercial activities; cost of traffic disruption; loss of tax revenue owing to migration of companies in the aftermath of disasters (and other long-term macroeconomic effects). *Indirect intangible damage*: psychological trauma; loss of trust in local and central government authorities; loss of reputation; social and political destabilization; reduction in the quality of life.

Once this classification is fixed, some first observations on disaster data are possible, starting from the basic consciousness that no single kind of data can autonomously describe the consequences of a disaster; consequently, this assumption implies that *many times (or nearly always) (single) data on disaster result as underestimated* in describing the effect of a disaster on a territory, or community, or city.

Moving on to aggregated statistics, difficulties multiply. The first problem is to establish criteria to select events to be included in the series of data. This is true, at the base of the pyramid of data, when one deals with records or the products of direct observations (Rowley and Hartley 2006), collects evidence or puts together the needed inputs, owing to a lack of thresholds which help in distinguishing what a disaster is—or a lack of indicators which set minimal requirements to recognize an event as a disaster. Again, even the more so, in the upper parts of the pyramid of data (the parts of information and knowledge), when one is in charge of the task of assembling data coming from different sources and needs to set additional selection thresholds to have a reasonable (i.e. not too large) amount of data to be handled. Concerning this issue, a classic and simple way to illustrate this type of selection, is that suggested also by the Centre for Research on the Epidemiology of Disasters (CRED), when dealing with the interpretation of disaster data. The reader of data on disasters must be "aware of the complex interactions between natural hazards and human vulnerabilities: the most violent storm over an uninhabited region will not be a disaster if no people are harmed, while even a small tsunami hitting a populous city with no early warning system can quickly become a major disaster" (CRED 2017).

Furthermore, when the goal is to aggregate data regarding a long period of time, something that is often desirable in this domain of study, the variability in reporting and recording data must also be taken into account, especially if the aim of a study is to extrapolate the trends or sizes of the phenomena. In reality, in the past no one systematically collected data on disasters, as is done today. Moreover, only in a woefully small number of countries is it possible to partially cover this lack of knowledge through the methods of historical enquiries, as, for instance, has been done in Italy: there are the works related to the *Italian Parametric Earthquake Catalogue* referring to the time period 1000–2014 (Rovida et al. 2016) and the database of landslides that occurred in Italy between 1279 and 1999 and caused deaths, missing people, injuries and homelessness (Guzzetti 2000).

By way of example, the case of the reinsurance company Munich Re is interesting: its collection contains about 26 catastrophes per year for the period 1900–1950, followed by 100 natural disasters added to the database every year for the period 1951–1980 (mainly in western countries), while starting from 1980, the number of records rises to about 700 per year (coming from all over the world) (Hoeppe 2016). Logically, at least the major part of this apparent so-heavy increase in the frequency of disasters is due to the recording method of data collection, rather than to real occurrence of phenomena and of their intensity and extent.

All those aspects considered, researchers and analysts suggest the years 1970 or 1980 as the years to be considered as an adequate starting point for the study of the dimensions of natural disasters at a global level, for instance from an economic point of view (IPCC 2012; CRED 2017; Boccard 2018; Swiss Re 2019).

When the goal is to analyze global data, which is the maximum level of aggregation, data sets face other additional kinds of complications. When the sources of the data are insurance companies (data generally collected to a high standard, thanks to the work of their agents), low rates of insurance coverage generate a distortion of geographical distribution of data and certain degrees of underestimation. When the sources of the data are international or intergovernmental agencies, the problem often may be the lack of information transmission (bottom-up process), due to higher or lower cooperative predisposition of local/national administrations, or, when the work is a data collection from a central service, the problem may be related to non-English speaking countries, even more so when its news networks are poorly developed. In addition, sometimes, mechanisms are influenced by the fact that some countries or territories voluntarily overestimate data on the consequences of disasters as a strategy, for example, in an attempt to obtain more financial aid, whereas some kinds of political regime may prefer to underestimate data on consequences of disasters for reasons of internal national consensus (Boccard 2018).

In order to overcome these kinds of problems, the United Nations Office for Disaster Risk Reduction (UNISDR), together with its partners, is working with governments to establish robust national disaster loss databases, as part of the jointly developed activities of the Sendai Framework for Disaster Risk Reduction 2015–2030: this derives from the awareness that improved record-keeping and standardized loss indicators will surely help in the capability to combat disasters (CRED 2017).

Having this delicate framework clear, one can enter in to contact with global data on disaster, keeping always in mind that handling economic temporal series of data, unavoidably, asks for at least two other factors to be taken into account, being: inflation, to be adjusted via country-specific consumer price index; fluctuations in exchange rates between local currencies and US$ or UE€. In addition, often, a normalization with corresponding local GDP helps in understanding the magnitude of each event.

Figures 4.1 and 4.2 show data on natural disaster events and data on losses, taken from four sources of high international reputation: the data of the Emergency Events Database (EM-DAT) for the Centre for Research on the Epidemiology of Disasters—CRED (www.cred.be) of the Institute of Health and Society (Université Catholique

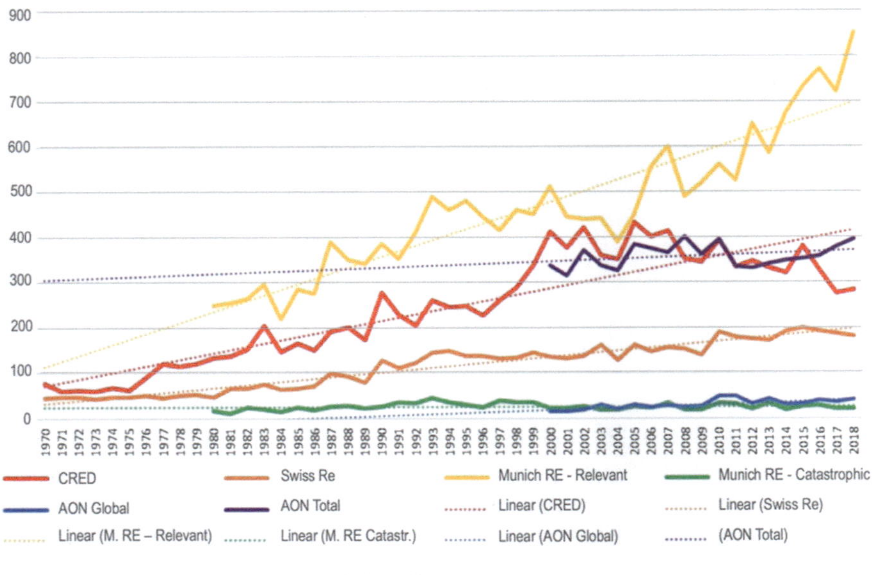

Fig. 4.1 Natural disaster events at global level

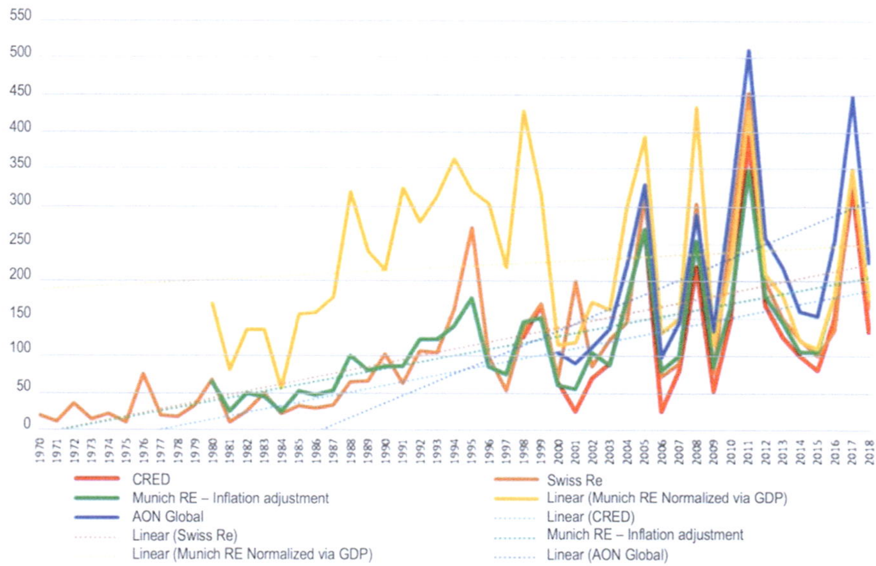

Fig. 4.2 Disaster global losses at a global level in the US dollars billion

de Louvain, Belgium) and the data of three international insurance/reinsurance and risk transfer companies such as Munich Re (www.munichre.com), Swiss RE (www.swissre.com) and AON (www.aon.com). Since events and losses are selected and estimated in partially or significantly different ways,[3] they are not always directly comparable, but they are to be considered as indications of the general orders of magnitude and trends, which can be taken to be statistically significative. It is important

[3]It is important to signal that in Figs. 4.1 and 4.2, data by sources are different also because different selection criteria were used. For instance, in Fig. 4.1, Munich Re Catastrophic series includes an event when it has caused ≥1000 fatalities and/or produced normalized losses ≥US$ 100 m, 300 m, 1 bn, or 3 bn, depending on the assigned World Bank income group of the affected country. Munich Re Relevant series includes an event when it has caused at least one fatality and/or produced normalized losses ≥US$ 100 k, 300 k, 1 m, or 3 m, once again depending on the assigned World Bank income group of the affected country (Munich Re 2019). In Fig. 4.2, AON Global series includes events which reached the billion-dollar-plus (USD) threshold after being adjusted for inflation based on the 2018 US Consumer Price Index and AON total events fixes the following criteria: economic loss: USD50 million; insured loss: USD25 million; fatalities: 10; injured: 50; homes and structures damaged or filed claims: 2000, economic data including only direct event impacts such as physical damage to property, infrastructure, and agriculture, net loss direct business interruption, and any mitigation or restoration costs for the event, not including secondary or tertiary losses, such as supply chain costs, monetary capital assets or values tied to loss of life (AON 2018). About Swiss Re data economic losses are all the financial losses directly attributable to a major event, i.e. damage to buildings, infrastructure, vehicles, etc. The term also includes losses due to business interruption as a direct consequence of the property damage but does not include indirect financial losses—i.e. loss of earnings by suppliers due to disabled businesses, estimated shortfalls in GDP and non-economic losses, such as loss of reputation or impaired quality of life.

to note that all linear interpolations of data series in Figs. 4.1 and 4.2 are growing, having positive angular coefficients, AON Global and Munich Re Catastrophic of Fig. 4.1 and Munich Re normalized via GDP of Fig. 4.2 series included. These three series are particularly significant because the first two are related to the largest events, the data of which are surely less affected by the recording methodology, and because the third corrects the series by taking into account the economic context of the stricken lands. That said, this overall tendency of the trends regarding events (Fig. 4.1) and losses (Fig. 4.2), offered by the data of high-reputation entities, can be considered as a creditable representation of reality.

Other data frequently reported in publications and reports on disasters are those related to: man-made disaster events (apparently decreasing); the number of victims (casualties or fatalities, without a recorded trend); affected population (injured, homeless, displaced or in need of emergency assistance); and insured losses (as percentage of the total). These kinds of data are very important for the analytical comprehension of events, but less so if one is interested in global tendencies, owing to the high level of connection with the local context (population density, buildings vulnerability, insurance market penetration, etc.).

4.2.3 Quantifying Risks

To operate with the aim of understanding the potential consequences of threats affecting a specific city, or community, or territory, or, in other words, to put into practice a risk analysis and assessment, independently of the extent of the piece of land considered, a green planner has the necessity to put in his DRM tool-box the fundamentals of quantifying risk in an unambiguous manner and on the basis of a wide methodological consensus.

With this aim, it must, first of all, be illustrated what is a *disaster risk* in order to allow the possibility of a specific consideration, by using the definition of the already cited expert working group on disaster risk reduction terminology of the General Assembly of the UN: a disaster is the potential loss of life, injury, or destroyed or damaged assets (that is: human, material, economic and environmental losses and impacts) which could occur to a system, society or a community in a specific period of time, determined probabilistically as a function of *hazard* (H), *vulnerability* (V), *exposure* (E) and *capacity* (C). Definition, for the use of which, in addition to what has been stated above it is necessary to fix these concepts: *hazard*, that is a process, phenomenon or human activity that may cause loss of life, injury or other health impacts, property damage, social and economic disruption or environmental degradation; *vulnerability*, that is the conditions determined by physical, social, economic and environmental factors or processes which increase the susceptibility of an individual, a community, assets or systems to the impacts of hazards; *exposure*, that is the situation of people, infrastructure, housing, production capacities and other tangible human assets located in hazard-prone areas (i.e. for instance, the number of people or types of assets in a given area); *capacity*, that is the ability of people and systems, using available skills and resources, to manage risk or disasters.

The simplest form of computing a risk is to multiply together the three components of hazard, vulnerability and exposure (Kron 2002), descending from the more complete integral form based on probability density function of a variable linked to the costs or losses caused by the phenomenon intensity, where costs or losses are linked to vulnerability and exposure.

Usually, the analytical calculation of the integral form is rarely possible, since often data and probability density function are available in poor quality, or incomplete, or even absent. Coherently, when using (1), a green planner must handle it carefully, since a hazard does not manifest itself in a single manner with a given probability of occurrence, but at least in several levels of intensity (also with the possibility of infinite variations).

$$R = H \times V \times E \tag{4.1}$$

To handle the (4.1) carefully means to scrupulously select the scenario of event to be considered in order to be ready with an action of defence or retrofit of an asset, with the purpose of efficiently and effectively obtaining the desired results or to reach the fixed objective of risk reduction by the strategy elaborated. In practice, the selection of a scenario of an event is the act of obtaining the measure of one impact among the many which may potentially strike the asset in question.

Some authors add to (4.1), albeit in different manners, the term of capacity (C), term able to emphasize and take account of the role of a community in reacting in a more or less effective manner to the disaster, as, for instance, in (4.2) (Roberts et al. 2009). In any case, to use the term C in a numerical form is no simple task, since it is multi-dimensional and is not easily quantified (Birkman et al. 2013).

$$R = \frac{H \times V \times E}{C} \tag{4.2}$$

4.3 At the Roots of the Matter and of the Connections of International Strategies

Among the various aspects of sustainable development (SD) of built environment to be considered when the aim is to provide a unique framework of reference both for practitioners and researchers of green urban planning, it is nowadays clear that it is impossible not to include those related to the disaster risk themes. Disasters are a transversal and serious concern, already integrated in the agenda of sustainable development since the World Summit on Sustainable Development held in Johannesburg, South Africa, 26 August–4 September 2002. That summit recognized the threat to communities deriving from disasters, during which, coherently, proactive vulnerability reduction was instituted as a strategy for disaster loss reduction. This pronouncement has later been proven correct not only by the work of the scientific community (Thomalla et al. 2006; McBean 2012) but also from the subsequent debates of intergovernmental and public institutions. This has been definitively been made clear after 2015, the year when three key agreements, having numerous parallels and connections among them (Mysiak 2016; Kelman 2017), were approved as UN and intergovernmental decisions: the Sendai Framework for Disaster Risk Reduction 2015–2030; the Paris Agreement 2015; the 2030 Agenda for Sustainable Development.

4.3.1 International Debate and up to Now Outlined Countermeasures to Face Disaster Risks at a Global Level

Joint international efforts to cope with disasters can be considered to have commenced on 14 December 1971, when the General Assembly of United Nations approved the resolution 2816 establishing the United Nations Disaster Relief Office (UNDRO), with the aim of mobilizing, directing and coordinating relief activities of the various organizations of the UN system and other donors, or intergovernmental/non-governmental, in response to a request of assistance from a State stricken by a disaster.

In subsequent years, international research and the resulting debates showed that an approach to the problem of disasters exclusively based on responding to their effects after phenomena had occurred was insufficient and extremely expensive in terms of losses of lives, material goods and cultural heritage. Today, the UNDRO no longer exists, having been incorporated, since April 1992, in what today is the United Nations Office for the Coordination of Humanitarian Affairs (UNOCHA). However, simultaneously, the UN General Assembly decided to launch the International Decade for Natural Disaster Reduction (IDNDR). This was the decade commencing 1 January 1990, with the aim, in cooperation with UNDRO/UNOCHA, both to promote the adoption and global utilization of advanced and universally applicable approaches and techniques to disaster risk reduction, and to demonstrate the positive connection between disaster risk reduction and economic plus social development.

At the end of the International Decade for Natural Disaster Reduction, in 1999, a United Nations' Office for Disaster Risk Reduction (UNDRR) was established, as a dedicated secretariat to facilitate the implementation of International Strategy for Disaster Reduction (ISDR). In 2000, the Millennium Declaration recognized, among many other critical issues, the specific threat to development caused by disasters, and, consequently, requested the global community to "intensify (our) collective efforts to reduce the number and effects of natural hazards and man-made disasters" (https://www.who.int/topics/millennium_development_goals/about/en/).

In 2005, the World Conference for Disaster Reduction, held in Kobe, Hyogo, Japan, produced the Hyogo Framework for Action 2005–2015 (HFA) "Building the Resilience of Nations and Communities to Disasters" (UNISDR 2005).

The HFA established set of indicators of disaster risk and vulnerability at national and subnational scales, with the aim of empowering decision-makers to assess the effects of disasters and to engage their societies on efforts towards disaster risk reduction. Advances were achieved in strengthening disaster preparedness, response and early warning systems, but advancement was inadequate in most countries for handling the trends of rapid urbanization and population growth in hazard-prone areas (CRED 2017).

Ten years later, at the Third UN World Conference on Disaster Risk Reduction in Japan in March 2015, another agenda of purposes was adopted, the Sendai Framework for Disaster Risk Reduction 2015–2030 (SFDRR). This new Framework outlines four *priorities for action* to prevent new and reduce existing disaster risks: understanding disaster risk; strengthening disaster risk governance to manage disaster risk; investing in disaster reduction for resilience; enhancing disaster preparedness for effective response and to "Build Back Better" in recovery, rehabilitation and reconstruction. On the basis of these priorities and to support the assessment of global progress, seven strategic Global Targets have been agreed upon and complemented by thirty-eight (38) indicators, conceived to align the implementation of the Sendai Framework with both the UN's global Sustainable Development Goals and its targets, for several of which baseline data were unavailable or did not have clear numerical targets, and with the Paris Agreement on climate change.

The Seven Global Targets (UNISDR 2015a) are:

(a) *Substantially reduce global disaster mortality by 2030, aiming to lower average per 100,000 global mortality rate in the decade 2020–2030 compared to the period 2005–2015.*
(b) *Substantially reduce the number of affected people globally by 2030, aiming to lower average global figure per 100,000 in the decade 2020–2030 compared to the period 2005–2015.*
(c) *Reduce direct disaster economic loss in relation to global gross domestic product (GDP) by 2030.*
(d) *Substantially reduce disaster damage to critical infrastructure and disruption of basic services, among them health and educational facilities, including through developing their resilience by 2030.*
(e) *Substantially increase the number of countries with national and local disaster risk reduction strategies by 2020.*
(f) *Substantially enhance international cooperation to developing countries through adequate and sustainable support to complement their national actions for implementation of this Framework by 2030.*
(g) *Substantially increase the availability of and access to multi-hazard early warning systems and disaster risk information and assessments to the people by 2030.*

Furthermore, the implementation of the Framework is guided by some principles to be adapted to national circumstances and laws as well as to international obligations. Among these principles, not integrally quotable here, two are noteworthy for the purposes discussed (UNISDR 2015b):

- *Coherence of disaster risk reduction and sustainable development policies, plans, practices and mechanisms, across different sectors;*
- *Addressing underlying risk factors cost-effectively through investment versus relying primary on post-disaster response and recovery.*

Among 38 indicators set up to monitor the Sendai Framework implementation progress (UNISDR 2017), some of them (nine) have been selected, definitively able to show to what extent disaster risk reduction is essential to achieve sustainable development, and, more specifically, sustainable urban planning.

The first group of selected indicators which should be familiar to a green urban planner is associated with target (b) which is that of reducing the number of affected people. Specifically, they constitute a sort of sub-group of those related to the people whose houses were damaged or destroyed. In detail they are:

Indicator B-3: Number of people whose damaged dwellings were attributed to disasters.

- *Sub-indicator B-3a: Number of dwellings/houses damaged attributed to disasters.*

Indicator B-4: Number of people whose destroyed dwellings were attributed to disasters.

- *Sub-indicator B-4a: Number of dwellings/houses destroyed attributed to disasters.*

Worthy of note is the definition on the basis of which damaged dwellings/houses differ from destroyed ones, being housing units which may continue to be habitable.

The second group of selected indicators with which a green urban planner should have familiarity is associated with target (c) which is that of reducing the direct economic loss of disasters. This is the sub-group of those indicators connected to the assessment of built environment losses, namely:

Indicator C-4: Direct economic loss in the housing sector attributed to disasters.
Indicator C-5: Direct economic loss resulting from damaged or destroyed critical infrastructure attributed to disasters.
Indicator C-6: Direct economic loss to cultural heritage damaged or destroyed attributed to disasters.

The assessment of direct economic losses in the housing sector (indicator C-4) is to be calculated on the basis of replacement costs and average sizes of stricken units. The assessment of direct economic losses to critical infrastructures (C-5) is obtained in different ways on the basis of rehabilitation and reconstruction costs in cases concerning buildings, linear infrastructures (among which roads, but also damaged flood protection walls/barriers) and the remaining critical elements (such as airports, ports, bridges). For the losses of cultural heritage (C-6), being intangible losses, in order to calculate, whenever possible, at least a portion of the direct economic loss, some sub-indicators is proposed. These are divided into two groups, being non-movable assets (buildings, monuments) and movable elements (art, historical artefacts). These computations are made on the basis of the cost of rehabilitating, recovering and restoring them to a standard similar to that of the pre-disaster situation or on the basis of the cost of replacing or acquisition of a new item with similar functions or characteristics. When the computation of direct economic loss, even if only partial, is impossible, the guidelines propose indicators of the magnitude of physical damage, of the number of movable and non-movable assets damaged or destroyed.

The third group of selected indicators that a green urban planner must know and employ is associated with target (d) which is that of reducing disaster damage to critical infrastructure and disruption of basic services (among which health and educational facilities). This sub-group consists of the following indicators:

Indicator D-2: Number of destroyed or damaged health facilities attributed to disasters.
Indicator D-3: Number of destroyed or damaged educational facilities attributed to disasters.
Indicator D-4: Number of other destroyed or damaged critical infrastructure units and facilities attributed to disasters.
Indicator D-8: Number of disruptions to other basic services attributed to disasters.

The computation of indicator D-4 includes protective infrastructure such as canals, dams, dykes and other water regulation mechanisms or cyclone and tornado shelter systems, whereas computation of indicator D-8 should comprise services that are needed for all societies to function effectively or appropriately, such as, for example, water supply, sewerage systems, electricity supply, telecommunications, waste management, emergency operations centres, the selection of which is left to Member States, who must insert the description thereof within the accompanying metadata.

A concrete way to use these collected data will be that of adopting and implementing *national and local disaster risk reduction strategies and plans (DRRP/S)*, to be made across different timescales, using cited targets indicators, with the aim to prevent the creation of risk, the reduction of existing risk and the strengthening of economic, social, health and environmental resilience (UNISDR 2015a).

National and local disaster risk reduction plans, when and where possible, are expressively indicated as the instrument to be linked to sustainable development and climate change adaptation plans (UNISDR 2015a).

4.3.2 Climate Change and Disasters (to Avoid Any Mismatch)

Climate change is a huge issue, about which it is beyond the scope of this chapter to aspire to a complete exposition. Nonetheless, it is useful to fix some elements in order to properly place the topic in relationship with the theme of disaster risk, considered from the viewpoint of the built environment. Indeed, climate change poses additional challenges and concerns to disaster risk management tasks (Blakely 2007), for at least two sets of reasons. The first set or group is well described by the assessments of the International Panel of Climate Change (IPCC),[4] which, in a specific work, highlighted how "climate change is very likely[5] to increase the occurrence and vary the location of some physical events, which in turn will affect the exposure faced by many communities, as well as their vulnerability" (IPCC 2012). The second group, once again as illustrated by IPCC, is related to the implication of climate change effects, that make it more difficult "to anticipate, evaluate and communicate both probabilities and consequences that contribute to disaster risk, in particular that associated with extreme events" (IPCC 2012).

As is well known, in order to avoid runaway climate change and with the objective of preventing "dangerous" human interference with the climate system,[6] the United Nations Framework Convention on Climate Change (UNFCCC), entered into force on 21 March 1994, has been suggesting since its initiation a strategy combining *mitigation* (the reduction of emissions of greenhouse gases, GHG, that contribute to climate change) and *adaptation* (the process of responding to actual or expected effects induced as a result of climate change) policies. In fact, as stated by UNFCCC, evidence of climate change impacts is strongest for natural systems, whereas some impacts on human systems have already been attributed to climate change and indeed detected, although the effects are still not always distinguishable from other influences, such as social, economic or environmental factors (i.e. pollution or change in land use).

The UNFCCC's first objective, concerning mitigation, is to stabilize greenhouse gas concentrations "at a level that would prevent dangerous anthropogenic (or human-induced) interference with the climate system", setting out that "such a level should be achieved within a time frame sufficient to allow ecosystems to adapt naturally to climate change, to ensure that food production is not threatened, and to enable

[4]The IPCC is the UN body for assessing the science related to climate change. It was set up in 1988 by the World Meteorological Organization and United Nations Environment Programme to provide policymakers with regular assessments of the scientific basis of climate change, its impacts and future risks, and options for adaptation and mitigation. The IPCC does not conduct its own research. It identifies where there is agreement in the scientific community (who works autonomously), where there are differences of opinion and where further research is needed.

[5]It is important to underline that the terms "very likely" indicate the assessed likelihood of 90–100% probability.

[6]It is interesting to note that in 1994, when the UNFCCC took effect, there was less scientific evidence available than now. More recent IPCC documents confirm challenges and concerns (IPCC 2014). Nowadays, the works for the sixth Assessment Report (AR6) are ongoing, foreseen to be presented in 2021/22.

economic development to proceed in a sustainable manner". The Kyoto Protocol is the international agreement adopted in 1997 (and updated in Doha, 2012) which fixes countries' emission reduction targets and mechanisms to stimulate green investments and the achievement of emission targets in a cost-effective manner.

On the basis of the third IPCC Assessment Report (AR3), also climate change adaptation (CCA) gained in importance, from the Cancun Agreements of 2010 until Paris Agreement 2015[7] (Cfr. Art. 7), where *national adaptation plans* (NAP) were identified as the means for setting medium- and long-term adaptation needs and implementing strategies to address them: this was based on the belief that the faster the climate will change, the longer, more difficult and expensive adaptation efforts will be. Every NAP is based on an initial assessment of impacts, vulnerability and risks (among which disaster risks). However, here, on the subject of risks, it is important to underline that Art. 8 of the Paris Agreement on loss and damage, highlighting the importance of enhancing understanding and actions to avert, minimize and address loss and damage due to the effects of climate change, includes extreme weather events and slow-onset events, as well as the role of sustainable development in reducing the risk of loss and damage (UNFCCC 2015).

That description provides the opportunity to specify, once more on the basis of the terminology of the UN working group on indicators and terminology relating to disaster risk reduction, that "*a slow-onset disaster is defined as one that emerges gradually over time. Slow-onset disasters could be associated with, e.g. drought, desertification, sea-level rise, epidemic disease*", whereas "*a sudden-onset disaster is one triggered by a hazardous event that emerges quickly or unexpectedly. Sudden-onset disasters could be associated with, e.g. earthquake, volcanic eruption, flash flood, chemical explosion, critical infrastructure failure, transport accident*". Slow-onset disasters are mainly coped with sustainable development and climate change mitigation/adaptation strategies; sudden-onset disasters are mainly covered by DRM. Nonetheless, the worsening of a slow-onset disaster can in turn worsen (increase) the probability of the occurrence of sudden-onset disasters—this is the case, for instance, of the rise in sea level caused by climate change (i.e. the melting ice caps) which is increasing the depth of the floodwaters in coastal floods, or of more frequent occurrence of drought and wildfire events, owing to consistently warmer temperatures (Swiss Re 2019).

Despite the different strategies to cope with slow and sudden-onset disasters, and considering the connections which have been shown, it is important to underline that CCA and DRM provide a series of complementary and combinable approaches for dealing with risks of climate extremes and climate-related disasters, thus opening the possibility of deriving benefit from the synergies in preparing and implementing their plans. Such a possibility is not without its difficulties (Birkman and von

[7]At the Twenty-first Conference of the Parties (COP) of UNFCCC held in Paris on 2015, the agreement (entered into force in 2016) charts new aims to strengthen the long-term global response to climate change, addressing the need to limit the rise of global temperatures, but including also the increasing of the ability of countries to deal with the impacts of climate change, starting from the consideration that adaptation is a global challenge to be faced with local, subnational, national, regional and international dimensions to protect people, livelihoods and ecosystems.

Teichman 2010), nowadays principally owing to two kinds of scales of mismatches, spatial and functional. The first mismatch is owing to the fact that climate change issues have been investigated mostly on a global scale, whereas disasters have been studied focusing on hazards, vulnerabilities and exposures at meso- or local/micro-scale. The second mismatch is due to the fact that, usually, in most countries, climate change issues have been tackled by the environment ministries and meteorological services, whereas disaster risk management often is under the responsibility of different entities, such as the ministry of the interior or defence and specialized agencies, mainly focused on post-disaster needs. These two differences are understandable since disasters occur when hazards—storms, earthquakes, industrial accidents, volcanic eruptions, etc.—impact on specific vulnerable people and assets. Whereas vulnerabilities and hazards increase for many causes, including, as already stated, population growth, urban development in risk-exposed areas, changes in land use, and also, but only as one among many causes, because of climate change. It seems to be an easy consideration, but far too often, despite the cited separation at the administrative level, at a political level, climate change and disasters risks are considered jointly as a unique problem, when, on the contrary, they are two separate issues which only partially overlap.

4.3.3 The Fruitful Link Between DRM and the 2030 Agenda for Sustainable Development

As already mentioned in the introduction to this volume, in September 2015, many countries adopted the document "Transforming Our World: the 2030 Agenda for Sustainable Development" (UN 2015), with its 17 Sustainable Development Goals (SDGs) and 169 associated targets to be developed as soon as practicable through national responses (NR).

In this document, the "Heads of State and Government and High Representatives" who decided to adopt the Agenda also "reaffirm" explicitly the outcomes, among others, of the Third United Nations World Conference on Disaster Risk Reduction of Sendai, and, when speaking of the new Agenda, they add that they are "determined" to promote resilience and disaster risk reduction.

Furthermore, examining this document from the point of view of this chapter, it is to be observed that DRM is directly recalled as theme connected to sustainability at different stages. Namely: in *Goal 1* (about poverty) and *Goal 2* (about food and agriculture), where disasters are factors which are aggravating the main problem to be faced[8]; in *Goal 11* (about cities and human settlements) and in *Goal 13* (about climate

[8]In *Goal 1. End poverty in all its forms everywhere*, target 1.5 underlines how for poor and vulnerable people the need to be protected from disasters is even more urgent, affirming namely as a target: "*By 2030, build the resilience of the poor and those in vulnerable situations and reduce their exposure and vulnerability to climate-related extreme events and other economic, social and environmental shocks and disasters*". In *Goal 2. End hunger, achieve food security and improved nutrition and promote sustainable agriculture*, target *2.4* fixes as a key problem also that of agriculture, linked to

change and its impact), where the fight against disaster is part of the main problem. Herein, the theme of this chapter is seen both when starting from consideration of the stricken areas and goods (the built environment), and when starting from consideration of the phenomena.

Goal 11. Make cities and human settlements inclusive, safe, resilient and sustainable explicitly addresses the topic in three targets, among which *11.5* is general, since it looks at the reduction of deaths and losses[9]; *11.b* is essentially a reaffirmation (or a promotion) of Sendai Framework[10]; *11.c* is specifically thought out in connection with the problem of buildings in poor countries, to be supported, technically and financially, in achieving sustainable and resilient buildings utilizing local materials.[11]

Goal 13. Take urgent action to combat climate change and its impacts, about which the document acknowledges "*that the UNFCCC is the primary international, intergovernmental forum for negotiating the global response to climate change*" sets two targets strictly connected with disasters. Namely, article *13.1*, in which the document unites the efforts to be made to fight climate-related hazards and natural disasters[12]; *13.3*, which underlines the need to improve education, awareness, human and institutional capacity also in climate change adaptation, impact reduction and early warning.[13]

Examining the other Sustainable Development Goals, one should note that DRM may benefit indirectly from an additional number of other aspects of an improved sustainability. First of all, in the author's opinion, *Goal 4*, about the quality of education,[14] is a key factor enabling and fostering the possibility of a community or a city to better act and fight against disasters in every aspect which must be considered. Furthermore, *Goals 6*, *7*, *9* and *15*, respectively handling water and sanitation,

---

disasters by the connection of the maintaining and the proper management of land and soil, fixing as a target: "*By 2030, ensure sustainable food production systems and implement resilient agricultural practices that increase productivity and production, that help maintain ecosystems, that strengthen capacity for adaptation to climate change, extreme weather, drought, flooding and other disasters and that progressively improve land and soil quality*".

[9] *11.5 By 2030, significantly reduce the number of deaths and the number of people affected and substantially decrease the direct economic losses relative to global gross domestic product caused by disasters, including water-related disasters, with a focus on protecting the poor and people in vulnerable situations.*

[10] *11.b By 2020, substantially increase the number of cities and human settlements adopting and implementing integrated policies and plans towards inclusion, resource efficiency, mitigation and adaptation to climate change, resilience to disasters, and develop and implement, in line with the Sendai Framework for Disaster Risk Reduction 2015–2030, holistic disaster risk management at all levels.*

[11] *11.c Support least developed countries, including through financial and technical assistance, in building sustainable and resilient buildings utilizing local materials.*

[12] *13.1 Strengthen resilience and adaptive capacity to climate-related hazards and natural disasters in all countries.*

[13] *13.3 Improve education, awareness raising and human and institutional capacity on climate change mitigation, adaptation, impact reduction and early warning.*

[14] *Goal 4. Ensure inclusive and equitable quality education and promote lifelong learning opportunities for all.*

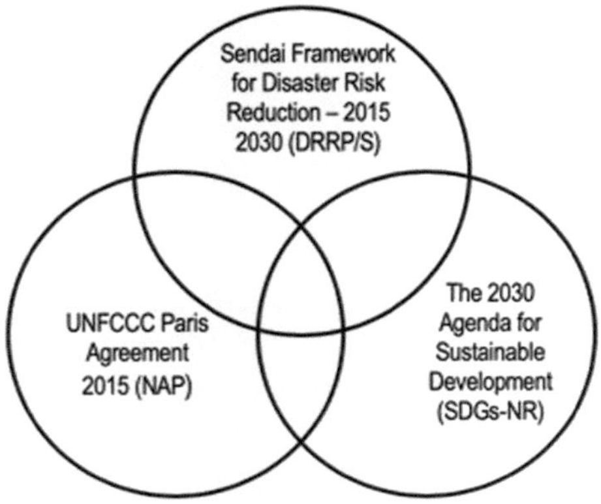

Fig. 4.3 Only partially overlapping relationship among Disaster Risk Reduction Plans and Strategies (DRRP/S), National Adaptation Plan (NAP) on Climate Change and Sustainable Development Goals National Responses

energy, resilient infrastructure and terrestrial ecosystems, forests, desertification and land degradation,[15] compose a physical framework for stronger environments, both natural and built, as basic preconditions to have and obtain cities and communities less susceptible to be stricken, wounded or deeply troubled by disasters.

Thus, SDGs show and definitely fix how DRM and CCA are crucial parts, among other elements, for planning sustainable urban development and for the quality of life of people in cities and communities, whose local authorities, together with each Government, must decide how Global Targets should be incorporated into planning preparation and assessment (see Fig. 4.3).

4.4 Improve the Ability to Plan Urban Spaces Able to Face Disasters

Despite the progress obtained thanks to the SFDRR 2015–2030, to the Paris Agreement and to the SDGs, the need of operational indicators and frameworks at local level to monitor progress in DRM plans and in planning and building-connected urban actions is still a key issue for fostering sustainable built environments (Wamsler 2004; Etinay et al. 2018). This goes together with the specific need of DRM to have more reliable data to consider such variables as hazard, exposure and vulnerability. The fulfilment of these needs represents a link between the development of risk-resilient communities (DrP at local level) and the connected DRM actions of implementing

[15]*Goal 6. Ensure availability and sustainable management of water and sanitation for all; Goal 7. Ensure access to affordable, reliable, sustainable and modern energy for all; Goal 9. Build resilient infrastructure, promote inclusive and sustainable industrialization and foster innovation; Goal 15. Protect, restore and promote sustainable use of terrestrial ecosystems, sustainably manage forests, combat desertification, and halt and reverse land degradation and halt biodiversity loss.*

DRA activities and DRR strategies. The issue is addressed by international agencies and researchers with many initiatives, about which some examples at different scales have been selected, with the aim of elucidating the complexity and evolution of the theme.

4.4.1 Taking into Account Resilience Issues

As shown in the sub-paragraphs dedicated to the terminology and about quantifying risks, the consequences of disasters are associated not only with the direct and instantaneous physical effects of phenomena on assets but also with the ability to cope with the effects of a disaster and to recuperate functionally and perform services, at least partially, despite the damage. In this regard, with the aim to illustrate the complexity of issues related to DrP, two kinds of approaches are illustrated, one economic (at macroscale) and one technical and operational (at micro-scale), both useful for a green planner aiming at taking into account these issues. Resilient urban spaces are not only about the retrofit or creation of physical assets in an appropriate way and in suitable locations but also about the increase in the capacity of the social, institutional, environmental and economic resources in order to make built environment operational and flexible (Bosher 2014).

Intuitively, one can imagine that rich countries or areas are more resilient than poor ones. Indeed, poor populations may invest less in preparedness and risk reduction, live in lower-quality buildings and often be exposed to food insecurity. More accurately, the World Bank (WB), in a recent report, affirms that one dollar "*in losses does not mean the same thing to a rich person and a poor person*", and, simultaneously, that a disaster can "*have a negligible impact on a country's aggregate wealth or production if it affects people who own almost nothing and have very low incomes*" (Hallegatte et al. 2017). Starting from these assumptions, WB examines the different abilities of poor and non-poor people in responding to the effects of disasters.

If a disaster causes one million dollars of losses in consumption, it is not the same if it is perfectly shared out across the population or when it is concentrated only in the poorest segment of the population, more susceptible to losing well-being, because proportionally more affected in income and connected consumption: this is particularly so if one also considers the hypothesis of equivalent damages on assets (buildings, means of production, etc.), because of less insurance, reserves and alternatives. That is why WB suggests a system of measurement to capture "*overall effects on poor and non-poor people, even if the economic losses of poor people are small in absolute terms*", in order for it to be used in the analysis of disaster risk management projects and to counterbalance the trend which systematically drives traditional disaster risk reduction investments towards wealthier areas and individuals while undervaluing the options of strengthening resilience, such as adopting adaptive social safety nets.

As seen above, the damages that a disaster causes on assets, usually evaluated on the basis of monetary replacement or repair costs, are an incomplete form of measurement. If the same losses are concentrated on fewer or poorer people, they have a higher impact than the equivalent losses touching richer people or being distributed across a wider population. In order to better compare these situations, at international, national or local level, a simple model has been suggested. If one considers this higher impact as the *risk to well-being* (R_{wb}), we can modify (4.2) using (4.3), in which (SER) is the *socio-economic resilience*

$$R_{\mathrm{wb}} = \frac{H \times V \times E}{\mathrm{SER}} \tag{4.3}$$

and where socio-economic resilience SER is a numerical term obtained as quantification of the ratio between losses on assets and losses on well-being, in its turn derived from calculation of the connected loss in consumption. When losses in consumption are greater than losses on assets, then SER is smaller than 1 and R_{wb} is greater than R, and vice versa. So that the ranking of a list of stricken or endangered regions could change, signalling those more in need of aids or DRR intervention.

All that considered, WB adds that well-being losses can be lessened by reducing asset losses, in this way preserving productivity and preventing loss of consumption and well-being, through the application of a set of actions as a priority. It is interesting, here, to point out that among these priorities, as stated by WB, are to be found: *land use and urbanization plans*, which need to be risk-sensitive; *more robust infrastructure*, to be designed taking into consideration uncertainty in investment decisions; *building regulations and standards* to rehabilitate buildings with low resistance to natural hazards; *climate-smart agriculture, access to markets, and efficient transport* to strengthen food security. Through the application of this set of priorities, all countries, the richest included, regardless of their context, geography or richness, can reduce risk and increase resilience.

Although such set of priorities has been conceived to be put in practice and not only to be used in a wide-ranging or descriptive sense, it is still not fully self-operational. This situation is not rare when using and applying the term and concept of resilience since it is neither easy nor intuitive to derive working (or field) instructions from its definition. Bearing in mind the goal of going deeper into the details of the application of such a set of actions, or, more in general, of the concept of resilience, in order to guide green urban planners in designing and building urban resilience, some additional tools, or key elements, can be proposed.

As previously stated, resilience is a quality to be promoted for systems, communities or societies. To make that a reality it is, first of all, important to bear in mind that nowadays systems are ever more complex, linked and interdependent and, consequently, often there exists the potential situation that a failure of one element of the system could trigger other cascading failures owing to their interconnection (Kirshen et al. 2008). The assessment and the retrofit design of this kind of system, put in place in order to reduce the possibility or the probability to fail, must safeguard minimum levels of functionalities and ensure quick restoration and return to satisfactory performances. Such goals are not achievable only through the strength or the resistance of each individual component and element of the system, but also demand flexibility and functional diversification.

A framework scheme to be given to green planners to set the specific resilience strategies to be applied for cities and communities could be inspired by these fundamental characteristics selected by the literature in the field (Godschalk 2003; Tyler and Moench 2012). More explicitly:

- *Flexibility and diversity in the system*: the ability to perform essential tasks under a wide range of conditions, and to convert assets or modify structures to introduce new ways of doing so (overlapping field of action and different solutions and options of essential services). A resilient system has key assets and functions physically distributed (or coming from diverse geographic areas) so that they are not all affected by a given event at any one time (spatial diversity) and can substitute for each other (e.g. multiple transportation modes) and have multiple ways of meeting a given need (functional diversity).
- *Autonomous and collaborative system*: the capability to operate independently of external support, at least for a time, and to give internal support to the parts of the community that are in need, or to their administrative bodies.
- *Redundancy and modularity of the components of the system*: spare capacity for contingency situations, to accommodate increasing or extreme surge pressures or demand; multiple pathways and a variety of options of systems for service delivery; or interacting components composed of similar parts that can replace each other, if one, or even many, fail (substitutable components). Redundancy is also supported by the presence of buffer stocks within systems that can compensate if flows are disrupted (e.g. local water or food supplies to buffer imports; backup energy generators or redundant energy connections; multiple access routes; various communication systems).
- *Safe failure of the components of the system*: ability to absorb sudden shocks (including those that exceed design thresholds, as retention zones or flood-proofing technics in case of inundation) in ways that avoid catastrophic failure. Safe failure also refers to the interdependence of various sub-systems, which support each other (limiting the use of hierarchic topologies of connections in favour of mesh topologies in which every node has a connection to every other node in the network) failures in one structure being unlikely to result in cascading impacts across other systems (one node can replace the malfunctioning of another).

4.4.2 Analytical Knowledge and Action Systems Options

To have analytical approaches to investigate and improve at least part of the components of a system, approaches to be understood to be detailed actions of DRA and DRR, would be ideal for the implementation of DrP policies for cities and communities and to support the work of green planners. Such instruments are sometimes part of specialist knowledge of professionals for designing critical infrastructures, but mostly they are too rarely available to urban and sustainability specialists.

Concerning essential functions, such as those allowed by residential, scholastic and office buildings, a contribution has been given to the green urban planner's community by the work, still experimental, in developing the *Predisposition of Resistance of Edifices to Disasters certification* (PRED certification, or CEPRED). PRED is suggested as a feasible new approach, to be applied on a voluntary basis, to make building owners and local administrations/authorities of communities more aware of the consequences of their choices related to the field of safety from disaster risk (Bignami 2014). The proposed certification (despite its experimental status) aims at working along lines similar to those of the international environmental certification protocols for buildings, neighbourhoods and cities.

The PRED methodology is, through a well-defined and structured set of indices, a way to summarize the disasters risk performance of a specific building, considering its specific functional situation as in the cases of other certifications applied to the building industry: e.g. the "energy" certifications of buildings (Dall'O et al. 2008), or the environmental certifications of buildings such as the US Leadership in Energy and Environmental Design (LEED®, www.usgbc.org/leed) or the UK Building Research Establishment Environmental Assessment Method (BREEAM®, www.breeam.org) systems. Logically, PRED faces not only a complex multi-dimensional problem requiring knowledge and experience from a wide range of disciplines, as in the previous cases, but also adds knowledge to be obtained in a more complex manner from a treatment of uncertainty methods and complies with the ISO 31,000 guidelines on risk management.

Concretely, the experimental proposal of PRED certification is a multi-risk assessment certification process, based on the idea of a voluntary system conceived to be managed by a third-party body providing results to the building's owners and, if so desired, to the local authorities or agencies in charge of DRA (or even of DRR and DrP). The system first analyzes the hazards of the territory and site of the buildings under examination, then estimates the vulnerability of buildings and, in the case of satisfying findings, eventually produces a result (assigned through the award of a special label) for the potential buildings' resistance (or strength) in the event of disaster. Such a procedure is made in order to direct spontaneously the building market towards a predisposition for better multi-disaster resistance performances and, in addition, to provide an element of rewarding the best properties, element to which attention is drawn by four or five Blue-shields (the Blue-shields being, in the cited label, the icons chosen to show the satisfactory resistance performances of buildings, measured from one to five).

The resulting system provides information useful to the various users of buildings, thanks to an analysis deriving from an objective and reproducible audit process. The methodology has been specifically developed to allow for subsequent improvement in land-use planning and building regulations on safety and related sustainability. Thanks to the PRED rating system, owners, residents and renters or buyers, or even local decision-makers, will know the specific condition of the buildings which interest them, having access to a wealth of detailed information. On the basis of the assessed conditions, they can decide whether or not to activate a risk reduction initiative.

Such a kind of actions or policies can lessen damage to buildings and inhabitants, showing how disaster risk reduction initiatives implemented in a systematic manner can give more and more help in protecting lives and the economic development of cities and communities.

The PRED certification, as a DRA certification, together with environment and energy certifications, may compose an integrated knowledge system contributing to the definition of overall sustainability performances of many kinds of buildings. Such kinds of synergies and combined effects are logically emphasized in many operational contexts, as, for instance, in France, to show the economical advantages of retrofit actions (Cfr. *Référentiel de travaux de prevention des risques technologiques dans l'habitat existant*—METL and MEDE 2014).

Such guidelines are fundamental instruments to disseminate the know-how regarding DRR for implementing DrP trough long-term investment (Bignami et al. 2019). To mention some instances, it is appropriate to list some of the extant guidelines. In the USA, one can cite, as an example, the handbook on *Reducing Flood Risk to Residential Buildings That Cannot Be Elevated* (FEMA 2015). For international agencies, one can cite the *guidelines on non-structural measures in urban flood management* (UNESCO 2001). In France, should be cited, once again on the subject of technological risks, the guide on the retrofitting of windows in case of explosion (Cfr. the guide *Renforcement des fenêtres dans la zone des effets de surpression d'intensité 20–50 mbar*—Leroux and Peron 2015), and, about flood risk, the retrofit guide in case of flooding (Cfr. *Référentiel de travaux de prevention du risque d'inondation dans l'habitat existant*—Fournier and Blas 2012). In Switzerland, one can cite the recommendations prepared by the insurances, to face landslides, floods and avalanches (Cfr. *Protection des objects contre le danger naturels gravitationnel*—Egli 2005).

4.5 Conclusion and Final Remarks

If green urban planning plays a fundamental role in the future of humanity, a decisive action towards the improving of DRM disciplines is a priority to be considered alongside the energy and environmental efficiency of buildings.

As shown in this chapter, the present-day development in hazard-prone locations such as, for instance, flood plains, vulnerable coasts and earthquake zones, together with climate change effects, continues to outweigh global disaster risk. Therefore, integrating disaster risk management into investment decisions is the most suitable way to contribute to developing sustainability for cities and communities.

Although DRM may not completely avert or eliminate the disaster threats, the involvement of potentially affected communities in disaster risk management activities at a local level goes in the right direction to achieve results. As has been clearly indicated in previous paragraphs, today the available tools to act are many—in summary: assessments of hazards, vulnerabilities and capacities; planning, implementation, monitoring and evaluation of risk reduction and of resilience promotion interventions are all necessary steps.

References

Amaratunga D, Malalgoda C, Haigh R, Panda A, Rahayu H (2018) Sound practices of disaster risk reduction at local level. Procedia Eng 212:1163–1170

AON (2018) Weather, climate & catastrophe insight 2018—annual report. http://catastropheinsight.aon.com

Bignami DF (2014) Towards a territorial multi-disaster buildings' resistance certification. Springer briefs in environmental science. https://doi.org/10.1007/978-88-470-5223-9

Bignami DF, Rosso R, Sanfilippo U (2019) Flood proofing in urban areas. Springer, Switzerland, p 268. https://doi.org/10.1007/978-3-030-05934-7. ISBN 978-3-030-05933-0

Birkmann J, von Teichman K (2010) Integrating disaster risk reduction and climate change adaptation: key challenges—scales, knowledge, and norms. Sustain Sci 5:171–184. https://doi.org/10.1007/s11625-010-0108-y

Birkmann J, Cardona OD, Carreno ML, Barbat AH, Pelling M, Schneiderbauer S, Kienberger S, Keiler M, Alexander D, Zeil P, Welle T (2013) Framing vulnerability, risk and societal responses: the MOVE framework. Nat Hazards 67:193–211. https://doi.org/10.1007/s11069-013-0558-5

Blakely JE (2007) Urban planning for climate change. Lincoln Institute of Land Policy, Working paper, Lincoln Institute Product Code: WP07EB1

Boccard N (2018) Natural disasters trends. SSRN Electron J. https://doi.org/10.2139/ssrn.3116231

Bosher L (2014) Built-in resilience through disaster risk reduction: operational issues. Build Res Inf 42(2):240–254. https://doi.org/10.1080/09613218.2014.858203

CRED (Centre for Research on the Epidemiology of Disasters) (2017) Economic losses, poverty and disasters 1998–2017. School of Public Health Institute of Health and Society (IRSS), Université catholique de Louvain

Dall'O' G, Galante A, Ruggieri G (2008) Guida alla valorizzazione energetica degli immobile—Metodi, strumenti e tecnologie per la riqualificazione del patrimonio edilizio esistente, Il Sole24Ore. ISBN-13:9788832470444

Egli T (2005) Protection des objects contre le danger naturels gravitationnel. AEAI—Association des établissements cantonaux d'assurance incendie. ISBN 3-003-00470-9

Etinay N, Egbu C, Murray V (2018) Building urban resilience for disaster risk management and disaster risk reduction. Procedia Eng 212:575–582

FEMA—Federal Emergency Management Agency (2015) Reducing flood risk to residential buildings that cannot be elevated. FEMA, p 1037

Fournier M, Blas M (eds) (2012) Référentiel de travaux de prévention du risque d'inondation dans l'habitat existant, Ministère de l'égalité des Terri-toires et du Logement Ministère de l'écologie, du Développement durable et de l'énergie, Republique Française. http://www.cohesion-territoires.gouv.fr/inondations. 26 Mar 2019

Geneletti D, La Rosa D, Spyra M, Cortinovis C (2017) A review of approaches and challenges for sustainable planning in urban peripheries. Landsc Urban Plan 165:231–243. https://doi.org/10.1016/j.landurbplan.2017.01.013

Godschalk DR (2003) Urban hazard mitigation: creating resilient cities. Nat Hazards Rev 4(3):136–143. https://doi.org/10.1061/(ASCE)1527-6988(2003)4:3(136)

Guzzetti F (2000) Landslide fatalities and the evaluation of landslide risk in Italy. Eng Geol 58(2):89–107. https://doi.org/10.1016/S0013-7952(00)00047-8

Hallegatte S, Vogt-Schilb A, Bangalore M, Rozenberg J (2017) Unbreakable—building the resilience of the poor in the face of natural disasters. World Bank Group, Washington, DC. ISBN (electronic): 978-1-4648-1004-6. https://doi.org/10.1596/978-1-4648-1003-9

Hoeppe P (2016) Trends in weather related disasters—consequences for insurers and society. Weather Clim Extremes 11(Suppl C):70–79

IPCC (2012) In: Field CB, Barros V, Stocker TF, Qin D, Dokken DJ, Ebi KL, Mastrandrea MD, Mach KJ, Plattner G-K, Allen SK, Tignor M, Midgley PM (eds) Managing the risks of extreme events and disasters to advance climate change adaptation. A special report of Working Groups I and II of the Intergovernmental Panel on Climate Change. Cambridge University Press, Cambridge, UK, New York, NY, USA, 582 pp.

IPCC (2014) In: Core Writing Team, Pachauri RK, Meyer LA (eds) Climate change 2014: synthesis report. Contribution of Working Groups I, II and III to the fifth assessment report of the Intergovernmental Panel on Climate Change. IPCC, Geneva, Switzerland, 151 pp.

Kelman I (2017) Linking disaster risk reduction, climate change, and the sustainable development goals. Disaster Prev Manag 26(3):254–258. https://doi.org/10.1108/dpm-02-2017-0043

Kirshen P, Ruth M, Anderson W (2008) Interdependencies of urban climate change impacts and adaptation strategies: a case study of Metropolitan Boston USA. Clim Change 86:105. https://doi.org/10.1007/s10584-007-9252-5

Kron W (2002) Flood risk = hazard × exposure × vulnerability. In: Wu M et al (eds) Flood defence. Science Press, New York, pp 82–97

Leroux B, Peron O (2015) Renforcement des fenêtres dans la zone des effets de surpression d'intensité 20–50 mbar—Plan de Prévention des Risques Technologiques, INERIS and Ministère de l'Écologie, du Développement durable, des Transports et du Logement, DRA-10-111924-09528c, Republique Française. www.developpement-durable.gouv.fr

McBean GA (2012) Integrating disaster risk reduction towards sustainable development. Curr Opin Environ Sustain 4(1):122–127. https://doi.org/10.1016/j.cosust.2012.01.002. ISSN 1877-3435

Merz B, Kreibich H, Schwarze R, Thieken A (2010) Assessment of economic flood damage (review article). Nat Hazards Earth Syst Sci 10:1697–1724

Messner F, Penning-Rowsell E, Green C, Meyer V, Tunstall S, van der Veen A (2007) Evaluating flood damages: guidance and recommendations on principles and methods. FLOODsite Project Deliverable D9.1, Contract No: GOCE-CT-2004-505420. Available at www.floodsite.net. Jan 2007

METL—Ministère de l'égalité des Territoires et du Logement, MEDE—Ministère de l'écologie, du Développement durable et de l'énergie (2014) Référentiel de travaux de prevention du risque d'inondation dans l'habitat existant, Republique Française. www.territoires.gouv.fr, www.developpement-durable.gouv.fr

Munich RE (2019) Relevant natural loss events worldwide 1980–2018. NatCatSERVICE, 06 Nov 2019

Mysiak J, Surminski S, Thieken A, Mechler R, Aerts J (2016) Brief communication: Sendai framework for disaster risk reduction—success or warning sign for Paris? Nat Hazards Earth Syst Sci 16(10):2189–2193. https://doi.org/10.5194/nhess-16-2189-2016

Norton J, Atun F, Dandoulaki M (2015) Exploring issues limiting the use of knowledge in disaster risk reduction. TeMA J Land Use Mobil Environ ECCA15:135–154. ISSN 1970-9870

Roberts NJ, Nadim F, Kalsnes B (2009) Quantification of vulnerability to natural hazards. Georisk: Assess Manag Risk Eng Syst Geohazards 3(3):164–173. https://doi.org/10.1080/17499510902788850

Rovida A, Locati M, Camassi R, Lolli B, Gasperini P (eds) (2016) CPTI15, the 2015 version of the Parametric Catalogue of Italian Earthquakes. Istituto Nazionale di Geofisica e Vulcanologia. http://doi.org/10.6092/INGV.IT-CPTI15

ISBN 978-0-7546-4431-6

Swiss Re (2019) Natural catastrophes and man-made disasters in 2018: "secondary" perils on the frontline, sigma n. 2/2019. Swiss Re Institute

Thieken AH, Ackermann V, Elmer F, Kreibich H, Kuhlmann B, Kunert U, Maiwald H, Merz B, Müller M, Piroth K, Schwarz J, Schwarze R, Seifert I, Seifert J (2008) Methods for the evaluation of direct and indirect flood losses In: 4th international symposium on flood defence: managing flood risk, reliability and vulnerability, Toronto, Institute for Catastrophic Loss Reduction, Ontario, Canada, 6–8 May 2008

Thomalla F, Downing T, Spanger-Siegfried E, Han G, Rockström J (2006) Reducing hazard vulnerability: towards a common approach between disaster risk reduction and climate adaptation. Disasters 30:39–48. https://doi.org/10.1111/j.1467-9523.2006.00305.x

Tiernan A, Drennan L, Nalau J, Onyango E, Morrissey L, Mackey B (2019) A review of themes in disaster resilience literature and international practice since 2012. Policy Des Pract 2(1):53–74. https://doi.org/10.1080/25741292.2018.1507240

Tyler S, Moench M (2012) A framework for urban climate resilience. Clim Dev 4:311–326. https://doi.org/10.1080/17565529.2012.745389

UNESCO (2001) In: Andjelkovic I (ed) Guidelines on non-structural measures in urban flood management—Paris

UNISDR (The United Nations Office for Disaster Risk Reduction) (2005) Hyogo framework for action 2005–2015: building the resilience of nations and communities to disasters. ISDR International Strategy for Disaster Reduction. www.unisdr.org/wcdr

UNISDR (The United Nations Office for Disaster Risk Reduction) (2015a) Sendai Framework for Disaster Risk Reduction 2015–2030

UNISDR (The United Nations Office for Disaster Risk Reduction) (2015b) Chart of the Sendai Framework for Disaster Risk Reduction 2015–2030. www.preventionweb.net/go/sfdrr

UNISDR (The United Nations Office for Disaster Risk Reduction) (2017) Technical guidance for monitoring and reporting on progress in achieving the global targets of the Sendai framework for disaster risk reduction—collection of technical notes on data and methodology

United Nations Development Programme—UNDP (2010) Guidelines for designing disaster risk assessment projects—methodology and tools. Global Risk Identification Programme (GRIP), Bureau for Crisis Prevention and Recovery (BCPR), United Nations Development Programme (UNDP), Chemin des Anémones, Châtelaine, Geneva, Switzerland, pp 11–13. http://www.gripweb.org

United Nations Framework Convention on Climate Change—UNFCCC (2015) Paris Agreement, Paris. https://unfccc.int/sites/default/files/english_paris_agreement.pdf

United Nations—UN (2015) The sustainable development goals report, New York

United Nations—UN (2016) General assembly, report of the open-ended intergovernmental expert working group on indicators and terminology relating to disaster risk reduction, A/71/644

Wamsler C (2004) Managing urban risk: perceptions of housing and planning as a tool for reducing disaster risk. Glob Built Environ Rev 4(2):11–28

Chapter 5
Circular Approach in Green Planning Towards Sustainable Cities

Monica Lavagna

Abstract Cities are like organisms, drawing in resources and emitting wastes. In nature, organisms are related to ecosystems characterized by a circular metabolism, so a green planning to move towards sustainable cities needs to adopt a circular approach to the way in which are managed the resources consumed (materials, food, energy, water and land) and the emissions produced (solid waste, airborne and waterborne) in cities and their related territories. Initially, the origins of the theme of circularity are illustrated, in particular, in relation to urban applications (urban metabolism) and territorial (industrial ecology) and current developments, with the analysis of a virtuous example. Thereafter, some possible strategic approaches to favour urban circularity (urban mining, building regeneration) are analysed, highlighting the importance of flow mapping and the creation of exchange platforms. Finally, some methods of assessing environmental impacts are shown to verify the effectiveness of circularity with respect to the objective of sustainability, and examples of benchmarks/targets adopted in some concrete cases experienced are presented.

5.1 Well-Being and Resource Reduction: A Challenge

The ultimate goal of green planning is to guarantee the well-being of the people who live in a city or territory. In Western societies, despite the high levels of access to consumer goods and services, a perception of a reduction in well-being is becoming more widespread. This is due to the deterioration of the quality of the environment in which people live (pollution, stress, social insecurity, etc.). Therefore, people's well-being is not based solely on economic stability, but also on the quality of social relations and on the healthiness and beauty of the environment that surrounds them. In particular, natural capital is a direct source of well-being for people (Bonaiuti 2011): drinking at a high mountain source, being able to swim in a clean sea, contemplating an alpine landscape are already free and available elements of well-being, which

M. Lavagna (✉)
Architecture, Built Environment and Construction Engineering, ABC Department, Politecnico di Milano, Via G. Ponzio, 31, 20133 Milan, Italy
e-mail: monica.lavagna@polimi.it

G. Dall'O' (ed.), *Green Planning for Cities and Communities*,
Research for Development, https://doi.org/10.1007/978-3-030-41072-8_5

must not be "produced", but must, however, be preserved (and not destroyed by the expansion of human activities).

"The environment was considered a sort of inert stage in which the actors, that is the organisms, played the game of natural selection. Now we recognize that the "stage" and the "actors" interact with each other constantly so that not only do organisms relate to the physical environment, but they also change the environment" (Odum 2003).

The pollution we create (heavy metals, chemicals, micro-plastics) enters the water cycle, contaminates the soils and the air and becomes elements that we breathe and eat, also thereby contaminating our organism.

Nature is therefore not simply a place to contemplate, but it is the source of the resources that sustain our life, so the changes that people make to the natural environment also affect their well-being in terms of health (toxicity) and survival (availability of resources, e.g., food due to infertility of contaminated soils).

It must, therefore, be considered that the natural capital is also the necessary basis for all human activities, even urban ones.

5.2 Urban Input and Output Flows

Cities are the expression of mankind's construction of an "artificial" environment, separate from nature: man is, in fact, a "deficient being" (Gehlen 1988), unfit to survive in the natural environment, so he must construct his conditions of survival through technical acts (building artefacts, building a refuge, building the city, etc.), which become elements of protection and mediation between mankind and nature.

At the same time, man (and his environment, like the city) continues to depend on nature: he needs natural resources to produce artefacts, to produce energy and to feed himself and he needs to be able to remove scrap, solid waste and polluting emissions, which nature is able to re-absorb. Cities are thus totally dependent upon natural ecosystems.

Cities are sustained by input flows of resources from nature (food, water, materials, energy) and output flows of emissions into nature (waste, air and water pollution). These flows lead to a close relationship between the city and its surrounding territory. Urban areas and cities occupy only a small part of the total land area (only 2%). However, although the surface area occupied by the city is relatively small, it is the part which represents the greatest threat to the rest of the land area (Fig. 5.1).

It is possible to identify a relationship between the city and the surrounding area, in terms of flows of resources and waste, in all periods of urban development; where the resource management has not been prudent and the territory has been degraded, civilisations have fallen (Agudelo-Vera et al. 2011). Today the problem is more pronounced: the more cities grow, the greater their dependence upon the surrounding areas. Attacks of robbery against the occupied territory, then lead to the infertility of the soils, to floods due to excessive waterproofing of the land, to water pollution, etc. Thus, it is necessary to establish a balance between the city and its territory.

Fig. 5.1 Relationship between man's artificial sphere and the ecosphere: nature "feed" the city with flows of resources in input and assimilate the waste in output, to support the human activities and life. *Source* Rees and Wackernagel (1996)

In addition, cities have direct and indirect impacts on a global scale (e.g., greenhouse effect due to emissions into the air released by cities and their own territories). So it is also necessary to find a balance between the city and the planet globally.

As long as the flows were small, nature was able to regenerate resources and dispose of waste and pollution. But in the last century, the exponential growth of the world's population, per capita consumption, urbanisation processes, industrial production have led to an unsustainable pressure on a planet limited in its resources. "Large-scale urbanization is a profoundly resource-demanding process, to build as well as to run cities (…). We can argue that we no longer live in a *civilization*. We live in a *mobilization*—of people, resources and products" (Girardet 2003). Today the cities, the territory and the entire planet suffer from environmental degradation, land use, scarcity of resources (raw materials, food, water, fuels), air pollution, water pollution, waste disposal (landfill, transboundary shipment of hazardous wastes, groundwater contamination), climate change, biodiversity loss, etc.

So we build cities to survive (as deficient beings), but the cities we build are the major cause of the destruction of natural ecosystems, which support our survival; consequently, we are putting our survival at risk.

In the face of climate change and the damage of anthropization, one talks of implementing strategies of resilience. However, it would be more appropriate to act to prevent changes rather than to act to adapt to devastating changes (living with a polluted sea, between destructive weather events, etc.). Hence, the need to rethink mankind's consumption and emissions on all scales, including the urban one. Currently, consumption and emission rates have become unsustainable, and to conserve the remaining natural capital (now irremediably eroded and deeply polluted), necessary for mankind to survive, it is essential to reverse the trend.

To trigger a reduction in mankind's consumption and emissions, a possible strategy is to extend the life of resources in two ways: extending the life of artefacts and reusing/recycling resources at the end of their useful life (closing the cycle). It is no coincidence that current policies are aimed at promoting the circular economy, which focuses precisely on these two objectives.

When referring to the circular economy, the aspect on which the most attention appears to be focused is end-of-life recycling (waste management), to reduce the transfer to landfills and reduce the withdrawal of new resources from nature (thanks to the availability of recycling resources).

However, the broader objective of the circular economy should in fact be that of the efficient use of resources, thereby reducing the resources necessary to guarantee the essential services for the functioning of mankind's activities and the extension of the useful life of the products in which resources are stored: all this is in order to avoid new waste streams and new withdrawals of resources (which occur also in the recycling activity).

Furthermore, the circular economy is a segment of the broader objective of environmental sustainability, so it is not only a question of using resources efficiently but also of guaranteeing a reduction in overall impacts and a better environment for human well-being (Fig. 5.2).

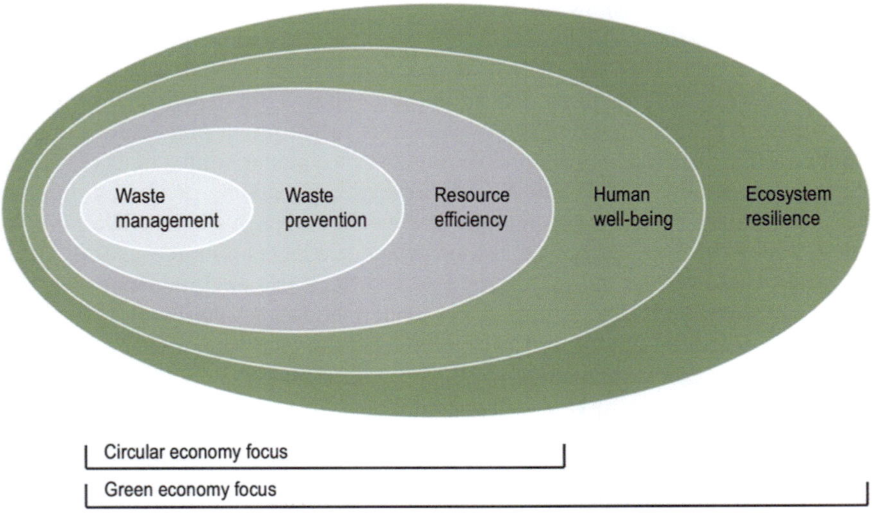

Fig. 5.2 Strategies of circular economy and green economy. *Source* EEA (2015)

5.3 Origins and Developments of Circularity Approach

The circular economy is currently focusing attention on issues that seem new since only now they find expression in political documents and roadmaps and therefore are expressed in concrete actions and virtuous applications. Actually, the themes are not new at all and originate from theories and principles developed as early as the 1960s, in parallel with, and within the reflections on sustainable development, developed in that period.

The first time the Earth was viewed by satellite in 1957, humanity became aware of the limited and fragile nature of the planet and the importance of the presence of man in the modification of the planet. It is precisely from that image that Kenneth Boulding, in his "The Economics of the Coming Spaceship Earth" (1966), recommends a change in the approach of the economy: from an attitude of conquest (cowboy economy), linked to the presumption of an infinite availability of territory and resources, to an attitude of prudent management (spaceman economy), linked to the effective limitation of territory and resources (the Earth as a spaceship). Then, in this context, he proposes the idea of a circular circuit of materials, water and energy (as happens in spaceships), taking into account that on Earth nothing enters except energy from the sun.

In their 1976 research report to the European Commission in Brussels, "The Potential for Substituting Manpower for Energy" (published as a book in 1982 with the title "Jobs of Tomorrow"), Walter R. Stahel and Genevieve Reday-Mulvey introduced the argument of extending the service life of buildings and such goods as cars and highlighted the waste inherent in disposing of old products instead of repairing them. In 1981, Stahel summarized these ideas in his article "The Product-Life Factor" and identified the sale of the use of goods (service) in place of the sale of goods as the sustainable economic model of a closed-loop economy: in this way, the industry adopts the reuse and service-life extension of goods as a business strategy (Fig. 5.3).

Extending product-life reduces the depletion of natural resources and consequently waste and decouples wealth from resource consumption.

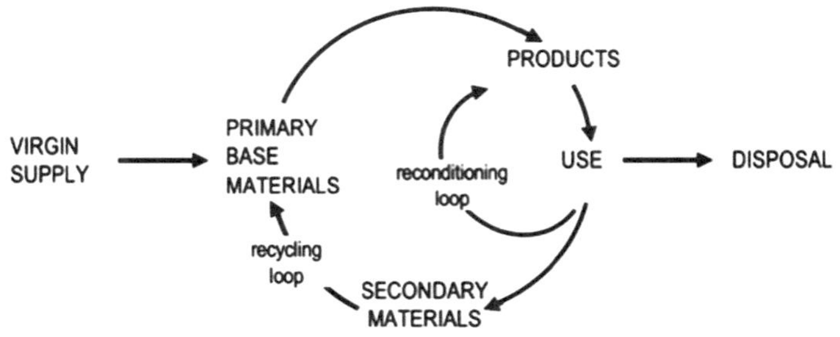

Fig. 5.3 Alternative life cycles of an industrial product (Stahel 1981)

Parallel to these theoretical developments, one can identify declinations of the circularity on the urban and territorial scale already in the 1960s, with urban metabolism and industrial ecology.

5.3.1 Urban Metabolism and Material Flow Analysis

Cities can be assimilated to living organisms. Cities absorb, transform and release various forms of energy and materials, such as food, water, heat and waste. In this, they recall the functioning of biological organisms and natural ecosystems. The set of "internal" chemical–physical energy transformations to these systems necessary for their support is called metabolism.

To manage urban ecosystems, it is necessary to study their metabolism.

In 1885 to create awareness of the massive flows of resources in cities, Geddes used the concept of urban metabolism and established an urban energy and material budget in physical input–output terms (Geddes 1885), but without any immediate follow-up and application.

In 1965, the health engineer, Abel Wolman, revised the concept of urban metabolism proposed by Geddes and developed the first wide-area environmental budgets, estimating the material flows that cross *cities* (Wolman 1965).

Starting from his studies of 1963 on energy flow and nature's metabolism, in 1975, Eugene Pleasants Odum highlighted the importance of urban metabolism and provided some data on the urban material balance (Odum 1975).

Howard Thomas Odum, the younger brother of Eugene, provided fundamental methodological bases for the analysis of energy flows in urban metabolism. In particular, he has developed the analysis of the energy necessary to produce the goods and services of anthropic systems (Emergia; Odum 1983). The energy balances are performed in units of energy on a temporal basis (powers) regulated by the equation: incoming flow rate = outgoing flow rate—consumption flow + stored flow rate. The same type of budget/balance can also be applied to materials.

In the early 1970s, the analysis of material flows through urban ecosystems was also promoted by UNESCO with a programme (Man and Biosphere, MAB) aimed at promoting studies on ecological approaches to urban systems and other human settlements [Unesco 2010].

Since the 1990s, the Material Flow Analysis (MFA) and the Input-Output Analysis (IOA) of anthropic systems have developed above all on a national scale in the USA and Europe (Wernick and Ausubel 1995; Matthews et al. 2000; European Communities and Eurostat 2001), but also on regional and urban scale, as in the case of Vienna (Daxbeck et al. 1997), Stockholm (Burström et al. 1998), Geneva (Faist Emmenegger and Frischknecht 2003), Hamburg, Vienna and Liepzig (Hammer et al. 2003), Paris (Barles 2009). These studies, carried out to understand urban metabolism, have become very important in supporting public sector policies and actions.

The urban ecologist, Herbert Girardet, (1992) has argued that the key lies in cities aiming at a circular "metabolism", where consumption is reduced by implementing efficiencies and where re-use of resources is maximized. Girardet significantly coined the concept and defined the difference between a "circular" and "linear" metabolism.

These scientific reflections and practical applications do not only affect the decisions of government and of the local authorities, but also the activity of designers and urban planners.

Richard Rogers, in his "Cities for a small planet" (1997), emphasizes that cities must be viewed as ecological systems and this assumption must provide information to our approach to designing cities and managing their use of resources. Rogers promotes the transition from a linear metabolism to a circular metabolism, considering it necessary to obtain sustainable cities, and considers it essential to introduce experts in urban ecology into city planning (Fig. 5.4). Rogers points out that the circular attitude must also concern land and buildings. "In cities, we cast off buildings, used land, industrial waste and millions of other damaged, used products directly into the urban environment and its wide hinterland.

Most of this is unnecessary. Careful recycling of land, buildings and waste could make a difference not just to the landscape and viability of both town and country,

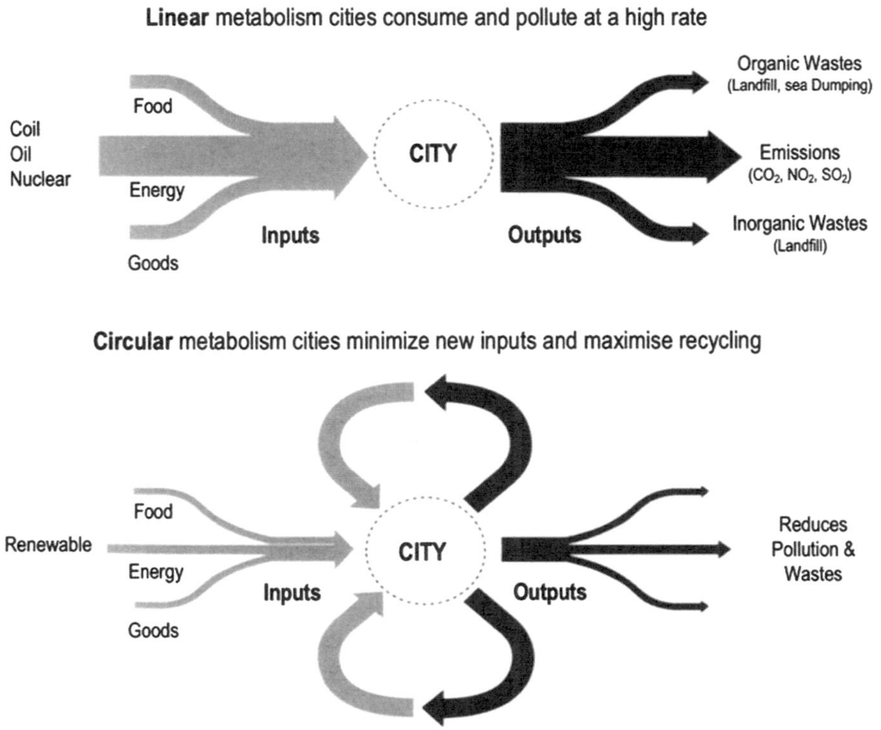

Fig. 5.4 From linear to circular metabolism of cities (Rogers 1997)

but also to our approach to cities in general. With care, buildings can last twice or three times their predicted life" (Rogers and Power 2000).

Tackling the analysis of urban flows in terms of metabolism allows one to move from the end-of-pipe approach, which has characterized the first attention paid to the environment, that is to say mainly controlling the environmental deterioration caused by emissions to water, soil and air, to an approach of prevention. It became understood that it is important to manage not only the outgoing polluting flows but also the incoming flows (consumptions) because decreasing consumption also reduces emissions. Hence, great importance is given to the closure of the cycles.

5.3.2 Industrial Ecology

The city should not be seen as an isolated system, but conceived in its relations with the surrounding land. Therefore, besides studying the flows that cross, enter and leave a city, it is also necessary to extend the analysis to the related territory.

In 1969, McHarg published the book "Design with Nature", in which he argued that cities should be planned as an integral part of natural systems. He proposed to use ecology to understand interactions between people and their environment and to use these as guiding principles for urban planning (McHarg 1969).

At the end of the 1980s, a scientific discipline was born, industrial ecology, which uses the methods of ecology to study material and energy flows in industrial systems. What for one industry is a waste for another can become a resource. The principle is based on the idea of closing the cycles of productive activities, connecting them in a network. This means not only analysing the individual activities but assuring a relationship between them thanks to their territorial proximity.

One of the first examples of industrial ecology is that of Kalundborg, in Denmark, where the waste heat of the thermoelectric plant is recovered and used to supply district heating to the city, the fly ash deriving from the combustion processes of the thermoelectric plant is reused in a cement works, and the washing sludge is reused in a plasterboard plant, etc. (Fig. 5.5).

There are many other examples. City rubbish, which is usually either dumped as landfill or incinerated, both with polluting effects, can be burned by local combined heat and power plants (CHPs) and supply a community's energy needs. Organic waste, which is currently discharged in such high concentrations that it poisons the environment (eutrophication), can instead be recycled to produce biofuel and fertilizers. Grey water can be filtered through natural systems on site and be reused for irrigation of the urban landscape or to restock local aquifers. Other possibilities along similar lines will be apparent to the reader.

This approach can solve not only the resource management but also the disruption of natural cycles, for instance the phosphorus and carbon-cycle, with benefit for the environment. It is not necessarily related only to industries, but can involve urban activities, creating a network with the urban metabolism.

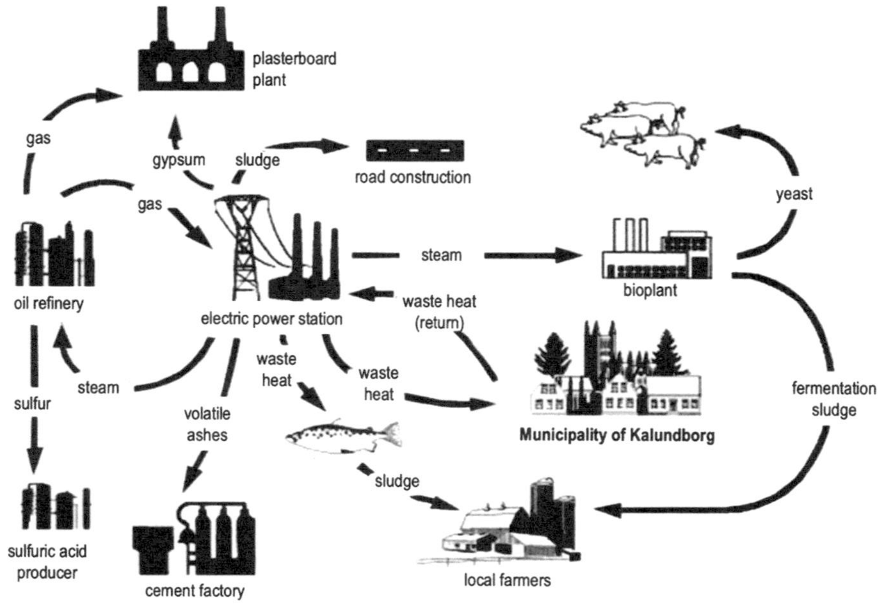

Fig. 5.5 Example of industrial ecology: relationship of waste as food for other industrial processes in Kalundborg (DK), 1989 (Kibert 2005)

5.3.3 *The Eco-Towns and the Hammarby Model*

Started in the 1980s, the eco-cities or eco-towns movement focuses on redeveloping the industrial centres of large towns and cities through industrial metabolism projects (Prendeville et al. 2018).

There are many new eco-towns which have been built in northern Europe (e.g. Vauban, Amersfoort, Malmö) based on the principles of sustainability. Particularly interesting is the case of Hammarby Sjöstad ("Hammarby Lake City"), in which the principles of industrial ecology were applied during the design of the city, connecting the city to its surrounding territory through a series of entering and exiting flows of resources (thus also taking into account waste and wastewater) (Fig. 5.6).

A first interesting aspect is the application of the circular approach already in the identification of the settlement, as a recovery of an abandoned industrial area. In the early 1990s, Stockholm's City Planning Administration acknowledged that the city's population was growing and would continue to grow into the next century. The 1999 city plan identified several areas across the city for development, most of which had once been industrial sites. Hammarby was developed on an old brownfield site. All contaminated soil was sanitized prior to development.

The most interesting aspects concern the strategies of optimisation of resources and circularities applied to energy, water, waste, in the so-called Hammarby model.

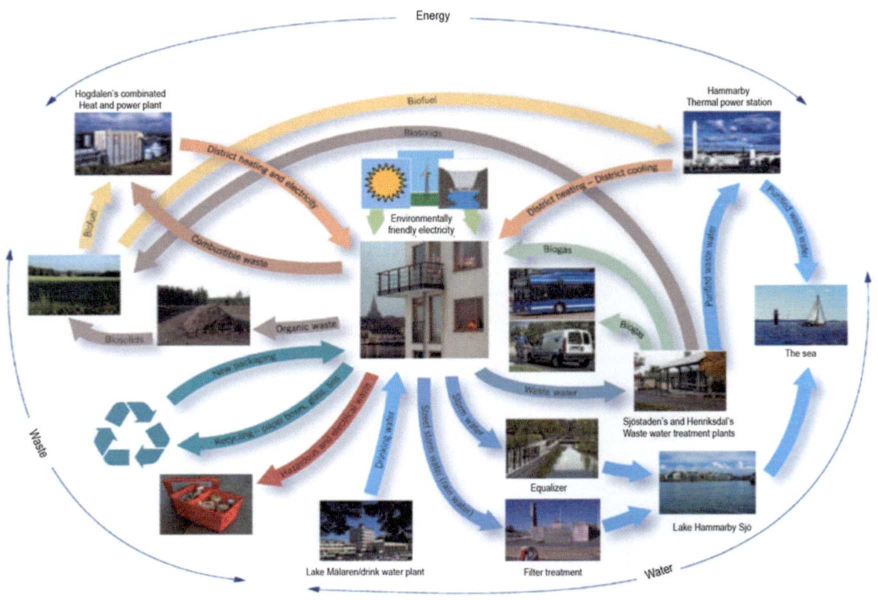

Fig. 5.6 Hammarby model. *Source* Lena Wettrén, Bumling AB, Hammarby Sjöstad Project

Energy self-sufficiency is achieved through energy-saving measures, the use of renewable energy sources (solar collectors, photovoltaic, hydroelectric and wind power plants), the use of bioenergy and incineration of local waste producing both locally generated heat and co-generated electricity (80% of the energy extracted from waste and wastewater). All apartments are connected to the district heating system, and the household waste supplies fuel for the district heating plant.

Separate waste collection takes place via underground channels with a vacuum system and is recycled. That waste which is not recyclable is taken to the waste-to-energy plant. 100% of waste is sorted and only 0.7% of waste goes to landfill, 50% of waste is recovered as energy, 16% of waste is turned into biogas, 33% goes into recycling of materials, and 1% is hazardous waste (Jernberg et al. 2019).

Wastewater is treated locally. The sludge produced by the treatment process is recycled and used for fertilizing farmland and forestry land. The waste releases biogas during processing that is used as fuel for vehicles such as buses, taxis and waste collection trucks, and to heat 1000 homes in the area. Heat is extracted from the treated water in the treatment plant, which is then used for district heating.

The Hammarby model was designed as a project that might be replicated in other cities. Hammarby Sjöstad has become an international example of sustainable urban planning. For example, it has inspired the design of Toronto's Waterfront, London's New Wembley and many cities in China and Thailand.

"Hammarby Sjöstad is a good example not only of focusing on the short-term aspects and obtaining short-term profits but also of investing for the future and increasing revenue. The cost for Hammarby was roughly 5% higher from a purely construction cost perspective, but in the end you get back roughly 25% more property value out on the site over time which shows how real value is created over time". (Henrik Svanqvist, Director of Communities, Skanska).

These experiences, however interesting, are linked to very particular situations, of small-scale intervention, with the ex-novo design of an urban reality (as generally associated with the regeneration of abandoned industrial areas), where it is easy to build relationships and virtuous activities. The intervention on existing cities is much more complex and with consolidated realities that are difficult to modify, where these models struggle to find an application. However, having such a model which demonstrates the advantages of its operation and functionality is a precious stimulus to try to apply it in a widespread manner.

5.3.4 Circular Economy

In the contemporary world, attention to the circular approach has been renewed as a result of the growing need for raw materials (increase in consumption) and to the growth in the costs of supply of raw materials. Finding alternative solutions, therefore, becomes essential to continue supporting the economy. This is why the circular approach is associated with the economy: it is necessary to change the approach of the economy (from linear to circular) to permit further economic development.

It is no coincidence that the productive and economic world then moved before the political world did so. Since 2009, the Ellen Mc Arthur Foundation (a US private foundation) has promoted the circular economy (in all sectors), so as to bring the themes back to the centre of the current debate and above all to favour their adoption at the political level. "A circular economy is an industrial system that is restorative or regenerative by intention and design. It replaces the "end-of-life" concept with restoration, shifts towards the use of renewable energy, eliminates the use of toxic chemicals, which impair reuse, and aims for the elimination of waste through the superior design of materials, products, systems, and, within this, business models" (Ellen Mc Arthur Foundation 2013).

The circular economy was introduced into European policies as of 2014, with the European Commission Communication "Towards a circular economy: A zero waste programme for Europe" (COM 398), that aims to be an incentive for investment in efficiency improvements or innovative business models, using the European Resource Efficiency Platform in order to identify opportunities for recycling business and exchange of by-products. In 2015, the European Commission Communication "Closing the loop—An EU action plan for the Circular Economy" (COM 614) highlights industrial symbiosis opportunities in production and consumption stages and supports innovative business models (sharing economy, supplying of services rather than the sale of goods) and information technology for efficiency platform, with the aim of maintaining the value of products, materials and resources for as long as possible.

According to Pomponi and Moncaster (2017), the circular economy in the built environment can be treated on three levels: at macro-level, which concerns the system of cities or urban agglomerations (such as eco-cities), at meso-level, that is at the scale of the buildings and at micro-level, focused on the materials.

Studies of the circular economy in the built environment have also been performed by the Ellen MacArthur Foundation and CE100 network (CE100 2016). They suggest the six actions called the "ReSOLVE framework" (Regenerate, Share, Optimise, Loop, Virtualise, Exchange). It is possible to apply the "ReSOLVE framework" at urban level: there "Regenerate" can be the regeneration of the building stock; "Share" can be car sharing but also building sharing (residential and office) and the infrastructure sharing (parking sharing, shared infrastructure areas, shared green areas) and shared water consumption (water treatment facilities); "Optimise" can be the promotion of industrial process (industrial ecology) and smart urban design (use inner-city vacant land, promoting compact urban growth, high-quality urban environments, integrated, sustainable and participative urban development); "Loop" can be the activity to close the cycle of urban flows (urban metabolism, urban mining); "Virtualise" can be a virtualisation of products (service than physical products) and tele-working; "Exchange" can be the development of platform for create networks of stakeholder and scrap trading.

In 2019, the Foundation, in collaboration with Arup, launched Circular Economy in Cities, a suite of easily accessible resources which provide a global reference on the topic. Its modules have been developed to respond to the growing interest in circular economy from city governments and mayors, and will offer insights to many other urban stakeholders. Circular Economy in Cities focuses on opportunities in three key urban systems (buildings, mobility and products) and looks at how city governments/councils can work to enable a transition to a circular economy.

5.4 Urban Resources Management Strategies

The scarcity and costliness of raw materials has led to the shifting of the focus from "fixed stocks", still, immobile materials in the natural matrix, to "anthropogenic stocks", real anthropogenic stocks of resources, imagined over the years in the cities (and neighbouring territories).

Hence, the birth of the concept of urban mining, which means to think of the city as a mine of materials, and the actions aimed at the regeneration and monitoring of the building stock, as a place of conservation and storage of the materials to be regenerated through maintenance or reuse/recycling. The mapping of materials within the cities, on the one hand, and the creation of platforms for the interchange of resources, to favour the creation of networks to keep resources alive (also keeping their economic value), on the other, become indispensable tools to promote these processes.

5.4.1 Urban Mining

The city is seen as a mine of materials that can be reused and hence the concept of Urban Mining (Brunner 2011), which is understood as the management of stocks of anthropogenic resources (as products, buildings, spaces) and waste, from urban catabolism (Baccini and Brunner 2012). It proposes long-term environmental conservation, conservation of resources and economic benefits (Cossu and Williams 2015).

For example, WEEE, waste electrical and electronic equipment, represent today real urban mines: it suffices to realize that from a ton of electronic circuit boards, one can obtain more than 2 t of copper, over 46 kg of iron, almost 28 kg of tin and aluminium and about 18 kg of lead, in addition to smaller quantities of silver, platinum and palladium. Indeed, it becomes less expensive to obtain them from waste management rather than from actual physical mining (Fig. 5.7).

This can be extended to the chain of construction and demolition waste, plastics, paper, metals and organic wastes. This opportunity to prolong the life of resources is important, but is orientated to waste management, with an end-of-pipe approach. Circular economy strategies should be orientated towards extending the useful life of products, and this aspect is particularly important for durable goods, such as buildings.

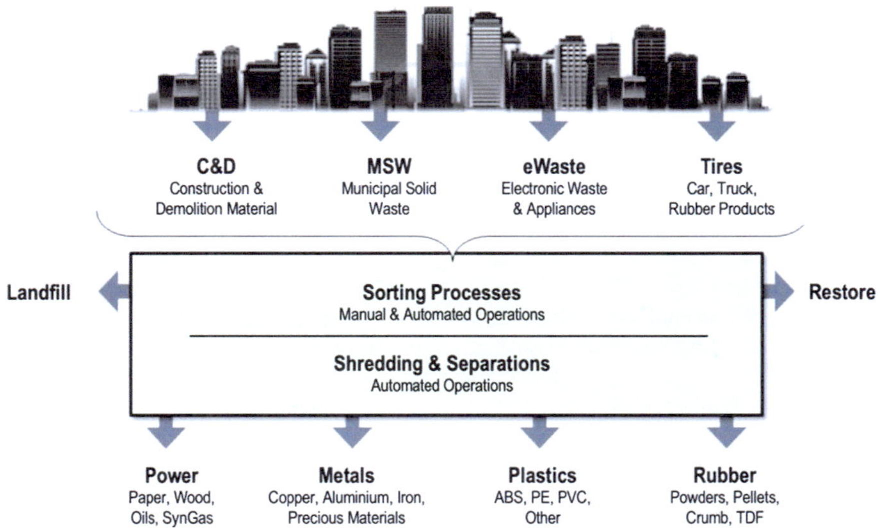

Fig. 5.7 Urban mining, recycle of urban waste streams and new value for resources (Vollmer and Soriano 2014)

5.4.2 Building Stock Regeneration and Building as Material Bank

Existing buildings constitute a precious asset of stored resources. These resources can be reused and recycled if the building is demolished, or they can be maintained, extending their useful life.

Urban regeneration is important first and foremost to avoid consuming new soil (using brownfield instead of greenfield sites) but also to avoid the presence of degraded and unsafe areas within the city. Furthermore, intervening on the existing stock means being able to manage the resources stored up in the building itself, reducing the extraction of new raw materials.

Urban regeneration works often aim at demolition and reconstruction, in order to construct a building with the performances required by contemporary standards and user expectations: in this case, it is important to try to intervene through selective demolition, recovery and reuse of materials (avoiding disposal in landfill) and construction of new flexible buildings (able to adapt to changes over time) and reversible buildings (with removable materials and components, so as to favour the reuse and recycling of the components).

Clearly when compared to demolishing and reconstructing, the possibility of extending the useful life of buildings and the materials stored therein is a more advantageous scenario since it is not necessary to reprocess the materials. So surely, the regeneration of the existing building stock must be encouraged and promoted (Giorgi et al. 2019).

The theme of the building as a place for storing materials and maintaining the value over time of the materials it contains has been the subject of European research BAMB (Building As Material Bank). Within this research, the role of selective deconstruction practices (to be applied in the case of demolition of existing buildings), with a view to reducing the use of natural resources (virgin quarry materials) through the recovery of construction and demolition waste has been emphasized. In addition, it has highlighted the role of design practices for new buildings (or renovation of the existing), attentive to the adaptability of the building over time (to prolong its useful life) and to the use of solutions of reversible construction (to facilitate the reuse of materials and components at the end of the building's life).

5.4.3 Madaster: The Cadastre of Urban Materials

"Waste is a material without an identity" (Rau and Oberhuber 2016). Starting from this assumption, the idea was to define a "cadastre" for materials stocked in the built environment, a real estate materials register, called Madaster.

Thomas Rau and Sabine Oberhuber, in the book "Material Matters" (2016), describe the transition to a new economic system where consumers are no longer owners, but temporary users of products and materials: "First, we must realize that with ownership comes responsibility. Nowadays, by necessity, all kinds of items end up in our possession for which, in the long term, we cannot bear responsibility. If the building we work in no longer meets our requirements, we are completely out of our depth when we are asked to consider its responsible use. We are not able to take care of all (raw) materials that were used to create the building—let alone reuse them".

The Madaster promotes the Material Passport: the passport ensures "identity" of the materials (as a guarantee), displaying also the value of materials and products during the life time of the building. Madaster also gives, for each building upload to the platform, a "circularity indicator" (CI). A building that was built with virgin materials and ends up as waste after a shorter average life is a completely "linear" building with a Madaster CI of 0%. On the other end, a building constructed from reused and/or rapidly renewable materials, that can be disassembled and easily reused at the end of the lifetime is a "fully circular" building with a score of 100%.

Madaster is the first developed example of a platform available to provide information about a building as material banks.

5.4.4 *Mapping of Flows and Exchange Platforms*

In addition to mapping the materials stored in buildings, the next step would be the ability to monitor the flows of materials and resources at the urban level. Through surveys and statistical analysis, it is necessary to analyse the (quantitative economic) flows of resources/waste in an urban area (district) in order to have a general mapping of the flows and be able to trigger controlled re-use strategies. The new digital technologies, ICT, can help the management of natural resources: through the use of "big data" and GIS, it is possible to manage and track waste, water, energy and the exchange of information on consumption between suppliers and users (Neirotti er al. 2014).

In addition to these tools, which should be in the hands of public administrations/local authorities, operators express the need to set up exchange platforms both for stakeholders to exchange information and good practices, and for producers/demolition companies/waste operators for the exchange of secondary raw materials.

The latter is particularly interesting, since they could help to build networks between processes, in a logic of industrial ecology.

Unfortunately, the waste involved is often fragmentary and impromptu (event), and this generates the problem of the location and the quantity of waste (the meaningfulness of the collected flows) to be reintroduced into a new production cycle. For the preliminary feasibility of creating new value chains, it is essential to be able to characterize the waste and verify the possibility of diffused collection in the territory, creating economies of scale. It is also necessary to identify significant and stable quantities: if the scrap is limited and occasional, it becomes difficult to activate recycling chains. If instead the waste being examined is significant and constant, stable relationships can be created between operators. For example, pre-consumer recycling is generally more easily characterized, because the origin tends to be stable and constant (Migliore et al. 2016). A good example is quarry waste which can become inert material for concrete.

In any case, mapping activities and interchange platforms can constitute an important scenario to facilitate the identification of forms of circularity.

5.5 Environmental Impacts at Urban Scale

The control of environmental pressures (consumptions and emissions) implies the need to effect environmental balances with a large area. It is not enough to quantify the flows of material in input and output (Material Flow Analysis), but it is also necessary to evaluate the environmental repercussions on, and consequently the implications for the natural environment.

Different approaches have been developed to study urban complexity and its impacts. Some examples of those approaches are to be found in Ecological Footprint and Life Cycle Assessment.

5.5.1 Ecological Footprint

The ecological footprint measures the environmental pressures determined by anthropic systems, expressing the "quantity of surface" necessary to support the metabolism of people, cities, regions, nations or anthropic systems in general. The methodology is based on the fact that many material and energy flows can be converted into land-area equivalents. Thus, the ecological footprint of a specified population is the area of land required to produce the resources consumed and to assimilate the wastes generated (Wackernagel and Rees 1995).

The "quantity of surface" measured with the ecological footprint includes both the natural resources necessary to support the communities (for example, the surface of the fields to produce wheat, the trees planted for paper, the space occupied by the buildings) and the territory needed to absorb the emissions and the waste generated (for example, the forest surface needed to absorb carbon emissions). It is expressed in "global hectares", being the extension of biologically productive territory, calculated with respect to the average of world productivity.

Some analysts have used the ecological footprint to quantify the metabolic nature of cities, thus clearly demonstrating that the territories necessary to support urban consumption have an amplitude hundreds of times greater than the directly constructed areas (Girardet 2004), and the dependence of cities not only on their surrounding territory and region, but also their dependence upon other nations if not indeed upon other continents.

This "visualisation" of the impacts makes it possible to become aware of one's dependence on foreign supplies of resources (raw materials) and the need to contain one's impacts within the limits of the bio-productive capacities of one's territory. Extending the reasoning globally, it possible to become aware of the need to contain the impacts of human beings within the limits of the bio-productive capacities of world's territory.

5.5.2 Life Cycle Assessment of Cities

Life Cycle Assessment (LCA) is a methodology typically used for the environmental evaluation of buildings, but can also be used in the assessment of a city or an urban region, even if there is no standardization for this level of application (Albertí et al. 2017).

LCA is a technique to assess environmental impacts of a system (a building or a city), compiling an inventory of relevant energy and material inputs and environmental releases and evaluating the potential impacts associated with identified inputs and outputs. In order to compile this inventory, it is necessary to define the spatial and temporal boundaries of the system.

LCA can support a policy and decision-making process, identifying the most effective environmental strategy among the various possible alternatives. The peculiar aspect of this evaluation is that it permits the verification of the effects with respect to all the phases of the life cycle (from the extraction of the raw materials to the disposal in landfills or the conferment to the recycling of all the resources necessary for the operation of the system). It also allows a broad picture of impacts to be considered, thus avoiding the possibility of burden-shifting and trade-off, actions which shift environmental problems from one phase of the life cycle to another or from one environmental impact to another. For example, moving towards the choice of nuclear energy can be viewed positively if one looks only at the reductions in CO_2 emissions during the usage phase due to the non-combustion of fossil fuels, but significant problems are created for waste disposal, thus with significant environmental impacts on other environmental indicators and in other phases of the life cycle. Strategic choices made without considering the complete set of environmental effects generated can cause significant consequences.

5.5.3 Sustainability Benchmarks at Urban Scale and Planetary Boundaries

Environmental assessments can allow one to know the environmental impact of a system (e.g. a city) and to make comparisons between alternative strategic actions, but they also provide information to make comparisons between alternative cities, to understand which is the most virtuous.

Being able to make comparisons between systems becomes very useful in establishing which are the best practices, trying to extend them to other situations. Through the comparison between systems, it is possible to define what is the "typical" impact of a city and therefore to establish environmental impact benchmarks (Lavagna et al. 2018; Frischknecht et al. 2019), and consequently also the targets, to be introduced in the policies.

Examples can already be found in which public administrations/local authorities act through the definition of targets, which in some cases are quite radical.

For example, the C40 Climate Leadership Group (more than 75 of the world's largest cities connected with the aim of developing and implementing policies and programmes which generate measurable reductions in both greenhouse gas emissions and climate risks) launched the call for "Reinventing Cities", with the aim of promoting the regeneration of disused urban areas through zero-emission redevelopment.

Another interesting example is the Swiss target 2000 W Society that the citizens of Zurich voted in 2008 to implement in their constitution. The long-term goal is to achieve a primary energy use of 2000 W per person and an emission of no more than 1 ton of CO_2 equivalent per person and year. The target addresses not only personal or household energy use, but the total for the whole society, including embodied energy, divided by the population. The typical uses considered are: living and office space (this includes heat and hot water), food and consumer discretionary items (including services such as transportation of these to the point of sale), electricity, automobile travel, air travel, public transportation, public infrastructure.

In both cases, however, the critical aspect remains that of taking into consideration only one/two indicators of environmental impact and therefore there is a risk of orientation towards choices that determine burden-shifting on to other types of impact.

Furthermore, defining benchmarks can be important to ensure that from the awareness of how much impact is made, one tries to move towards reducing impacts, but if one wants to find a balance between one's activities and the natural environment, it is necessary to take a step further. It is not enough to try to reduce the current behaviours that are now out of bounds, but it is necessary to respect the limits of the planet (Meadows et al. 1972): being sustainable means considering the carrying capacity of our planet, that is nature's ability to provide us with resources and re-absorb our emissions.

In the case of the Ecological Footprint, it has already been shown that mankind's activities are beyond the limits of the planet and that three planets would be required to support growth at current rates.

In the case of LCA, one would need to know what are the environmental boundaries at a planetary level for each of the different calculated environmental indicators, but it is not an easy task to define them.

In 2009 an attempt was made on some indicators by a group of Earth system and environmental scientists led by Johan Rockström from the Stockholm Resilience Centre and Will Steffen from the Australian National University. The group wanted to define a "safe operating space for humanity" for the international community.

They identified nine "planetary life support systems" essential for human survival, attempting to quantify how far seven of these systems had been pushed already. Estimates indicated that three of these boundaries (climate change, biodiversity loss and the biogeochemical flow boundary) appear to have already been crossed.

This type of study is of paramount importance to be able to understand how far we are from sustainability and to address the issue of sustainability not as a small adjustment of behaviour, but as a radical inversion of our production and consumption models.

5.6 Conclusion and Suggestions

The application of the circular economy at the urban level is still to be clearly defined, overcoming the current limitative approach of waste management, aimed simply at solving the landfill problem and at ensuring a continuous supply of resources (without reducing consumption). "The concept of the Circular Economy in itself is over-hyped, scarcely investigated and therefore as yet ill-defined. It is so far dominated by a business-focused narrative for competitive advantage, raising questions about the placement of the Circular Economy within a broader urban sustainability agenda" (Prendeville et al. 2018).

Circular economy strategies must not be reductively interpreted as closing the outgoing flows (and therefore in a perspective of improving waste management), but as an overall reduction in flows, as it was in the original interpretations of the theme.

It is not enough to close the cycle, and it is also necessary to aim at reducing consumptions. The consumption process involves the entropic degradation of the materials and energy used: a certain amount of energy and matter will no longer be usable at the end of the process.

It is necessary also to avoid the rebound effect, that is the risk that people, trusting in the fact that so much the products are recycled, would increase its consumption.

These could also be the "hidden" objectives of the current interpretation of circular economy by economists, still focused on economic growth and the GDP indicator, elements that conflict with a scenario of stationary consumption of resources.

The most interesting strategies of the circular economy are those aimed at an economic paradigm shift: decoupling economic well-being from the growth of produced goods, aiming at other business models (car sharing), shifting the economy from materials-based manufacturing to knowledge-based industries (Chung and Gillespie 1998).

It should also not be forgotten that circularity must be a sustainability-oriented objective, so verification of the effectiveness of circular strategies is required. We must pay attention to strategies that proceed in parallel on different themes because optimizing single-issue actions can cause displacements of environmental problems: if I maximize energy efficiency, maybe increase the materials used and the waste produced (burden-shifting); if I maximize the recycling, maybe the throw-away increases (rebound effect). For example, the incineration of urban solid waste combined with district heating is an effective circularity process; however, if waste were perfectly differentiated, there would not be much to burn, and in addition the burning of waste still creates harmful substances (PM2.5). For this reason, it is important to act on the reduction and the extension of the useful life of the products (and of the resources stored in them).

It is also important that the theme of reuse and recycling does not become a way to solve the problem of landfills, putting waste in the house, or even on the plate. The issue of people's health must be put in the first place, and the recycling of waste must be strictly controlled.

Finally, it is important that politics, industry and research work together to succeed in achieving the ambitious goal of effective environmental improvement, combined with the development and well-being of society.

The policy can be the mover or the obstacle. It is important to act on regulations, regulatory impositions (limits on the consumption of land, prohibitions on the opening of new quarries, etc.) and economic levers (taxes on landfill costs, etc.), as driving elements for effective modification of the market (Giorgi et al. 2018). An appropriate market's use of natural capital is the most effective actions that a public administrator can do. It is also important to identify and remedy barriers, without obviously compromising health and safety.

The industry can be the proponent of change. Even if it is said that it is consumers who orient the market, in reality, today it is industry and the market that build people's induced needs and orient consumption. If the market went from a disposable logic to a circular logic, from a logic of obsolescence programmed to a prolongation of the life of the resources, from a profit based on the sale of the quantity to a profit based on access to quality services, this would reorient society's behaviour, untying profit from resource consumption.

Research can open scenarios (vision and objectives) and provide solutions. It is no coincidence that all the major issues (environmental sustainability, energy efficiency, circularity) were first addressed by the research world and then applied in the market and in politics after at least twenty years. The world of research has the ability to anticipate future scenarios and to identify the technical solutions to follow them.

Many of the themes dealt with in this text are not new, as has been demonstrated, but they are not even current since they are not yet fully applied: it took decades for the policy to welcome them, and years will be needed for the market and society to receive them. The hope is that the change will be more radical than that seen so far. It is not enough for us to adjust our behaviour, but it is necessary an effective change of mentality and objectives, which puts the environment and our health as focus.

References

Agudelo-Vera CM, Mels AR, Keesman KJ, Rijnaarts HHM (2011) Resource management as a key factor for sustainable urban planning. J Environ Manage 92:2295–2303

Albertí J, Balaguera A, Brodhag C, Fullana-i-Palmera P (2017) Towards life cycle sustainability assessment of cities. A Rev Backgr Knowl Sci Total Environ 609:1049–1063

Baccini P, Brunner PH (2012) Metabolism of the anthroposphere: analysis, evaluation, design. The MIT Press, Massachusetts Institute of Technology, Cambridge

Barles S (2009) Urban metabolism of Paris and its region. J Ind Ecol 13(6):898–913

Bonaiuti M (2011) Introduction. In: Bonaiuti M (ed) From bioeconomics to degrowth, N. Taylor and Francis, Georgescu-Roegen Bioeconomics in Eight Essays, Routledge

Boulding, Kenneth (1966) The Economics of the Coming Spaceship Earth

Brunner PH (2011) Urban mining. A Contrib Reindustrializing City J Ind Ecol 15(3):339–341

Burström F, Brandt N, Frostell B, Mohlander U (1998) Material flow accounting and information for environmental policies in the City of Stockholm. In: Bringezu S, Fischer-Kowalski M, Kleijn R, Palm V (eds) Analysis for action: Support for policy towards sustainability by material flow accounting. In: Proceeding of the conaccount workshop. Wuppertal Institute for Climate, Environment and Energy, Wuppertal, Germany, 11–12 Sept 1997 pp 136–145

CE100 (2016) Circularity in the built environment: case studies a compilation of case studies from the CE100. Ellen MacArthur Foundation. Available at: https://www.ellenmacarthurfoundation.org/assets/downloads/Built-Env-Co.Project.pdf

Chung C, Gillespie B (1998) Globalisation and the environment: new challenges for the public and private sector. In: Globalisation and the environment: perspectives from OECD and dynamic non-member economies. OECD, Paris

Cossu R, Williams ID (2015) Urban mining: concepts, terminology, challenges. Waste Manag 45:1–3

Daxbeck H, Lampert C, Morf L, Obernosterer R, Rechberger H, Reiner I, Brunner PH (1997) The anthropogenic metabolism of the City of Vienna. In: Bringezu S, Fischer-Kowalski M, Kleijn R, Palm V (eds) Regional and national material flow accounting: From paradigm to practice. In: Proceeding of the conaccount workshop. Wuppertal Institute for Climate, Environment and Energy, Leiden, Wuppertal, 21–23 Jan 1997, pp 249–254

EEA European Energy Agency (2015) Green economy. https://www.eea.europa.eu/downloads/a138a0fa46574c9e80d23c791e361e52/1528363382/green-economy.pdf

European Communities and Eurostat (2001) Economy-wide material flow accounts and derived indicators: a methodological guide, Luxembourg

Faist Emmenegger M, Frischknecht R (2003) Métabolisme du canton de Genève. Phase 1, ESU Service, Uster

Frischknecht R, Balouktsi M, Lützkendorf T, Aumann A, Birgisdottir H, Ruse EG, Hollberg A, Kuittinen M, Lavagna M, Lupišek A, Passer A, Peuportier B, Ramseier L, Röck M, Trigaux D, Vancso D (2019) Environmental benchmarks for buildings: needs, challenges and solutions—71st LCA forum, Swiss Federal Institute of Technology, Zürich, 18 June 2019. Int J Life Cycle Assess 24(12):2272–2280

Gehlen A (1988) Man: his nature and place in the world

Geddes P (1885) An analysis of the principles of economics. In: Proceedings of the royal society of Edinburgh. Williams and Norgate, London

Girardet H (1992) The gaia atlas of cities. Gaia Books

Girardet H (2003) Cities, people planet. In: Vertovec S, Posey DA (eds) Globalization, globalism, environment, and environmentalism: Consciousness of connections. Oxford University Press, New York, pp 87–102

Girardet H (2004) The metabolism of cities. In: Wheeler SM, Beatley T (eds) The sustainable urban development reader. Routledge, London, New York, Canada

Giorgi S, Lavagna M, Campioli A (2018) Guidelines for effective and sustainable recycling of construction and demolition waste. In: Benetto E, Gericke K, Guiton M (eds) Designing sustainable technologies, products and policies. Springer, Berlin, pp 211–221
Giorgi S, Lavagna M, Campioli A (2019) Circular economy and regeneration of building stock in the Italian context: policies, partnership and tools. In: SBE 2019 BAMB-CIRCPATH: buildings as material banks—a pathway for a circular future, bruxelles, IOP conference series: earth environmental science, vol 225
Hammer M, Giljum S, Bargigli S, Hinterberger F (2003) Material flow analysis on the regional level: questions, problems, solutions. NEDS Working Paper 2, Hamburg. http://alt.seri.at/en/publications/other-working-papers/2009/09/20/material-flow-analysis-on-the-regional-level-questions-problems-solutions/
Jernberg J, Hedenskog S, Huang CC (2019) Hammarby Sjostad. An urban development case study in Sweden, Stockholm. https://hammarbysjostad20.se/wp-content/uploads/2019/06/Hammarby-Sjostad_report_eng.pdf
Kibert Charles J (2005) Sustainable construction: green building design and delivery. Wiley, Hoboken
Lavagna M, Baldassarri C, Campioli A, Giorgi S, Dalla Valle A, Castellani V, Sala S (2018) Benchmarks for environmental impact of housing in Europe: definition of archetypes and LCA of the residential building stock. Build Environ 145:260–275
Matthews E et al (2000) The weight of nations: material outflows from industrial economies. World Resource Institute, Washington, DC
McArthur E, Foundation (2013) Towards the circular economy: economic and business rationale for an accelerated transition. https://www.ellenmacarthurfoundation.org/assets/downloads/publications/Ellen-MacArthur-Foundation-Towards-the-Circular-Economy-vol.1.pdf
McHarg Ian L (1969) Design with Nature. Garden City, N.Y., published for the American Museum of Natural History [by] the Natural History Press
Meadows Dennis L, Meadows Donella H, Randers J, Behrens III William W (1972) The limits to growth. Potomac Associates Universe Books, Falls Church
Migliore M, Lavagna M, Talamo C (2016) Circular economy in the building sector through the innovation and the development of new industrial strategies: the role of the information in the management of by-products and waste. In: 41st IAHS world congress sustainability and innovation for the future. Albufeira, Algarve, Portugal
Neirotti P, De Marco A, Cagliano AC, Mangano G, Scorrano F (2014) Current trends in smart city initiatives some stylised facts. Cities 38:25–36
Odum Eugene P (1975) Ecology, the link between the natural and the social sciences. Holt, Rinehart, and Winston, New York
Odum, Howard T (1983) Systems ecology: an introduction, Wiley, New York
Odum Eugene P (2003) How ecology has changed. In: Vertovec S, Posey DA (eds) Globalization, globalism, environment, and environmentalism: consciousness of connections. Oxford University Press, New York, pp 9–15
Prendeville S, Cherimb E, Bockenb N (2018) Circular cities: mapping six cities in transition. Environ Innov Soc Transit 26:171–194
Pomponi F, Moncaster A (2017) Circular economy for the built environment: a research framework. J Clean Prod 143:710–718
Rau T, Oberhuber S (2016) Material Matters: the alternative to our society of overexploitation. Bertram en De Leeuw Publishers, Haarlem
Rees William E, Wackernagel M (1996) Urban ecological footprints: why cities cannot be sustainable—and why they are a key to sustainability. Environ Impact Assess Rev 16(4–6):223–248
Rogers R (1997) Cities for a small planet. In: Gumuchdjian P (ed) Faber and Faber, London
Rogers R, Power A (2000) Cities for a small country. Faber and Faber, London
Stahel Walter R (1981) Product life factor. http://www.product-life.org/en/major-publications/the-product-life-factor

Unesco (2010) Urban connections of biosphere reserves. http://www.unesco.org/new/en/media-services/single-view/news/urban_connections_of_biosphere_reserves/
Vollmer C, Soriano P (2014) A new approach to urban mining, materials reclamation and business/job creation. https://studylib.net/doc/18369093/smart-biometrics-to-tsa
Wackernagel M, Rees William E (1995) Our ecological footprint: reducing human impact on the earth. Gabriola Press New Society Publishing, Gabriola Island, BC, Canada
Wernick Iddo K, Ausubel Jesse H (1995) National materials flows and the environment. Annu Rev Energy Env 20:463–492
Wolman A (1965) The metabolism of cities. Sci Am 213:179–190

Chapter 6
Universal Design in Sustainable Urban Planning

Alberto Arenghi

Abstract There is a *fil rouge* connecting Universal Design and sustainability which within the Brundtland Report can be recognized in the pillar of social sustainability. The demographic and epidemiological aspects that are affecting western countries, the definitions of health and healthy cities given by the World Health Organization and the international documents dealing with sustainable development oblige building designers and urban planners to reconsider their social role and become "health operators". The strict link between human beings and the built environment underlined both by Universal Design and International Classification of Functioning is the reasons why our cities and settlements need high quality urban spaces in order to enhance everyday life social dimension. Three urban projects in Manhattan, Den Haag and Copenhagen will be presented to show how sustainable urban planning can promote social interaction and inclusion, cohesion of communities, human health and well-being, cultural expression and dialogue among a wide diversity of people and cultures.

6.1 Demography, Epidemiology and Health

While planning cities, designing buildings and objects, thinking communications, we should always ask ourselves who will deal with them.

Some useful information can be obtained by analysing both demographic trends and epidemiological data. Actually, by considering the demographic situation in Europe and the USA related to 2017 and 2037 (Fig. 6.1), it is clear how the ageing

A. Arenghi (✉)
Department of Civil, Architectural, Environmental Engineering and Mathematics, University of Brescia, Via Branze 43, 25123 Brescia, Italy
e-mail: alberto.arenghi@unibs.it

G. Dall'O' (ed.), *Green Planning for Cities and Communities*,
Research for Development, https://doi.org/10.1007/978-3-030-41072-8_6

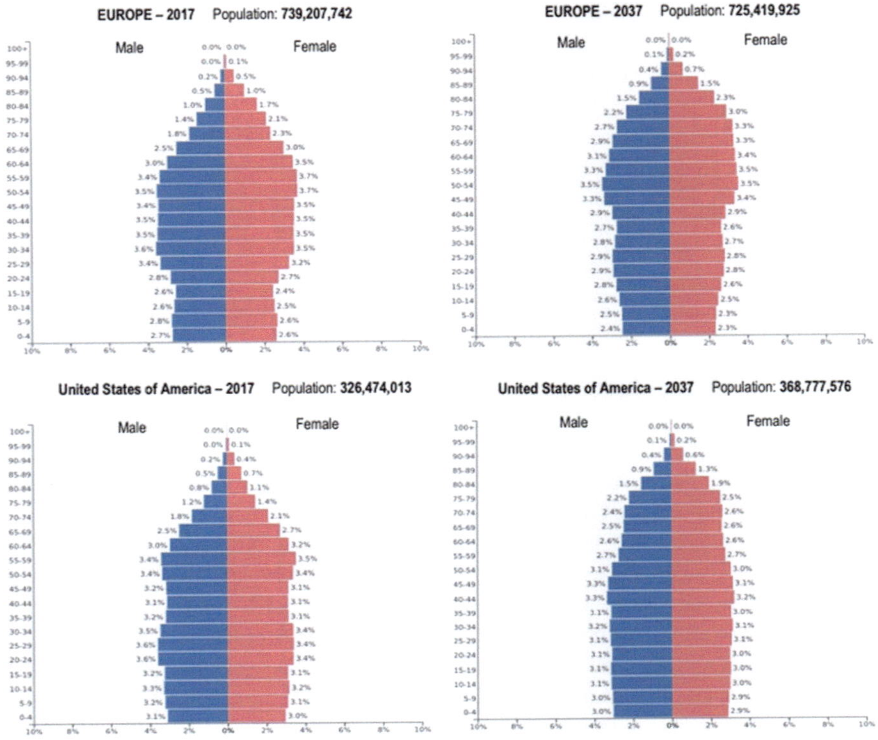

Fig. 6.1 Demographic trends in Europe and the USA related to 2017 and 2037 (https://www.populationpyramid.net)

effect is getting increasingly worrying (above all in Europe). Moreover, around 55% of the world population lives in urban areas today, increasing to 68% by 2050[1] (U.N. 2018).

Epidemiology tells us that in the same countries over the last century, the percentage of deaths attributed to infectious versus chronic diseases has completely changed. At the beginning of the twentieth century, most of the deaths were due to infectious diseases. Today, on the contrary, chronic diseases are the main cause of death.[2]

[1] According to John Wilmoth "The increasing concentration of people in cities provides a way of more economically providing services We find that urban populations have better access to health care and education" (The Guardian 2018).

[2] If we consider the trends related to New York in 1880 deaths for infectious diseases accounted for 57% and chronic diseases for 12%, in 1940, the scenario completely changed, and deaths for infectious diseases accounted for 11% and chronic diseases for 64%. This is due both to the improved hygienic conditions within the city and to the introduction of antibiotic. The trend has increased, and in 2005, the figures are 9% for infectious diseases and 75% for chronic diseases (City of New York 2010).

Most of these diseases are lifestyle related, driven primarily by the choices we make. People are constantly adopting unhealthy habits, whether intentionally or not, that may eventually lead to coronary artery disease, ischemic stroke, diabetes, obesity and some cancers. Some of the contributors are tobacco and junk food, together with decreased physical activity (Patra 2018).

By considering the demographic trends and the epidemiological data, the scenario we have to face is characterized by a population, mainly living in urban areas, which is becoming increasingly older and living with chronic diseases. Most of the time, above all when considering elderly people, the comorbidity is also present.

In this perspective, some interesting questions arise about the role of urban planner and building designer. Can they improve the everyday life of the people who live in our society within urban contexts? Have they any responsibility in keeping people in good health?

Firstly, we need to understand what is health, what we mean by it; secondly, who is mainly dealing with it.

According to the World Health Organization in 1948, "health is a status of complete physical, psychological and social well-being and not just the absence of illness" (WHO 1948). The definition is a bit cryptic even if it introduces an element, "social well-being" that gives to health a social dimension so that the same person can have good or bad health depending on social environments. In 1986, the World Health Organization has stated that "health is created and lived by people within the settings of their everyday life, where they learn, work, play and love" (WHO 1986).

This clarifies that health is a status which refers to our everyday life, and it is not only a medical issue.[3] Moreover, it has mainly a social dimension that recalls, indirectly, the environmental characteristics of the place(s) where everyday life has to be spent.

Once again, in 1998, the World Health Organization has provided the following definition: "A healthy city is one that is continually creating and improving those physical and social environments and expanding those community resources which enable people to mutually support each other in performing all the functions of life and in developing to their maximum potential" (WHO 1998).

From this evidence, it is quite clear that urban planners and building designers have a strong responsibility for people's health by acting on the environment where they spend their everyday lives. Urban planners and building designers have a prominent social role, which is often forgotten, and must become 'health operators'.

The focus must move from 'curing' to 'taking preventive care' since our society is characterized by chronic diseases which people must face in environments which mitigate their negative consequences.

More recently, the above-mentioned concepts have been stated in 2016 by the Shanghai Declaration on promoting health in the 2030 Agenda for Sustainable Development[4] (WHO 2017).

---

[3]In this perspective, Hans George Gadamer has written: "Wanting to define medical science in the best way, it can be qualified as a science of disease" (translation by the author) (Gadamer 1994).

[4]The 2030 Agenda for Sustainable Development will be considered subsequently.

In particular, the declaration says:

- "[we] united in the knowledge that health and sustainable urban development are inextricably linked and steadfastly committed to advancing both. We also recognize that health and well-being are at the core of the United Nations Development Agenda 2030 and its Sustainable Development Goals";
- "We commit to a Healthy Cities programme of action";
- "We recognize that creating Healthy Cities requires a comprehensive approach—it can never be the responsibility of one sector alone";
- "We also recognize that there is a powerful link between SDG 3 (Good Health for All) and SDG 11 (Make Cities and Human Settlements Inclusive, Safe, Resilient and Sustainable): unlocking the full potential of our cities to promote health and well-being and reduce health inequities will help to deliver both these goals";
- "Cities are at the front line of sustainable development, and we are convinced that mayors have the power to make a real difference. We must and will be ambitious in localizing the 2030 agenda, and we will set health targets to hold ourselves accountable. We recognize that everyone in the city will need to do their part to work towards these ambitious priorities".

6.2 Accessibility, Universal Design and ICF

In light of what has been said in the previous paragraph, the concept of accessibility and the cultural references related to Universal Design and International Classification of Functioning have to be analysed.

6.2.1 Accessibility

The concept of accessibility, to be understood as the ability of a social environment to be usable with the maximum ease and utility possible for any type of user, has historically and erroneously been associated only with the concept of disability, as well as the need to remove as many as possible architectural barriers in public and private spaces and services.

In this perspective, the definitions of accessibility to be found in the building standards of different countries have, more or less, the same meaning.[5]

Iwarsson and Ståhl have written that "Accessibility is a relative concept, implying that accessibility problems should be expressed as a person–environment relationship. In other words, accessibility is the encounter between the person's or group's

[5] According to the Italian law, "Accessibility is the possibility, even for people with partial or total motor or sensory impairments, to reach the building or its parts, to get in easily and to use all the spaces and the furniture safety and in autonomy".

functional capacity and the design and demands of the physical environment. Accessibility refers to compliance with official norms and standards, thus being mainly objective in nature.

Whenever using the concept of accessibility, statements must be based upon valid and reliable information gathered in three steps:

(1) The personal component (description of functional capacity in the individual or group being targeted, based on knowledge on human functioning).
(2) The environmental component (description of barriers in the environment being targeted, in relation to the norms and standards available).
(3) An analysis juxtaposing the personal component and the environmental component (description of accessibility problems)" (Iwarsson and Ståhl 2003).

Besides the meaning related to the person–environment relationship (functional aspect), accessibility has also an important legal value referring to the individual rights (socio-humanistic aspect) expressed by the Universal Declaration of Human Rights (UN 1948). These rights are present directly in the Constitutions or in the legal frameworks of different countries and deal not only with the access to the environment but also to the means of transportation, information, services, culture, etc., by considering people with disabilities. The legal value expresses the equal opportunity that should be insured for every citizen, being synonymous of empowerment and enabling action.

The definition[6] given the UN Convention on the Rights of Persons with Disabilities (UN 2006) has underlined this meaning.

In this perspective, Mia Mingus has recently observed that accessibility cannot be treated "as a logistical interaction, rather than a human interaction" (Mingus 2017).

Iwarsson and Ståhl in their work have introduced and defined the concept of usability which differs from accessibility. In particular "the concept of usability implies that a person should be able to use, i.e. to move around, be in and use, the environment on equal terms with other citizens. Accessibility is a necessary precondition for usability, implying that information on the person–environment encounter is imperative. However, usability is not only based on compliance with official norms and standards; it is mainly subjective in nature, taking into account user evaluations and subjective expressions of the degree of usability. Usability is a measure of effectiveness, efficiency and satisfaction. Most important, there is a third component distinguishing usability from accessibility, viz. the activity component.

---

[6]Art. 9.1 says: "To enable persons with disabilities to live independently and participate fully in all aspects of life, States Parties shall take appropriate measures to ensure to persons with disabilities access, on an equal basis with others, to the physical environment, to transportation, to information and communications, including information and communications technologies and systems, and to other facilities and services open or provided to the public, both in urban and in rural areas. These measures, which shall include the identification and elimination of obstacles and barriers to accessibility, shall apply to, inter alia:

(a) Buildings, roads, transportation and other indoor and outdoor facilities, including schools, housing, medical facilities and workplaces;

(b) Information, communications and other services, including electronic services and emergency services".

Whenever using the concept of usability, statements must be based upon valid and reliable information gathered in four steps:

(1) The personal component (description of functional capacity in the individual or group at target, based on knowledge on human functioning).
(2) The environmental component (description of barriers in the environment at target, in relation to the norms and standards available, but also based on user evaluation).
(3) The activity component (description of activities to be performed by the individual or group at target, in the given environment).
(4) An analysis integrating the personal, environmental and activity components (description of usability problems, i.e. description of the extent to which human needs, based on individual or group preferences, can be fulfilled in terms of activity performance in the environment at target)" (Iwarsson and Ståhl 2003).

To avoid any confusion or dogmatic position, two important assumptions made should be pointed out:

- accessibility of a place, a goods item or a service is not a fixed and final product, a result achieved once and for all, but is a dynamic notion subject to constant checks and audits based on the evolution of knowledge, sensitivity, social transformations and technological innovations: it is a never-ending process;
- accessibility of a place, a goods item or a service cannot be expressed with a 'yes' or 'not', but rather as a level of satisfaction on a scale of values, which we can define as 'accessibility degree'. This assessment is temporary and 'uncertain', and its margin of error is very wide, depending on many factors and first of all, on the accuracy with which these groups are defined (Arenghi et al. 2016).

6.2.2 Universal Design

In the early 1980s, Ronald Mace fully expressed the meaning of Universal Design by defining it as "the design of products and environments to be usable by all people, to the greatest extent possible, without the need for adaptation or specialized design. The Universal Design intends to simplify life for everyone by making products, communications, and the built environment more usable by as many people as possible, at little or no extra cost. Universal design benefits people of all ages and abilities" (Story et al. 1998). The definition does not strictly refer to "all" but "to as many people as possible" and therefore has a sense of limit: both with respect to the solution (each solution may present difficulties for a specific user) and with respect to the situation (humanity's complexity cannot be traced back to immutable patterns: there will always be particular situations that require customized solutions).

Now, the new integrated approach including Universal Design aims to respond equally to the needs of as many people as possible. Everyone should be able to enter and use any part of the built environment as independently and naturally as possible.

Table 6.1 Seven principles of Universal Design

Principle	Definition
1. Equitable use	Usable and marketable to people with diverse abilities
2. Flexibility in use	Accommodates a wide range of individual preference and abilities
3. Simple and intuitive use	Easy to understand, regardless of experience, knowledge, language skills or current concentration level
4. Perceptible information	Communicates necessary information effectively, regardless of ambient conditions or sensory abilities
5. Tolerance for error	Minimizes hazards and adverse consequences of accidental or unintended actions
6. Low physical effort	Can be used efficiently and comfortably, with a minimum of fatigue
7. Size and space for approach and use	Appropriate size and space for approach, reach, manipulation and use regardless of body size, posture or mobility

The criteria defining "normality" should be enlarged to ensure that the construction of the built environment is based upon Universal Design principles (Table 6.1). A new awareness of design and construction is needed.

To explain the process which led to the definition of Universal Design, a Gaussian distribution can be considered for the representation of the whole of humankind with all its differences (Fig. 6.2a). By trying to represent all, it is designing for a standard that does not fit almost anyone (Fig. 6.2b) which emerges. When disability is considered, another standard (designing for disability) (Fig. 6.2c) results. With a holistic approach which considers as many potential users as possible, Universal Design arises (Fig. 6.2d). In this case, somebody is still missing. A grey area of population needs some adjustments, some dedicated or particular solutions (Fig. 6.2e).

This approach has also been expressed by the UN Convention on the Rights of Persons with Disability (UN 2006) which defines at article 2: "Reasonable accommodation" means necessary and appropriate modification and adjustments not imposing a disproportionate or undue burden, where needed in a particular case, to ensure to persons with disabilities the enjoyment or exercise on an equal basis with others of all human rights and fundamental freedoms.

Universal Design has an anthropocentric approach and from a certain point of view, considers the user as a passive subject whom the designer has to pay his/her attention and direct his/her effort to.

Even though the term accessibility and one of its meanings is related to persons with disabilities and the Universal Design approach has too been developed to solve the problems caused by architectural barriers to persons with disabilities, nowadays, it is clear that both refer to a consistent variety of people (elderly, pregnant women,

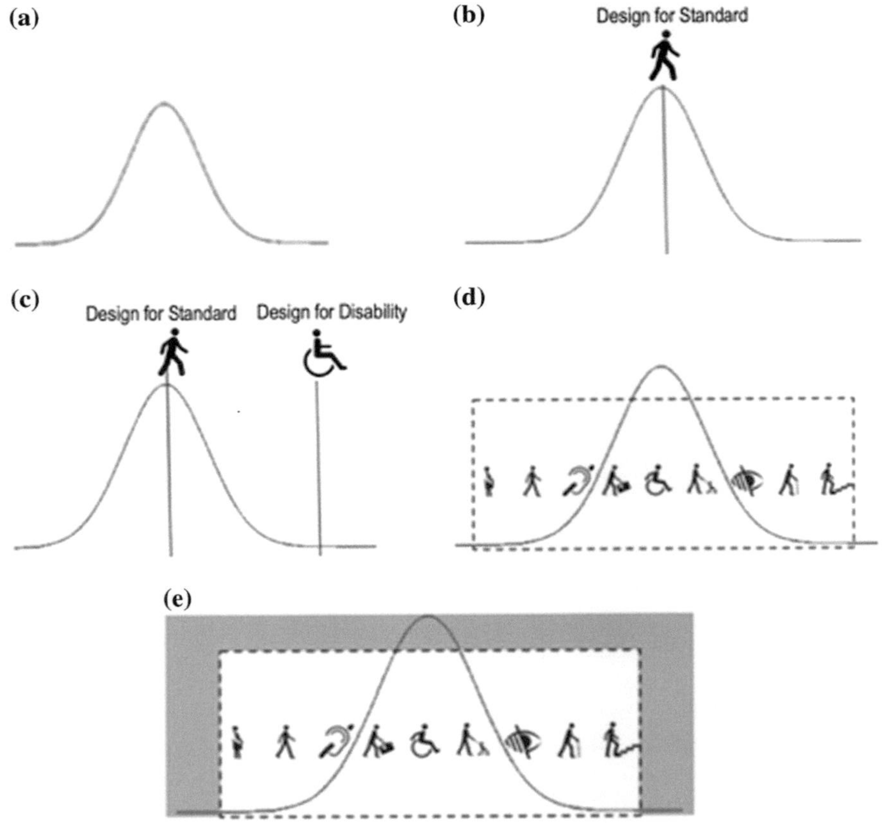

Fig. 6.2 Process which led to the definition of Universal Design (scheme by the Author)

children, people with temporary impairments, people moving heavy loads and persons with disabilities). This is particularly important when dealing with the scenario which has been presented in the first paragraph.

6.2.3 International Classification of Functioning

The International Classification of Functioning (ICF) introduced by World Health Organization in 2001 has favoured the evolution of the concept of disability, from the medical model to the bio-psycho-social model, calling attention to the possibilities of participation, denied or favoured by environmental conditions. This has induced public and private operators to identify and to conduct positive actions to ensure the effective and full integration of all citizens and their participation in social life in all its aspects, including the cultural one. This new anthropological approach has

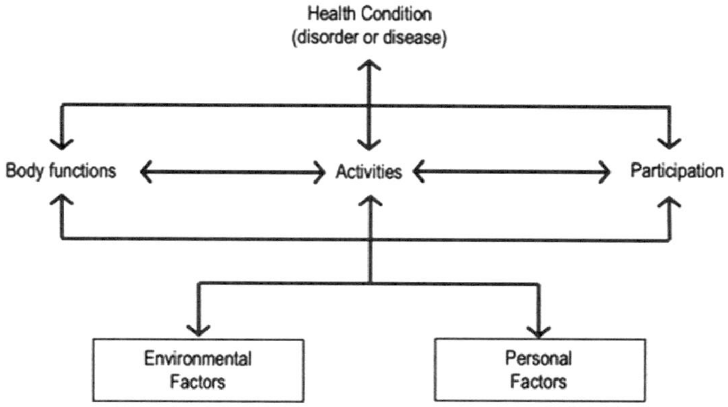

Fig. 6.3 Interaction between the components of ICF (WHO 2001)

strongly moved attention to user participation so that the user is an active subject and the environment must be designed accordingly: this is not only in order to reduce the experience of disability but above all to enhance everyone's experience and performance.

The ICF describes the person in his/her inseparable value body–function–environment, such that:

- good health could become illness if placed in an unsuitable environment,
- on the opposite manner, a well-designed environment makes a biological suffering condition acceptable if not liveable.

The ICF has facilitated the evolution of the concept of disability, from the medical model to the bio-psycho-social one; drawing attention to the possibility of participation, denied or favoured by environmental conditions (Fig. 6.3).

6.3 Sustainable Urban Planning

It is widely recognized that the three pillars of sustainable development are economic, environmental and socio-cultural sustainability.

However, over the years, it has been observed that the social dimension of sustainability, in comparison with the others (and particularly the economic aspect), has received less attention and importance (McKenzie 2004), only slightly affecting the national and international strategies.

Nevertheless, it has been noted that more subjective key themes emerge nowadays, complementing and/or substituting the traditional one for social sustainability: the sense of the places, social participation, inclusion, quality of life amongst others. Moreover, it has been acknowledged (Syrazwani and Mariam 2013) that the built environment—which provides space to grow and involve the community—can foster the new criteria for social sustainability, amongst which accessibility, equity, empowerment, participation, cultural identity and institutional stability play a critical role in allowing a socially correct distribution of benefits (and costs) coming from the management of the environment (Khan 1995).

In its broader meaning of "process for creating sustainable, successful places that promote well-being, by understanding what people need from the places in which they live and work", social sustainability fosters the design of human habitat (private and public living spaces and facilities) to allow active participation and inclusion (Syrazwani and Mariam 2013).

Social sustainability and healthy urban spaces also have an important impact on economic sustainability. Gro Harlem Brundtland, director of the World Health Organization in 2001 defined "discomfort" (the disease) as subtraction of resources from development, while the investments made for health can be a concrete resource for economic development. Only healthy people, with the support of an efficient health service, can ensure the sustainable development of their societies. A loss of health involves the person who suffers but also his/her family and society as a whole, affecting emotional and psychological resources with economic implications which are often dramatic and irreversible.

The UN 2030 Agenda for Sustainable Development (UN 2015) is one of the first documents which try to encompass as many people as possible similarly to the approach proposed by the Universal Design.

As above-mentioned goal 11 and goal 3 are those that better fit the current discussion:

- **Goal 11—Sustainable Cities and Communities: Make cities inclusive, safe, resilient and sustainable**. "Cities are hubs for ideas, commerce, culture, science, productivity, social development and much more. At their best, cities have enabled people to advance socially and economically (…). There needs to be a future in which cities provide opportunities for all, with access to basic services, energy, housing, transportation and more".[7]
- **Goal 3—Good Health and Well-being: Ensure healthy lives and promote well-being for all at all ages**. "Ensuring healthy lives and promoting the well-being at all ages is essential to sustainable development (…)".

The New Urban Agenda "share(s) a vision of cities for all, referring to the equal use and enjoyment of cities and human settlements, seeking to promote inclusivity and ensure that all inhabitants, of present and future generations, without discrimination of any kind, are able to inhabit and produce just, safe, healthy, accessible, affordable,

[7] In particular consider the following targets:
11.1 By 2030, ensure access for all to adequate, safe and affordable housing and basic services and upgrade slums;
11.2 By 2030, provide access to safe, affordable, accessible and sustainable transport systems for all, improving road safety, notably by expanding public transport, with special attention to the needs of those in vulnerable situations, women, children, persons with disabilities and older persons;
11.3 By 2030, enhance inclusive and sustainable urbanization and capacity for participatory, integrated and sustainable human settlement planning and management in all countries.

resilient and sustainable cities and human settlements to foster prosperity and quality of life for all" (UN 2017).

These documents stress their attention to the social dimension of a sustainable city. The planning and the design of streets, sidewalks pavements and cycling lanes, squares, waterfront areas, gardens and parks must be safe, inclusive, accessible, green and have high quality in order to promote social interaction and inclusion, human health and well-being, economic exchange and cultural expression and dialogue among a wide diversity of people and cultures. They must ensure human development and build peaceful, inclusive and participatory societies, as well as to promote living together, connectivity and social inclusion.

6.4 Designing Sustainable and Healthy Urban Spaces

Jan Gehl is a master reference when dealing with the relationship between urban spaces and the inhabitants' behaviour. Gehl identifies three main outdoor activities: necessary activities, optional activities and social activities. If the first is almost independent of the exterior conditions, optional and social activities are strictly related to the quality of the outdoor spaces. In a good urban environment, people are encouraged to go for a walk, stop for a pause, sit, eat, play (optional activities) and thus meeting other people and sharing time for greetings and conversations, communal activities of various kinds, "and finally—as the most widespread social activity—passive contacts, i.e. simply seeing and hearing other people" (Gehl 2011).

"Just as it is possible through choice of materials and colours to create a certain palette in a city, it is equally possible through planning decisions to influence patterns of activities, to create better or worse conditions for outdoor events and to create lively or lifeless cities" (Gehl 2011).

In particular, "the city with reasonably low, closely spaced buildings, accommodation for foot traffic and good areas for outdoor stays along the streets and in direct relation to residences, public buildings, places of work and so forth. Here, it is possible to see buildings, people coming and going and people stopping in outdoor areas near the buildings because the outdoor spaces are easy and inviting to use. This city is a living city, one in which spaces inside buildings are supplemented with usable outdoor areas, and where public spaces have a much better chance of working well" (Gehl 2011).

Gehl's lesson not only explains how important is the quality of the urban space to enhance its social dimension, but his words recall the definition of health and healthy cities by the World Health Organization. When thinking back to the scenario presented at the beginning of this Chapter, it is clear that if a city offers spaces which provide freedom, permitting dialogue, allowing coming and going, affecting the sense of belonging to a community, then that city can be defined as healthy and sustainable.

An interesting approach for sustainable and healthy urban planning has been developed by the City of New York as of the beginning of the century. The study had been conducted to face up to the increasing problem of obesity that affects an important part of the American population. Mayor Bloomberg had asked his Technical Office whether it was able to contribute to tackling the issue by using architecture and urban planning.

They came up with Active Design whose definition is "(Active Design is) an approach to the development of buildings, streets and neighbourhoods that uses architecture and urban planning to make daily physical activity and healthy foods more accessible and inviting" (City of New York 2010).

According to Active Design, five design qualities are critical to a good walking environment. "These characteristics were defined qualitatively and then related to the physical features of the street environment:

- **Imageability** is the quality of a place that makes it distinct, recognizable and memorable. A place has high imageability when specific physical elements and their arrangement capture attention, evoke feelings and create a lasting impression;
- **Enclosure** refers to the degree to which streets and other public spaces are visually defined by buildings, walls, trees and other vertical elements;
- **Human scale** refers to a size, texture and articulation of physical elements that match the size and proportions of humans and equally important correspond to the speed at which humans walk;
- **Transparency** refers to the degree to which people can see or perceive objects and activity—especially human activity—beyond the edge of a street;
- **Complexity** refers to the visual richness of a place. The complexity of a place depends on the variety of the physical environment.

The researchers found that the presence of these five qualities enhances the public realm" (City of New York 2010).

In the following sub-sections, three urban projects are presented. Although they come from different contexts and have been elaborated starting from different requirements and demands, all of them explain what is the meaning of urban health, inclusion and accessibility: in two words urban sustainability as described above.

With regard to Universal Design principles, **flexibility in use** appears to be the common characteristic: people (in particular inhabitants, tourists and children) live these spaces as a function of what they feel both physically and psychologically, according to how their natural capabilities permit their interaction with those spaces and dependent upon the people who are experiencing them.

6.4.1 The High Line Park, Manhattan

The High Line was constructed in the early 1930s as part of New York's West Side railway. Part of it was taken out of service in the 1960s, and the last train ran in 1980, carrying "three boxcars full of frozen turkeys" for Thanksgiving.

In 2003, an open competition was held to convert the existing infrastructure into a public park. The winning proposal by James Corner Field Operations with Diller Scofidio + Renfro includes over a dozen access points to the elevated park.

Starting in 2006, work began on the redevelopment of the line, creating an elevated park, a strip of green at a height of 10 m, right above the busy metropolitan streets. With the last segment opened in 2014, a 2.4 km long panoramic garden with 210 different species of plants connects Gansevoort Street with 34th Street.

The High Line is now a green path inspired by post-industrial ruins, where nature takes over the infrastructures in decline: the design of the greenery is in fact inspired by the vegetation itself which, after years of neglect, takes over and many plants which had grown spontaneously between the tracks have been incorporated into the landscape project, giving rise to a disorderly beauty that alternates wild places with urban gardens.

"Through a **strategy of agri-tecture** (Fig. 6.4)—part agriculture, part architecture—the High Line surface is digitized into discrete units of paving and planting which are assembled along the 1.5 miles into a variety of gradients from 100% paving to 100% soft, richly vegetated biotopes. The paving system consists of individual precast concrete planks with open joints to encourage emergent growth such as wild grass through cracks in the sidewalk.

The long paving units have tapered ends that comb into planting beds, creating a textured, "pathless" landscape where the public can meander in unscripted ways. The park accommodates the wild, the cultivated, the intimate and the social" (Diller Scofidio + Renfro 2019).

Fig. 6.4 High Line park: "agri-tecture". Photo by Iwan Baan, Courtesy of Diller Scofidio + Renfro

Fig. 6.5 High Line park: the path through the skyscrapers. Photo by Gianluca Di Rosario

Panoramic seats and chaise longues, rigorously made of a biocompatible material, alternate with greenery, overlooking the Hudson River or the street below and create suggestive resting and social gathering spaces and allow the users to enjoy fabulous views of the Manhattan skyline. The route has become the venue for alternating events: visitors can experience t'ai chi, walking tours, meditation, comedy, music, gardening and even boxing (Fig. 6.5).

6.4.2 *Melis Stokepark, Den Haag, Nederlands*

"In 2007, Carve was asked by the municipality of the Hague to design two 'integrated play facilities', playgrounds suitable for children with and without disabilities. But how does one design a playground where the difference in play between children with and without disabilities is eliminated? A playground that offers challenges and appeals to all? In Carve's vision, **'playing together' doesn't mean playing next to each other**" (Carve 2019).

The playground, located inside a park, is not fenced. It is not a set of equipment because the equipment creates an integrated project composed of natural and constructed elements. It has a ring shape that constitutes a path with many changes in altitude and is, at the same time, a permeable boundary in several points. The external

Fig. 6.6 Melis Stokepark: internal part. Courtesy of Carve

walls of the ring are covered in wood with supports for climbing, while the internal walls, covered with blue anti-shock material, are inclined surfaces to climb, to slide down, depending upon the user's abilities, and can be used in different ways (Figs. 6.6 and 6.7). Several passages lead to and from this inner area and offer their seating and playing possibilities. The curved shape generates a cosy environment, which is helpful for those children who have difficulty in dealing with large open spaces.

Inside, around and on the ring, there are wide spaces available for challenging active games, repetitive movement games (turning, sliding, jumping and swinging), constructional play (sand) and for games of fantasy (tunnels, platforms, shelters).

6.4.3 Superkilen, Nørrebro, Copenhagen, Denmark

The 750 m long urban area of Superkilen was created in the city of Copenhagen at a key meeting place for different cultures and neighbourhoods, clearly identifiable by their strong heterogeneity.

Superkilen is a park that supports diversity. In this way, it is possible to define the project, in which every user of any origin and nationality can identify with different objects present on the surface of the public space. The objects come from 57 different nations, and each of them characterizes a specific area of the square.

Fig. 6.7 Melis Stokepark: external part. Courtesy of Carve

Apparatus for physical exercises coming from the beaches of Los Angeles, elements from the sewerage systems of Israel, palm trees of China and luminous signs from Qatar are some of the objects of common use which, if placed in a square, become elements of furniture and decoration of a multi-ethnic square.

Superb is the result of the creative collaboration between BIG, Topotek 1 and SUPERFLEX, which constitutes a rare fusion of architecture, landscape architecture and art from early concept to the construction stage.

Superkilen is the meeting of different cultures and vegetation which, thanks to large sectoral flowerbeds, leads to a system of different seasonal blooms which follow one another, allowing the square to "be alive" in every period of the year.

The area is subdivided into three communicating and distinct parts, each identified by a different dominant colouration. The red zone continues the extended area dedicated to outdoor sports activities, with a wide range of recreational and group games. At one extremity of the longitudinal spread of this project, the red square includes a piece of gently rising ground, creating a local market and a parking system (Fig. 6.8).

The black square is identified by the dark tone of the pavement and the presence of deformed levels of the surface, thus recalling desert dunes. This area of the square is dedicated to the East with the presence of Moroccan fountains and references to Turkish culture with benches and chairs. This area is dotted with Japanese cherry trees, which have been inserted so that during the suitable flowering period, they colour the otherwise totally black space (Fig. 6.9).

Fig. 6.8 Superkilen: the red area. Courtesy of BIG Architects

Fig. 6.9 Superkilen: the black area. Courtesy of BIG Architects

Fig. 6.10 Superkilen: the green area. Courtesy of BIG Architects

The green area is identified with the "Green Park" which, starting from the black surface of the square, extends over a long strip through the districts of the city. The environment is endowed with a few curved paths and is identified by the strong spontaneity of nature itself, with different vegetation and overlapping contours, as if to point out the uncontrollability of natural elements. When following the green path, the visitor can use the spaces without limits or boundaries and in complete safety. This is a place designed for families who want to spend an afternoon on the lawn for a picnic or to sunbathe (Fig. 6.10).

6.5 Conclusions

Most industrialized countries are experiencing important and unavoidable demographic changes (an increasingly older population flanked by a steady decline in the birth rate) and a significant change (underway since the beginning of the last century) of an epidemiological nature in which today the mortality rate is attributable to chronic diseases rather than infectious diseases. If on the one hand, all this must be read as a success story (longer life expectancy), and on the other hand, it requires important and urgent answers on the understanding of the needs of a society that already today is increasingly constituted by elderly people with chronic diseases.

Moreover for these very reasons, the theme of accessibility of cities, buildings and services can no longer be addressed by thinking that it concerns a niche of people (persons with disability) but more properly by seeing therein a modality, which becomes necessity, to allow a large part of the population to lead a life in an "enabling" environment, (healthy cities), that foster participation and healthy lifestyles.

Universal Design principles and International Classification of Functioning are useful references to conceive the design of buildings and products and the planning of urban spaces which work not only for as many users as possible but also enhance everyone's experience. This approach fits tangibly social sustainability which, over the last few years, has been gaining ever more attention in the international documents dealing with sustainable development.

The actual planning of accessible and sustainable urban spaces means considering the prominent social role that a designer has to take into account in order to promote social interaction and inclusion, communities' cohesion, human health and well-being, economic exchange and cultural expression and indeed dialogue between a wide diversity of people and cultures. It must ensure human development and build peaceful, inclusive and participatory societies, as well as promoting living together, connectivity and social inclusion.

This result can be attained only by planning high quality urban spaces which people can use in different ways, depending on their attitudes and therefore enhancing their health.

A cultural leap is, therefore, necessary because sustainability underlies the complexity which today can no longer be addressed sectorally but with an "ecosystemic" approach where the ecosystem is the citizen in everyday life. From this point of view, those who today, in the minds of many, appear only to represent costs to meet the needs of a population niche should instead be taken and acted upon as an added value, even in economic terms.

References

Arenghi A, Garofolo I, Lauria A (2016) On the relationship between 'universal' and 'particular' in architecture. In: Petrie H et al (eds) Universal design 2016: learning from the past, designing for the future. studies in health technologies and informatics. vol 229. IOS Press, Amsterdam, Berlin, Washington DC, pp 31–39

Carve (2019) Melis stokepark. available at: http://www.carve.nl/en/item/21. Accessed July 2019

City of New York (2010) Active design guidelines: promoting physical activity and health in design

Diller Scofidio + Renfro (2019) The high line. Available at: https://dsrny.com/project/high-line. Accessed July 2019

Gadamer HG (1994) Dove si nasconde la salute, Raffaello Cortina Editore, Milano [original title: Gadamer, H.G. (1993), Über die Verborgenheit der Gesundheit, Suhrkamp, Frankfurt a. M.]

Gehl J (2011) Life between buildings. Using Public Space, Island Press, Washington DC

Iwarsson S, Ståhl A (2003) Accessibility, usability and universal design-positioning and definition of concepts describing person-environment relationships. Disabil Rehabil 25(2):57–66

Khan MA (1995) Sustainable development: the key concepts, issues and implications, keynote paper given at the international sustainable development research conference. Wiley, Manchester, UK, Sustain Dev 3(2):63–69

McKenzie S (2004) Social sustainability: towards some definitions. University of South Australia, Adelaide, Australia
Mingus M (2017) Forced intimacy: an ableist norm. available at: https://leavingevidence.wordpress.com/2017/08/06/forced-intimacy-an-ableist-norm/. Accessed July 2019
Patra S (2018) Public health: infectious vs chronic diseases. J Biomed Sci 7(4):15
Syrazwani AK, Mariam J (2013) Universal design as a significant component for sustainable life and social development. In: ASEAN conference on environment-behaviouras studies proceedings. Hanoi, Vietnam, 19–22 Mar 2013, pp 179–190
Story MF, Mueller JL, Mace RL (1998) The universal design file: designing for people of all ages and abilities. Center for Universal Design, NC State University
The Guardian (2018) Two-thirds of world population will live in cities by 2050, says UN. available at: https://www.theguardian.com/world/2018/may/17/two-thirds-of-world-population-will-live-in-cities-by-2050-says-un. Accessed July 2019
UN (1948) Universal declaration of human rights, Paris. available at: https://www.un.org/en/universal-declaration-human-rights/. Accessed July 2019
UN (2006) Convention on the rights of persons with disabilities (UN-CRPD) (A/R/61/106), New York
UN (2015) Transforming our world: the 2030 agenda for sustainable development (A/R/70/1), New York
UN (2017) New Urban Agenda. HABITAT III (A/RES/71/256), New York
UN (2018) World urbanization prospects 2018. Available at: https://esa.un.org/unpd/wup/. Accessed July 2019
WHO (1948) Constitution of the World Health Organization, Geneva
WHO (1986) Ottawa Charter for Health Promotion, Geneva
WHO (1998) Health Promotion Glossary, Geneva
WHO (2001) International classification of functioning, Disability and Health (ICF-DH), Geneva
WHO (2017) Promoting health in the SDGs, Report on the 9th Global conference for health promotion. all for health, health for all, Geneva, Shanghai, China, 21–24 Nov 2016

Chapter 7
Green Energy Planning of Cities and Communities: New Paradigms and Strategies for a Sustainable Approach

Giuliano Dall'O'

Abstract In green planning, the correct management of energy in towns and cities is a critical issue and is the topic of this chapter. In the first part, a general overview is provided, identifying the new paradigms to be pursued according to the needs of reducing, and possibly eliminating, the emission of greenhouse gases and those gases that have an impact on the environment, also considering the technological evolution currently in progress and opportunities offered by the market of energy generation and management market. In the second part of the chapter, a methodological approach to Green Energy Planning on a territorial scale is proposed and discussed. The chapter is completed by an in-depth analysis of the European project which promotes an approach to bottom-up energy planning, the Covenant of Mayors: its methodologies are discussed and a case study is analysed.

7.1 New Paradigms for Energy Planning on an Urban Scale

7.1.1 Affordable and Clean Energy

When discussing energy, one of the major causes of environmental pollution in cities and metropolitan areas and indeed the production of greenhouse gases (GHG) on a global scale, a paradox becomes apparent. On the one hand, the consumption of fossil energy must be reduced, with a drastic change of approach, which is not only technological but also economic and social, but on the other hand, access to energy is not yet guaranteed to a significant portion of the world's population: overcoming this deficiency means a potential increase in energy consumption.

One of the 17 Sustainable Development Goals, namely the SDG 7 aims to "Ensure access to affordable, reliable, sustainable and modern energy for all".

G. Dall'O' (✉)
Architecture, Built Environment and Construction Engineering, ABC Department, Politecnico di Milano, Via G. Ponzio, 31, 20133 Milan, Italy
e-mail: giuliano.dallo@polimi.it

G. Dall'O' (ed.), *Green Planning for Cities and Communities*,
Research for Development, https://doi.org/10.1007/978-3-030-41072-8_7

The objectives to be achieved in 2030, established by the UN, are very ambitious but absolutely necessary[1]:

- ensure universal access to affordable, reliable and modern energy services;
- increase substantially the share of renewable energy in the global energy mix;
- double the global rate of improvement in energy efficiency;
- enhance international cooperation to facilitate access to clean energy research and technology, including renewable energy, energy efficiency and advanced and cleaner fossil fuel technology and promote investment in energy infrastructure and clean energy technology;
- expand infrastructure and upgrade technology for supplying modern and sustainable energy services for all in developing countries, in particular least developed countries, small island developing states and land-locked developing countries, in accordance with their respective programmes of support.

The SDGs are measured every year through indicators, and in this way it is possible to check the progress. As for SDG 7, some results have been obtained (UN 2018):

- the global electrification rate rose from 83% in 2010 to 87% in 2015, with the increase in accelerating to reach 89% in 2017. The global share of the population with access to clean cooking fuels and technologies reached 61% in 2017, up from 57% in 2010; despite this progress, close to 3 billion people still rely primarily on inefficient and polluting cooking systems;
- the renewable energy share of total final energy consumption gradually increased from 16.6% in 2010 to 17.5% in 2016, though much faster change is required to meet climate goals. Even though the absolute level of renewable energy consumption has grown by more than 18% since 2010, only since 2012 has the growth of renewables outpaced the growth of total energy consumption;
- global primary energy intensity ratio of energy used per unit of gross domestic product (GDP) improved from 5.9 in 2010 to 5.1 in 2016, a rate of improvement of 2.3%, which is still short of the 2.7% annual rate needed to reach target three of Sustainable Development Goal 7;
- international financial flows to developing countries in support of clean and renewable energy reached $18.6 billion in 2016, almost doubling from $9.9 billion in 2010.

Despite the progress made, the targets to be reached are still very distant. Amongst the many critical points, a particularly worrying one emerges: around 800 million people still remain without electricity.

A problem that is increasingly common also in the industrialised countries is of an economic nature: even when energy is available, sometimes its cost limits its usage (so-called fuel poverty[2]). If human life without energy is impossible, what attributes

[1]The data are taken from: https://sustainabledevelopment.un.org/sdg7.

[2]In the UK, fuel poverty is defined by the Warm Homes and Energy Conservation Act as: "a person is to be regarded as living "in fuel poverty" if he is a member of a household living on a lower income in a home which cannot be kept warm at reasonable cost" (source: http://www.legislation.gov.uk/ukpga/2000/31/contents).

should energy have? The answer is to be found in SDG 7: energy must be affordable, reliable, sustainable and modern for everybody.

The energy planning of our cities, which we call Green Energy Planning, must start from this very objective, trying to provide a concrete answer, through its actions, to what is required to meet the SDG 7.

7.1.2 Towards a Green Energy Transition

Without the presence of human beings, the Earth's ecosystem would be self-sufficient in terms of energy: plants have always exploited solar energy and have become the basic elements of the food chain. The discovery of energy from fossil sources, initially at low cost, has allowed humanity to live better, for example in comfortable environments both in winter and in summer or to achieve absolute mobility through public transport, or indeed private vehicles.

The anthropization of the land masses, also through the use of energy which has grown exponentially over the years, has generated a conflict between the built environment (by human beings) and the natural one. Between 1971 and 2017, the world total primary energy supply (TPES) increased by more than 2.5 times (from 5.519 to 13.972 Mtoe) (IEA 2019). Comparing the world population between those two historical thresholds 3.76 trillion in 1971 and 7.9 trillion in 2017 one observes an increase of about two times: this means that over the years per capita energy consumption has increased. In addition to the environmental problem, or in parallel with them, one should consider the critical issue of the availability of fossil energy sources.

Renewable energy sources, whose contribution has been considered marginal over the years, show an interesting growth trend. According to Eurostat, the share of energy from renewable sources in gross final energy consumption has almost doubled in recent years, going from about 8.5% in 2004 to 17.0% in 2016.[3] Not only the difference expressed in percentage terms between the two reference thresholds but also the linear growth trend is interesting.

The use of renewable energy sources offers significant possible advantages, including a reduction in GHG emissions, diversification of energy supply and less dependence on fossil fuel markets (in particular those of oil and gas). The development of the use of renewable energy sources could (and indeed already has) also stimulate employment by creating jobs in new "green" technologies.

The "green energy" transition, supported by the energy policies expressed by many countries (e.g. the European Directives on energy efficiency and innovative energy sources) and by technological innovation characterised by interesting growth trends, should be one of the paradigms of Green Energy Planning.

Its success will be even greater if at the same time maximum efforts are made to reduce energy requirements by promoting action on energy efficiency. In the

[3]Source: https://ec.europa.eu/eurostat/databrowser/view/t2020_31/default/line?lang=en.

energy planning of towns and cities, all efforts must be made to promote priority position renewable energy sources, leaving behind what was their traditional role as an integrative source.

As far as the energy transition is concerned, an interesting scheme has been proposed by the Energy Atlas (Heinrich Böll Foundation 2018) which defines in 12 points the actions that can support an energy transition in Europe (Table 7.1).

Table 7.1 Twelve brief lessons on Europe's energy transition (Heinrich Böll Foundation 2018, graphic designer Bartz/Stockmar CC BY 4.0)

#	Lesson	#	Lesson
1	Energy has historically been a key driver of European COOPERATION. But current EU proposals are not enough. To comply with the Paris Climate Agreement, we MUST GIVE UP fossil fuels altogether by 2050	7	Digitalisation can make this transformation more DEMOCRATIC AND EFFICIENT and can reduce the bill for the end consumer
2	A 100% renewable energy system in Europe is now technically possible using existing STORAGE and DEMAND RESPONSE technologies	8	The European energy transition promises to increase PROSPERITY in a sustainable way (creating more local jobs) and boost Europe's global LEADERSHIP in green innovations
3	Stronger INTERCONNECTIONS of markets and infrastructure across Europe will make the energy transition cheaper for all Europeans	9	Since 2013, renewables have helped SLASH Europe's import bill for fossil fuels by more than a third, CUTTING ITS DEPENDENCY on unstable and unpleasant regimes
4	The biggest potential lies in INCREASING EFFICIENCY. Europe-wide we could reduce our energy demand by half by 2050	10	A SOCIALLY JUST TRANSITION is both essential and viable: all over Europe, the renewables sector already provides more well-paid, secure local jobs than the coal industry
5	A switch to 100% renewables in Europe will trigger SYSTEM CHANGE—away from centralised, monopolistic utilities to decentralised, community power projects and innovative business models	11	ENERGY POVERTY is being tackled by pioneering community power projects, acting in solidarity with those in their own community addressing this challenge
6	Framed by smart strategies and legislation, this system change can be driven by CITIZENS, CITIES AND ENERGY COOPERATIVES, leaving much more wealth in communities	12	Europe's Neighbourhoods Policy should INSPIRE AND SUPPORT other countries to decarbonise their economies. A socially just energy transition in Europe's neighbouring regions can stimulate their progress and stability

7.1.3 Making Connections for Smart Energy

Livio De Santoli, a professor of Università La Sapienza—Rome, argues that "Making connection represents today the path of change. Connections between the various disciplinary fields, to replace the image of the hierarchy amongst knowledge that of a shared knowledge. Connections between operators to limit the excess of specializations and disciplinary isolation, connections between individuals to foster communities, connections that determine bi-directional relationships between the nodes of a network. Connections of a smart network, an intelligent network" (De Santoli 2016).

Connection, therefore, a concept completely opposite to a hierarchical vision of energy management in large towns and cities adopted up to now, is a fundamental paradigm in Green Energy Planning.

In green cities and in green communities, **distributed generation of electricity** is the new energy paradigm. From a single generation point, for example, a thermoelectric power plant, there is a move to an architecture based on multiple generation points distributed throughout the territory, for example **solar photovoltaic power plants**, **wind power plants** or CHP/**cogeneration systems** (combined production of electricity and heat).

Distributed generation is conceptually similar to a computer network in which several terminals connected to a single network contribute to producing and exchanging information that can be used by the entire system. While in a centralised generation, the production of electricity is ensured by large-scale power plants, and in distributed generation the power plants are more numerous, each with a considerably lower power output and can be used locally.

Distributed generation paves the way for the diversification of energy carriers and the **intelligent use of renewable energy sources** which, due to their characteristic of diffuse sources, respond to the needs of small–medium-sized generation systems and favour the creation of micro-networks of transmission.

The network infrastructure compatible with distributed energy generation is the **smart grid** or intelligent network (Fig. 7.1).

Innovation allows electricity to travel between multiple nodes, rendering the network able to respond promptly to the demand for more or less consumption by one or more users and making its management as immediate and optimal as a real intelligent organism.

This characteristic is fundamental when we consider that many of the small electricity production plants that are connected to the grid are powered by renewable energy sources, i.e. random energy sources that generate and inject energy into the grid independently and cannot be controlled with respect to the request from users.

In the distributed generation scheme, moreover, the consumer can become a supplier and vice versa, and the new neologism "**prosumer**" has this meaning. The task of the smart grid is also to manage energy flows in directions that can vary continuously, unlike what happens in traditional centralised electricity generation systems.

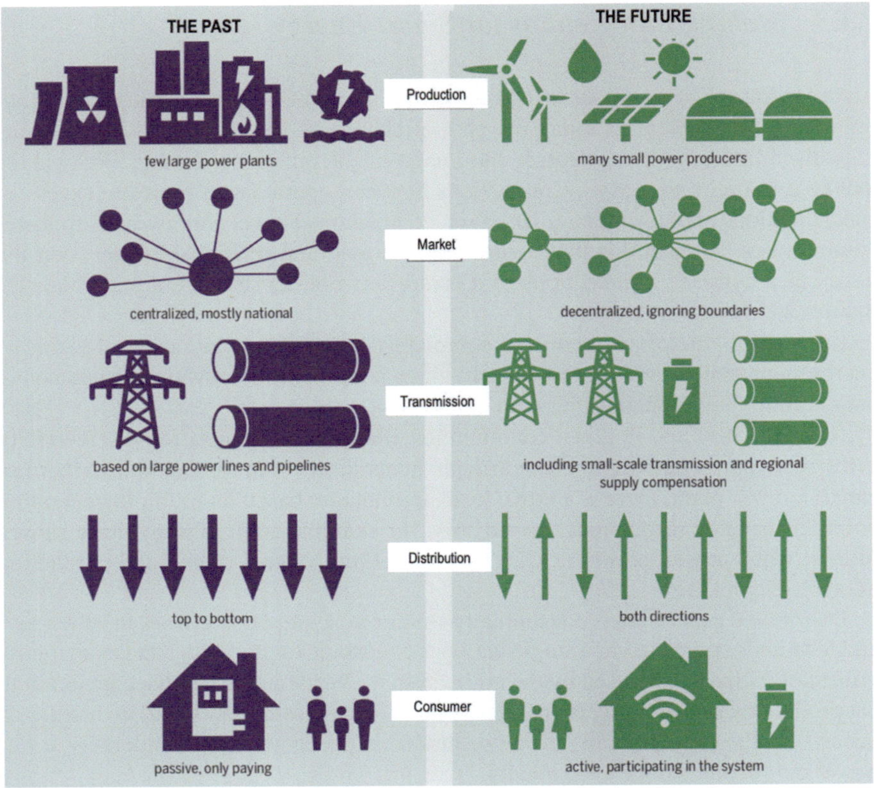

Fig. 7.1 Expected structural changes in energy management due to smart grids (Heinrich Böll Foundation 2018, graphic designer Bartz/Stockmar CC BY 4.0)

The intelligence of a smart grid stems from the fact that, in practice, there is an overlap or a juxtaposition between two technologies: that of electricity distribution and that of Information and Communication Technologies (ICT). Thanks to the latter, the self-production stations distributed throughout the territory or zone in question are able to communicate with the power plants, exchanging information with them on the energy produced and consequently regulating the dispatching of energy.

The management of smart grids is regulated by software capable of guaranteeing remote management and control functions.

As already happens in the mobile telephone sector, in the near future, it will become possible for a user to decide to make several supply contracts and activate them in different time slots, depending upon the related economic convenience. All using the same network and the same meter, which obviously must be set up for this function (**smart metering**).

A smart grid can respond to various technical needs, for example:

- balancing energy supply and demand (dispatching);

- integrating with the home or building automation system of the building;
- opening to the possibility for the user to use external services for optimised energy management.

The diffusion of energy production plants from renewable sources has stimulated applied research for the future execution of smart grids.

Some smart functions are already present in our networks, and renewable energy production systems can supply excess energy to the network and, thanks to the new generation of electronic meters, remote reading, fault detection and management of the contractually engaged power.

However, the current situation can be considered to be a transition phase in which the networks created with traditional architecture acquire, thanks to the technological development of ICT, a partial intelligence and thereby the capability for example to manage the energy introduced by the self-producers (Dall'O' 2014).

7.1.4 Energy Storage Systems

Energy storage is an essential factor in all applications where there is a mismatch between production and use. Storage systems can have a dual function: in the case of conventional energy sources, they can compensate for temporary drops in production or limit the number of generator starts, or indeed reduce peak values, while still guaranteeing the expected demand level and thus improving the level of efficiency of the plant to which it is connected.

In the case of the use of renewable energy sources, i.e. those of an intermittent nature such as solar energy or more particularly wind energy, energy storage systems make it possible to compensate for the mismatch between production and demand and are increasingly showing themselves to be indispensable in ensuring greater control of the production phases and improving performance, management and quality. They thereby reduce the costs of the electricity produced and used, a necessary ingredient for an open energy market.

As far as technologies are concerned, the storage systems which are currently prevalent can conserve energy in the following different forms (Piterà 2016):

- electrochemistral: batteries with aqueous electrolyte, high temperature, lithium and electrolyte circulation;
- mechanics: hydroelectric pumping, compressed air storage and flywheels;
- electrical: supercapacitors and magneto-electric superconductors;
- chemical (production): hydrogen and syngas (currently these are under intense development);
- thermal: molten salts, solar pond, solid accumulations and phase change materials (PCM).

The use of storage systems can greatly improve the performance of smart grids inserted in systems that use renewable energy sources.

7.1.5 Smart Metering

The term "smart metering" defines control systems based on sensor networks for real-time monitoring of light, gas and water consumption. Thanks to the possibility of computer and communication technology interfaces, it permits intervention on plants, in terms of regulating the exchange of both energy and information in plant operation, also offering the possibility to intervene remotely in case of problems or failures, without having to resort to actual on-site action.

Smart metering is a technology of great utility in the field of energy efficiency because it is a tool that can be used to measure the savings which can be achieved following efficiency improvements. Its application, in fact, makes it possible to accompany every intervention from the design phase, with the measurement and evaluation of the energy consumption and energy losses of a plant or a building before the redevelopment works, passing through monitoring during the phase of carrying out the works and finishing with the measurement and control of consumption in tele-management of post-redevelopment operation.

The technological items of which it is composed, in particular the sensors, involve technologies that are already mature, widely available on the market and accessible at reasonable prices. Therefore, the use of smart metering is desirable at every level of the distribution network of energy resources (Dall'O' 2014).

7.1.6 The Prosumer as a New Player in the Green Energy Strategy

The diffusion of plants powered by renewable energy sources, such as the solar PV, in the "grid connected" configuration required the connection of the systems to the public electricity grid. In cases where the electrical energy produced by the plant is higher than the user's needs, the excess is sold to the electricity system (grid), otherwise the user purchases energy from the same system. The user is therefore no longer just a consumer of energy but is also a producer of energy.

The neologism "prosumer" refers to an electricity consumer producing electricity for his/her own consumption (and possibly for feed-in into the grid).

The word is derived from a combination of "producer" and "consumer".

The Renewable Energy Directive (EU 2016) gives the following definition:

> Renewable self-consumer means an active customer or a group of customers acting together as defined in Directive (EU, 2016) who consume and may store and sell renewable electricity which is generated within their premises, including a multi-apartment block, residential area, a commercial, industrial or shared services site or in the same closed distribution system, provided that, for non-household renewable self-consumers, those activities do not constitute their primary commercial or professional activity… (Lettner et al. 2018).

The availability of smart grids, or in any case of devices that allow the use of electric current networks in a smart way, opens the energy market to prosumers, as individuals or organised in structures, allowing to obtain both economic and functional advantages also for the community. It is up to the prosumer, in fact, to carry out the initial economic investment on the energy production plant, and this avoids the construction of new power plants. The prosumer, on the other hand, achieves payback on the investment in less time by selling the produced energy.

When discussing prosumers, the reference is generally to the production of solar photovoltaic energy; however, other technologies can also be used, such as combined heat and power/cogeneration. In this case when there is a demand for heat (e.g. for winter heating), any excess electricity produced by the CHP plant is transferred to the grid.

7.1.7 Energy Communities as a Driver for Local Development

The concept of an Energy Community refers to a set of energy users who decide to make common choices in terms of satisfying their energy needs, in order to maximise the benefits deriving from this collegial approach, thanks to the implementation of technological solutions for distributed energy generation and intelligent management of energy flows.

In the context of the evolution of the electrical system towards the smart grid, the Energy Communities—although in general not limited to the electrical energy carrier—represent one of the constituent elements, being typically connected to the public electricity grid, although there are cases of so-called off-grid applications, in contexts where the electricity grid is not present in a widespread manner (e.g. isolated rural areas and islands).

There are many categories of energy users potentially interested in being part of an Energy Community. In particular, it is possible to identify:

- residential users, such as residential properties and complexes;
- industrial users, such as industrial districts;
- tertiary users, such as shopping/logistics centres and hospital complexes.

In particular, the aggregations of users that form an Energy Community can be either homogeneous, if they belong to the same category, or mixed, if instead they belong to different categories.

The creation of an Energy Community makes it possible to achieve a series of benefits for the energy users present within it, which range from improving the quality and reliability of the energy supply—understood as the possibility of guaranteeing the energy users of the Energy Community high power quality—to optimising energy expenditure—understood as the possibility of guaranteeing energy utilities a lower cost of energy supply than that of traditional procurement methods.

The different categories of energy users, based on their specific peculiarities, attribute a different relevance to the potential benefits, the relative weight of which allows to characterise the aggregations of energy utilities and, consequently, to identify the configuration of the Energy Community (Energy & Strategy Group 2014).

7.1.8 New Paradigms for Urban Mobility

The relationship between density and the viability of public transportation, with consequential benefits for the related energy consumption, has long been recognised, while the wider relationship between urban form and energy consumption has been less clear. However, new data is emerging which demonstrates a complex but significant relationship between the latter pair.

One study which drew upon data from 274 cities found the "economic activity, transport costs, geographic factors and urban form explain 37% of urban direct energy use and 88% of urban transport energy use". Applying these results to projected growth in urbanisation, the study found that there was a potential greenhouse gas savings "wedge" available from the application of appropriate urban planning and transport policies. The wedge represented a reduction of around 26% of greenhouse gas emissions compared with business as usual scenario (Creutzig et al. 2015).

The European Environmental Agency (EEA) has adopted an intervention strategy based on three pillars, the ASI strategy (Avoid, Shift, Improve) which challenges the current transport system (EEA 2016).

The first pillar of *Avoid* (reducing the need for mobility) refers to all actions aimed at improving the overall efficiency of the transport system, to avoid or reduce the formation of passenger and freight transport demand through the use of smart technology such as ICT and intelligent transport system (ITS).

The second pillar of *Shift* (moving to a more efficient transport model) refers to all actions aimed at improving the efficiency of the journey (more energy efficient, lesser emissions, safer).

The third pillar *Improve* refers to all actions aimed at improving vehicle and fuel efficiency through technological development or even by improving lifestyles. Alternative fuels (e.g. biofuels and electricity) should be the main energy carriers for reducing greenhouse gas emissions in the mobility sector (Meneghetti 2018).

The green energy transition of the cities cannot neglect the new paradigms of sustainable mobility that affect not only the energy aspects but also the infrastructural aspects of the city.

The technological evolution in the mobility sector is very rapid, and the infrastructures associated with the new models have difficulty in adapting to this evolution.

First of all, cities will have to develop public transport infrastructures which will increase the number of users, being both amongst residents of the urban area and citizens from extra-urban areas.

The densification of the cities will privilege the short routes: the network of pedestrian and bicycle paths will therefore become strategic transport infrastructures that will have to be equipped and integrated with the bike sharing services. The spread of power-assisted bicycles will require an increase in the associated electrical charging points.

The automotive sector has long been moving towards fully electric vehicles. Urban vehicle charging infrastructures must increase by an order of magnitude compared to the current situation. All this will involve a search for not only suitably equipped parking areas within the urban space but also electricity recharging infrastructures (Meneghetti 2018). The city of Lucerne in central Switzerland is a good example of successful implementation.

The transition from mobility powered by endothermic engines to electric mobility will involve the definition of an urban energy balance in which a portion of the electrical uses will have to take this transition into account. Energy to make vehicles move will no longer be an individual problem or a fuel service station problem but will become a problem for the city and its infrastructure. Electric cars better if not owned but used with shared modes (e.g. car sharing) will significantly contribute to the reduction of harmful airborne gases (and indeed of background noise). However, in a vision that goes beyond the city, an important aspect which cannot be overlooked is how this electricity is produced.

The decision to produce electricity from renewable sources, starting with solar PV, should be privileged.

Electric vehicles play a vital role because they can interact with the power grid. A fleet of electric vehicles can flexibly store and deploy power en masse. BMW's charge forward programme, running in California, is an example: the vehicle owners are paid an incentive if they agree to flexible battery recharging. This allows for "de-peaking", where cars are not recharged immediately, but at times of day when grid loads are at their peak.

Transferring power from vehicle batteries to the grid is also possible and desirable. Cars act as "batteries on wheels": the grid taps their batteries if it is short of power. Each vehicle needs only to return a small percentage of its stored energy to have an overall impact on the grid. However, the equipment required is still too costly: prices need to come down significantly if this approach is to be widely accepted (Heinrich Böll Foundation 2018). In Europe, the electrical infrastructure situation is such that this still appears as a somewhat distant possibility for private users.

So far, we have considered the energy and environmental aspects of mobility. In a truly green vision of mobility, we must also consider the aspect of accessibility: "*equity and inclusivity are at the heart of Universal Access. This objective accounts for distributional considerations and places a minimum value on everyone's travel needs, providing all, including the vulnerable, women, young, old, and disabled, in both urban and rural areas, with at least some basic level of access through transport services and leaving "no one behind… omissis … Urban transport systems and services need to be upgraded—and in some cases planned from scratch—in an integrated way, that ensures the balanced access of urban residents regardless of income, mode of travel, gender, or disability status. Urban mobility should foster and enable cities to flourish, without creating over-dependence on any particular mode of travel*" (SuM4All 2017). Indeed reduction of vehicle density and hence urban traffic congestion will represent an improvement in the quality of urban living.

7.2 A Methodological Approach to Green Energy Planning on a Territorial Scale

7.2.1 *Defining a Green Energy Planning Strategy*

In energy planning on an urban or regional scale, there is no valid reference model for all the realities of a given zone; specific situations can complete the study and definition of intervention projects according to priorities that vary from place to place. For example, the energy planning of a city like Helsinki requires a concept that is substantially very different from the model that could be applied to a city of considerable historical significance (having associated structures to be carefully preserved) such as Rome.

A peculiarity of Green Energy Planning (GEP) is to consider with great attention, through an in-depth analysis, the urban context not only from energy or environmental point of view but also from a social, historical and cultural point of view.

The energy balance of a city arises from a comparison between energy supply and energy demand. The reduction of the negative effects on the environmental impact due to the use of traditional energy sources (fossil sources) can take place on both fronts: on the supply side of energy by replacing energy from fossil sources with renewable energy and on the demand side by reducing consumption through strategies of efficiency and savings.

Acting on the demand for energy is important and appropriate but requires considerable economic inversions, furthermore, there may be physical impediments (e.g. limited spaces for the installation of solar systems).

A concept applied in urban energy planning is that of "integrated resources planning" or "least cost planning". The methodological approach outlines a series of actions of improvement that can be applied to the energy demand side.

According to these theories, it is more interesting and useful to make substantial and coordinated investments mainly for users or infrastructures that use energy resources.

Resources can be defined on the demand side as "demand side resources" and are those technological or operational opportunities which allow an increase in the efficiency of the supply of energy services.

Demand side management (DSM) concepts and methods of energy management have been developed in the USA since the 1980s, in response to problems encountered by companies producing and distributing electricity in relation to the high costs, environmental problems and layout constraints for new production plants.

DSM actions, conducted directly or indirectly by energy companies, can be considered as structured energy-saving actions, or energy-saving incentives, based on non-traditional economic assumptions (Chiesa and Dall'O' 2006).

The definition of strategies to reduce environmental factors is complex and must obviously take into account the specificity of the territorial context. The green energy transition of cities, however, may be more feasible and economically viable if demand side needs are reduced through energy retrofit improvement works.

Cities and large towns are expected to play a key role in achieving the ambitious energy targets, and Green Energy Planning is the tool to reach these targets. Urban planners can contribute to shape energy and efficiency and low-carbon cities. However, the complexity involved in such a broad task impedes the realisation of any simple solution. The paper of Cajot et al. (2017) has the aim of examining in detail the many interrelated challenges and obstacles which hinder efficient urban energy planning.

The Green Energy Green Planning strategy is divided into the following operational phases, some of which are developed in sequence while others can proceed in parallel:

- definition of the working group;
- data gathering;
- analysis of the sectors;
- examination of the energy infrastructures already present in the territory;
- sample energy audits;
- definition of the indicators;
- baseline definition (reference energy balance);
- definition of the action plan;
- checking of the consistency with regional, national and international legislation;
- economic evaluation;
- implementation of the Green Energy Projects;
- updating of the database;
- monitoring.

A typical flow chart of a Green City Planning Strategy is illustrated in Fig. 7.2 that defines four macro-phases: audit phase, design phase, implementation phase and monitoring phase.

7.2.2 *Audit Phase*

The audit phase deals with two aspects: the definition of the working group and the analysis of the current situation (reference energy balance) that constitutes the baseline, starting point and reference point on the basis of which to implement the actions (projects). It is important to create an operations team involving different individuals characterised by clearly defined positions of leadership, and thus able to take decisions.

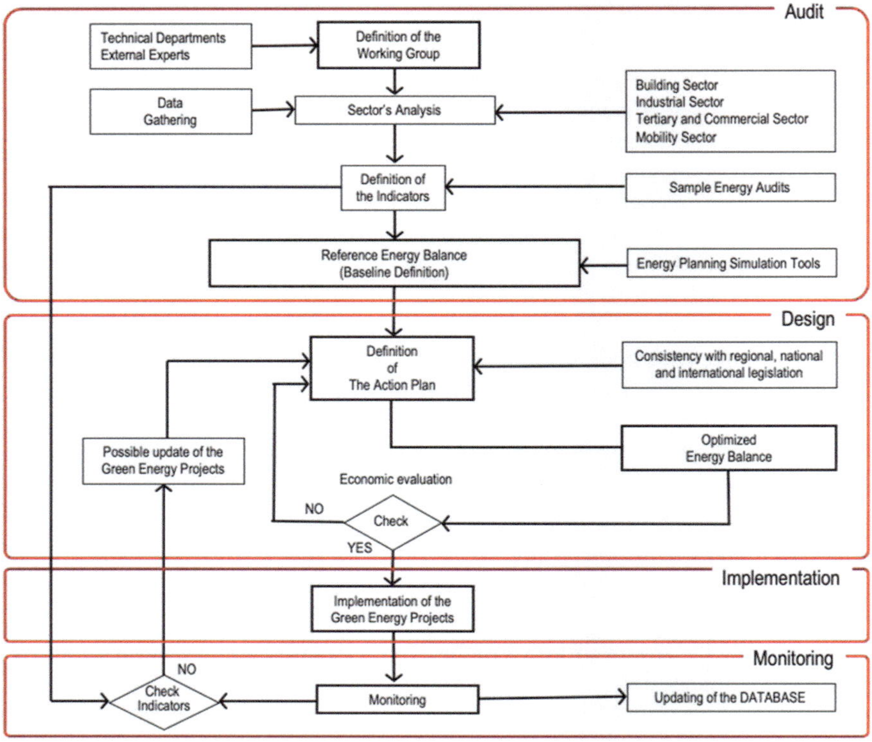

Fig. 7.2 Typical flow chart of a green energy plan for cities and communities: the actions are divided into three macro-phases: audit, design, implementation and monitoring

The managers in charge of the Technical Departments involved (e.g. City Planning Department, Building Department, Environmental Department, Mobility Department, City Manager, Mobility Manager) could be supported by external experts.

A Planning Secretariat should also be appointed, with the responsibility of managing and coordinating day-to-day operations. The chair of the Steering Committee should be entrusted to a high profile person, for example the city's energy manager.

Once the Steering Committee has been established, the audit phase should consider the data gathering for each sector (e.g. Building, Industrial, Tertiary and Commercial, Mobility). The information to be collected may concern:

- the physical and geographical characteristics of the territory (series of maps of various scales), possibly on digital support (e.g. a Geographic Information System (GIS) platform or a City Information Modelling (CIM) platform);
- general urban planning data (total surfaces, urbanised areas, number of inhabitants, population density, industrial and artisan craft development areas);

- the data relating to the changes in the urban layout (new buildings, new infrastructures, abandoned areas to be redeveloped, etc.);
- data relating to energy infrastructure in the area (e.g. district heating networks, district cooling networks, gas distribution networks, plant configurations for the supply of electricity, data relating to urban lighting systems, etc.);
- data relating to plants for renewable energy sources (solar thermal systems, PV solar systems, wind farms, etc.);
- the data relating to the thermophysical characteristics of the building stock;
- vehicle traffic data.

The data on energy consumption can be found at the utilities which supply the energy carriers (electricity companies, gas companies). The sample survey to define the detailed structure of the application can be carried out and must include the following activities:

- quantification and disaggregation of energy flows that affect the area in question;
- historical reconstruction of energy flows within the various sectors with territorial and sectoral breakdown;
- definition of energy indicators (kWh/inhabitant, kWh/m^3 of building, kWh/unit of product, etc.);
- territorial and sectorial analysis of flows to determine the consumption scenario;
- reconstruction of the theoretical energy requirement for each sector based on end uses and the energy carrier used.

The survey phase concludes with the elaboration of the reference energy balance which constitutes the baseline. The knowledge of how much energy enters the city (broken down by renewable energies and non-renewable energies) defines the supply of energy while the evaluation of how this energy is consumed defines the energy demand.

The reconstruction of energy flows divided up per sector is not simple and may require complex processing generally supported by energy planning simulation tools.

The supply of energy can be obtained from the data provided by the utility companies which provide the energy services while the demand for energy requires estimates on users and user equipment and therefore are based on theoretical assessments. Since it is not possible to perform detailed energy audits on all the users, it is advisable to define a suitable scheme of sampling amongst the users.

The reference energy balance is described with a diagram of the type shown in Fig. 7.3, which shows the subdivision of energy sources in various sectors of usage. The diagram takes into account not only conventional (non-renewable) energy sources but also renewable energy sources.

For the study of territorial energy reorganisation policies, it is necessary to reproduce a similar diagram for each of the sectors examined.

The territorial energy balance is the fundamental operational support to the planning activity and to the definition of the precise actions to be implemented, and therefore, it must also contain analytical and interpretative tools of the situation investigated.

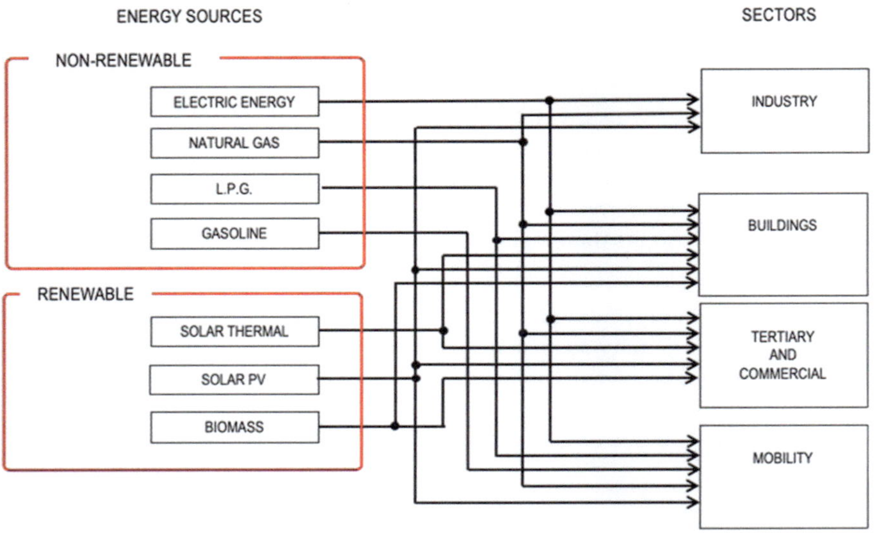

Fig. 7.3 Typical structure of a reference energy balance with the breakdown of energy sources by sector

7.2.3 Design Phase

The design phase defines the actions that can be implemented in the short, medium and long term.

At this stage, it is important to:

- define realistically achievable goals;
- define objectives that are not in conflict with legislative instruments and with guidelines approved by higher-level administrations (e.g. regional or national);
- define strategies consistent with the Sustainable Development Goals (SDGs) defined by the United Nations.

In choosing the actions to implement, the working group should ask itself the following questions:

- Is the proposed action strategic with respect to the needs identified in the audit phase?
- Is the implementation time compatible with the needs identified during the audit phase?
- Are there technical skills available within the Technical Departments capable of managing the action both from a technical and from an economic point of view?
- Is it feasible from an economic point of view? In other words, are the financial resources available to cover the costs not only of implementation but also of management and maintenance?

If the objectives to be achieved can be clear, the paths to follow, and therefore the actions to be implemented in order to attain those objectives, are various. For an examination of the actions and activities which can be implemented to reduce energy consumption, see Sect. 7.3 of this chapter.

7.2.4 Implementation Phase

In this phase, the projects are implemented, long-term actions, for example, the creation of an extended district heating or cooling network, may take several years. Long-term strategic planning makes it possible to allocate the available economic resources more effectively.

7.2.5 Monitoring Phase

By having a monitoring phase, it is possible to verify whether the actions implemented were truly effective in achieving the objectives: this is assisted by a verification over time of the partially implemented actions.

The Green Planning Strategy, although it may include the implementation of several sectoral projects, cannot be considered a macro-project, with a beginning and an end, but rather defines a continuous path of improvement that must be constantly verified through the monitoring of the results achieved.

The monitoring is performed by means of a comparison between the indicators periodically detected (e.g. every year) and the indicators obtained from the baseline that becomes the reference base. It is thus the presence of a monitoring phase which makes the difference between a project and a process.

Through a precise, detailed control, it is possible to identify the partial or global ineffectiveness of some actions, identify their causes and make corrections. All is done according to the typical quality systems approach which, using the Deming cycle (PDCA),[4] are recursive processes.

The use of an energy management system compliant with ISO 50001 (ISO 2018) makes it possible to manage all Green Energy Planning phases efficiently and effectively.[5]

---

[4]PDCA, an acronym for Plan–Do–Check–Act, is a four-step iterative management method used for the control and continuous improvement of processes and products.

[5]Standard ISO 50001:2018 provides a framework of requirements for organizations to: develop a policy for more efficient use of energy, fix targets and objectives to meet the policy, use data to better understand and make decisions about energy use, measure the results, review how well the policy works and continually improve energy management.

7.3 Covenant of Mayors for Climate and Energy

7.3.1 The Covenant of Mayors Project

The Covenant of Mayors (CoM), launched by the European Commission after the adoption, in 2008, of the 2020 EU Climate and Energy Package, is a unique bottom-up movement that succeeded in mobilising a great number of local and regional authorities to develop action plans and direct investments towards climate change mitigation measures. The objective of the project is to endorse and support the efforts deployed by local authorities in the integration and in the implementation of sustainable energy policies.

The CoM Project has had remarkable success over the years: to-date (www.covenantofmayors.eu, September 2019) 9664 local government and public administrative entities (municipalities, networks of municipalities, provinces/counties and regions), covering a population of 326,671,680 inhabitants, joined the pact of the mayors by signing the commitment.

The Covenant of Mayors for Climate and Energy is open to all local authorities democratically constituted with/by elected representatives, whatever their size and whatever the stage of implementation of their energy and climate policies.

Neighbouring small and medium-sized local authorities/councils can also, under certain conditions, decide to join as a group of signatories.[6]

The CoM Project, born within the European Union, has found international success worldwide. "*Since its launch in 2008, the CoM initiative has grown into the world's largest city movement. It has secured the commitments of thousands of cities across six continents and more than 120 countries, representing almost 10% of the world's population. The initiative was first extended to counties neighboring Europe to the East and the South and later to the countries of sub-Saharan Africa. In 2016, the Covenant of Mayors entered a major new phase of its history by joining forces with another city initiative, the Compact of Mayors.*

The resulting Global Covenant of Mayors for Climate & Energy now also covers North America, Latin America and the Caribbean, and much of Asia including countries in the South-East as well as China, India and Japan. Thus far, a total of nine regional and national Covenant Offices have been set up.

These regional and national Covenants operate as local chapters of a global alliance that is capitalizing on the experience gained in Europe and beyond".[7]

As of October 2015, Covenant signatories committed themselves to adopting an integrated approach to climate change mitigation and adaptation. They are required to develop Sustainable Energy and Climate Action Plans with the aims of cutting CO_2 emissions by at least 40% by 2030 and increasing resilience to climate change within the first two years of adhesion (Dall'O' et al. 2018). The CoM Project is developed in three steps:

---

[6]Source: https://www.eumayors.eu/about/covenant-community/signatories.html.

[7]Source: http://www.climatealliance.org/activities/covenant-of-mayors/global-covenant.html.

Step 1: with the signature of the Covenant of Mayors that involves the creation of adequate administrative structures, development of a Baseline Emission Inventory, Risk Vulnerability Assessment and Action Plan (within two years);

Step 2: with the submission of the Sustainable Energy and Climate Action Plan (SECAP);

Step 3: signatories commit themselves to report progress every two years.

As illustrated in Fig. 7.4, the project develops and progresses in a cyclical manner which drives towards a continuous improvement of the sustainability level (Bertoldi 2018).

The SECAP is based on a Baseline Emission Inventory (BEI) and a Climate Risk and Vulnerability Assessment(s) (RVAs) which provide an analysis of the current situation. These elements serve as a basis for defining a comprehensive set of actions that local authorities then plan to undertake in order to reach their climate mitigation and adaptation goals.

SECAP monitoring must be done every two years while an update of the Inventory of Emissions is required every four years.

Fig. 7.4 Covenant of Mayors for Climate and Energy step-by-step process (Bertoldi 2018)

The monitoring phase, independently verified by the European operational support structure for project management, the Joint Research Centre (JRC) of Ispra, guarantees the quality of the project and is therefore strategic for its success.

By means of the monitoring, it is possible to check whether the actions provided by SECAPs have been implemented and have permitted the attainment of the programmed targets and whether the strategies need to be revised by modifying the actions or planning other actions.

Feedback derived from in-depth analysis of monitoring results can help to understand what the best strategies are for promoting a sustainability-driven development policy at local and regional level. The ideas that may emerge can be useful for public administrators but also for all stakeholders in the sustainability chain, such as designers, construction companies, contractors, sub-suppliers, energy and environmental service providers.

The interest of the scientific community for the theme of the European project Covenant of Mayors, and more generally for its implementation tool, the SEAP (or SECAP), is demonstrated by the presence of interesting papers on these matters in indexed databases of the scientific journals of reference.

The topic of monitoring the SEAP is critically addressed by Delponte et al. (2017). In the paper, the authors use as test case the implementation of the SEAP of the city of Genova (Italy) to verify the efficacy of the adopted approach. The authors consider the strengths and weaknesses of the tool, testing particularly cost-benefit analysis, bankability, peer review and participatory level. The aim of the paper is to propose some recommendation to better outline the "Monitoring and Evaluation" methodology and to help other cities to define a strategy for SEAP monitoring and fulfilment.

Berghi (2017) focuses on energy use in the urban transport sector within the SEAPs in three large Italian cities, Milan, Palermo and Rome, which are characterised by completely different cultural, economic and environmental contexts. The originality of this paper consists in having developed, through the comparison of different SEAPS, a preliminary multidimensional framework for their harmonisation at metropolitan level.

Pablo-Romero et al. (2016) analyse the effects of Energy Action Plans on electricity consumption in CoM signatory municipalities in Andalusia, the region of Spain with the most signatories. The study demonstrates that the Covenant implementation has a positive effect on reductions in electricity consumption, since the municipalities involved show greater rates of reduction in electricity consumption after signing it.

Interesting, finally is the contribution on the CoM topic provided by Croci et al. (2017) which analyses the emission reduction strategies of a sample of 124 cities with more than 100,000 inhabitants, signatories of the CoM, which delivered a SEAP by February 2014. In the study, more than 5500 actions planned by the sample cities are analysed and categorised, and the most relevant actions in terms of recurrence and mitigation impact are assessed.

7.3.2 Definition of the Baseline Emission Inventory (BEI)

The CoM focuses on reducing the energy consumption in the local territory, i.e. area administered by the signatory but also matching energy demand with the supply of sustainable energy by improving energy efficiency and promoting the use of local renewable energy resources.

The emission inventory represents a fundamental phase for the design of a SEAP because it defines the baseline values from which to elaborate the planning strategies of the actions that will lead to environmental improvement.

For the signatories, the BEI is an important opportunity to build an energy–environmental balance on a local scale, based on real data that is often not available beforehand, or was available but only partially and in a disaggregated manner. The construction of the EIB also becomes a strategic opportunity to stimulate cooperation and dialogue between the different departments of a municipality (e.g. urban planning, environment, transport, mobility, etc.) which often operate independently.

A complete guide for the preparation of the BEI, which can become a complex activity, is reported in Bertoldi (2018).

The geographical boundaries of the "local territory" are the administrative boundaries of the entity (municipality, region) governed by the local authority which is a signatory to the CoM.

The BEI must cover the CoM key sectors. While the baseline CO_2 inventory will essentially be based on final energy consumption, some non-energy-related activity sectors may also be included. The three main types of GHG emissions to be potentially included in the BEI are as follows:

- Direct emissions due to final energy consumption split into the macro-sectors of "Buildings, equipment/facilities and industry" and "Transport"(mandatory).
- Indirect emissions related to grid supplied energy (electricity, heat or cooling) that are consumed in the local territory.
- Non-energy-related direct emissions which occur in the local territory.

Greenhouse gas direct and indirect emissions are calculated for each energy carrier by multiplying final energy consumption by the corresponding emission factor (see Bertoldi 2018).

In line with the European Union Energy strategy, the new target for the reduction of GHG emissions proposed by the CoM for Climate and Energy is at least a 40% reduction by 2030. The reduction target, to be achieved through the implementation of the actions for those areas of activity relevant to the local authority's mandate, is defined in comparison with a baseline year, which is set by the local authority. While the emissions in the BEI and the reduction per action have to be calculated and reported as absolute emissions, the local authority can decide to set the overall CO_2 emission reduction target either as an "*absolute*" or a "*per capita*" reduction.

7.3.3 *Definition of the Climate Risk and Vulnerability Assessment (RVA)*

The Covenant of Mayors for Climate and Energy supports an integrated approach to local climate action. 2030 Covenant signatories have committed to the common goals of accelerating the decarbonisation of their territories, strengthening their capacity to adapt to unavoidable climate change impacts and ensuring their citizens access secure, sustainable and affordable energy.

Taking measures to reduce GHG emissions (mitigation) and adapting to the unavoidable climate change risks are complementary sets of actions addressing two aspects of climate change. One cannot be fully successful without the other, and there are numerous co-benefits between adaptation and mitigation measures (CoM 2018).

Assessing Climate Change Risk and Vulnerability (RVA) has gained significance since 2010, given its central role in the Cancun Adaptation Framework,[8] the EU Adaptation Strategy and the countries development of National Adaptation Plans.[9]

European cities are heavily vulnerable to the impacts of climate change. Heat, flooding, water scarcity and droughts (amongst others) can have an impact on health, infrastructure, local economies and the quality of life of the inhabitants.

The reduction of the impacts in the cities is a topic considered with great attention within the CoM Project that foresees the preparation of a Climate Risk and Vulnerability Assessment (RVA).

Before analysing this topic, it is useful to take on board two important concepts: *adaptation to climate change* and *resilience*.

The European Environmental Agency defines Adaptation to climate change as "Adjustment in natural or human systems in response to actual or expected climatic stimuli or their effects, which moderates harm or exploits beneficial opportunities" and Resilience as "The ability of a social or ecological system to absorb disturbances while retaining the same basic structure and ways of functioning, the capacity for self-organization, and the capacity to adapt to stress and change".[10]

"*All levels of government, including cities and regions, need a sound understanding of the climate risks and vulnerabilities on their territory to guide their decision making and policy shaping.*

Assessing climate risks and vulnerabilities is one of the first steps in the adaptation cycle which provides the necessary information (What? Where? Why?) supporting tailored proactive measures for each site-specific context (How?)" (Bertoldi 2018).

On defining the RVA, one should consider the following steps:

1. Preparing the basis;

[8]UNFCCC (2010) CANCUN ADAPTATION FRAMEWORK (CAF)—Adopted at the 2010 Climate Change Conference in Cancun, Mexico (COP 16/ CMP 6). In the Agreements, Parties affirmed that adaptation must be addressed with the same level of priority as mitigation.

[9]https://www.eea.europa.eu/publications/climate-change-impacts-and-vulnerability-2016/key-findings.

[10]https://climate-adapt.eea.europa.eu/help/glossary.

2. Assessing risks and vulnerabilities;
3. Identifying adaptation options;
4. Selecting adaptation options;
5. Implementation;
6. Monitoring and evaluation.

For each step of the adaptation cycle, signatories should go through a set of key actions, as illustrated in the following paragraphs. Of particular importance, and indeed the heart of the matter, is the assessment of the risks and vulnerabilities which is the second step.

The issues related to risk assessment and vulnerability are dealt with in Chap. 4 of this book, while for a discussion of the themes involved in drafting the documents supporting the CoM Project; see Bertoldi (2018).

The "Urban Adaptation Support Tool" jointly developed by the CoM—The European Office and the European Environment Agency, is the main adaptation resource for the Covenant community. It is a tool for the implementation of an adaptation strategy and has been tested and approved by Covenant signatory towns and cities.[11]

7.3.4 Definition of the Sustainable Energy and Climate Action Plan (SECAP)

In the Covenant of Mayors Project, the action planning tool to be promoted to reduce energy consumption, and consequently greenhouse gas emissions, and to reduce the risks associated with climate change is the Sustainable Energy and Climate Action Plan (SECAP).

This programming tool can be considered a green energy plan (GEP) and its structure, therefore, can be defined on the basis of the concepts already expressed in the previous sections of this same chapter.

The Covenant of Mayors adds two important elements to the basic outline of the GEP: the first is the climate risk and vulnerability assessment itself and the second is the fact that this is a voluntary instrument. Rather than depending upon an obligation determined by a law, it is a voluntary choice that the mayor makes in committing the zone administered by his local authority to achieve the objectives of reducing greenhouse gases, before 2020 and more recently, through to 2030. The characteristics of a SECAP are shown schematically in Table 7.2.

SECAP defines actions that, if implemented, can reduce greenhouse gas emissions, allowing targets to be reached by 2030.

Actions in a SECAP can be divided into three major categories: direct actions, indirect actions and regulatory actions.

Direct actions are those that the municipality carries out through its own funding: for example, a school's energy re-qualification or the refurbishment of a public

[11] Available on: https://www.covenantofmayors.eu/support/adaptation-resources.html.

Table 7.2 Features of a Sustainable Energy and Climate Action Plan (SIMPLA Project 2018)

Issue	Description
Time span	To 2030
Fields of actions	Municipal buildings equipment/facilities; residential buildings; tertiary (non-municipal) buildings equipment/facilities; mobility (private and public transportation); public lighting; green public procurement (GPP); local electricity production; local heat/cold production (district heating/cooling); others (e.g. industry, agriculture, forestry, etc.) land use planning; environment and biodiversity
Relevance of a local authority's territorial size	Urban contexts of such a size where a balanced development of all transport modes is feasible and realistic
Process steps	Political commitment; start of stakeholder involvement process; planning; baseline definition; adapting administrative structure; establishment of a long-term vision with clear objectives; SECAP elaboration; implementation of actions; monitoring and reporting process
Objectives	At least 40% of CO_2 emissions reduction by 2030 and climate adaptation
Definition of baseline	Comprehensive overview of energy generation and consumption. Risk and vulnerability assessment
Indicators	% reduction of CO_2 emissions; data regarding energy use, generation from renewable energy sources (RES) and saving indicators for each action (MWh); energy delivered by electrical vehicles charging stations (kWh/year); litres of water delivered by public water houses (litres); vulnerability-related indicators (i.e. length of transport network located in areas at risk, number of consecutive days/nights without rainfall, etc.) Impact-related indicators (i.e. % of habitat losses from extreme water events, % of livestock losses from pest/pathogens, etc.); outcome-related indicators (i.e. of transport, energy, eater, waste, ICT infrastructure retrofitted for adaptive resilience, % of coastline designed for managed realignment, % of forest restored, etc.)

(continued)

Table 7.2 (continued)

Issue	Description
Elaboration of scenarios	Limited relevance: initial and final (2030) scenarios and optional "long-term scenario" beyond 2030
Cost and benefit analysis	Recommended but not mandatory
Report	Monitoring Emission Inventory every four years, standardised and mandatory report submitted every two years

lighting installation. For these actions, success depends solely on political will and financial availability.

Indirect actions are the actions that the municipal administration encourages through information, technical support or coordination: for example, an information campaign at schools, the creation of an energy office to inform citizens or the planning of public meetings with citizens and stakeholders. The success of these actions is more difficult to estimate as initiatives depend upon the third parties involved.

However, the contribution of the municipality is important as demonstrated by the photovoltaic systems discussed in the previous section.

Regulatory actions. For these actions, the municipality establishes the mandatory rules that must be applied, such as a new building code. The success of these actions is certain if they are cogent actions. Their application, and the actual results measured with the reduction of energy consumption, depends on the evolution of the building market.

Public awareness and social engagement play a pivotal role for successful climate action. Measures to induce changes in behaviour and to provide education contribute significantly to the decrease in energy consumption, employing social and non-technological approaches that must be included in policies made in support of energy efficiency and energy savings. The creation of the Local Energy Agency to support citizens is a very important support action.

The key elements of a successful SECAP are reported below (Bertoldi 2018):

- build support from stakeholders and citizen participation: if they support the SECAP, nothing should hold it back;
- secure a long-term political commitment;
- ensure adequate financial resources;
- do a proper GHG emissions inventory: this is truly vital;
- make a Climate Change RVA, based on an analysis of the local/regional trends of various climate variables and city socio-economic and biophysical specificities;
- integrate the SECAP into the everyday management and administrative processes of the municipality: it should not be just another nice document, but part of the corporate culture;
- ensure proper management during implementation;
- assure that staff has adequate skills for the purposes and if necessary offer training;

- learn to devise and implement projects over the long term.

On the official website of the CoM (www.covenantofmayors.eu), all SEAP and SECAPS of the signatories of the pact are available as an example.

7.3.5 *Case Study of SEAP Implementation*

The SEAP implementation and monitoring case study described in this subsection are the result of a research study project promoted by the ABC Department of the Politecnico di Milano. The aim of this research was to evaluate the potential as well as the critical aspects of this energy and environmental planning tool in real cases (Dall'O' et al. 2018).

The territorial context in which the analysis was carried out concerns some municipalities located in the Lombardy Region, Northern Italy, within the provinces of Bergamo and Lodi and the Metropolitan City of Milan. Although the municipalities considered in the study are 12 (Table 7.3), it must be considered that the municipalities which have joined the Covenant of Mayors in these two provinces are numerous. The reason for this massive participation in the CoM scheme derives from the very great activity of the provinces of Milano and Bergamo in promoting the adherence of their municipalities to the Covenant of Mayors right from the initial stages of the project.

Table 7.3 General information of the municipalities under study (Dall'O' et al. 2018)

Municipality	Cod.	In. 2015	Area (km^2)	Density 2015 (in./km^2)	DD_W (°)	Date of signature CoM
Pioltello	# 1	36,912	13.11	2816	2404	05/03/09
Melzo	# 2	18,710	9.66	1937	2404	22/04/09
Bareggio	# 3	17,293	11.29	1532	2563	28/10/10
Lainate	# 4	25,708	12.93	1988	2505	30/04/10
Cinisello Balsamo	# 5	75,078	12.70	5912	2404	11/03/10
Cernusco S/Naviglio	# 6	33,436	13.32	2510	2404	20/05/09
Lodi	# 7	44,945	41.43	1085	2592	19/11/08
Canegrate	# 8	12,511	5.30	2361	2617	29/07/09
Cesano Boscone	# 9	23,792	3.99	5963	2404	31/03/09
Albino	# 10	18,171	31.32	580	2543	12/02/10
Villa Cortese	# 11	6,213	3.56	1745	2617	24/11/09
Garbagnate M.se	# 12	27,175	8.86	3067	2449	01/12/11

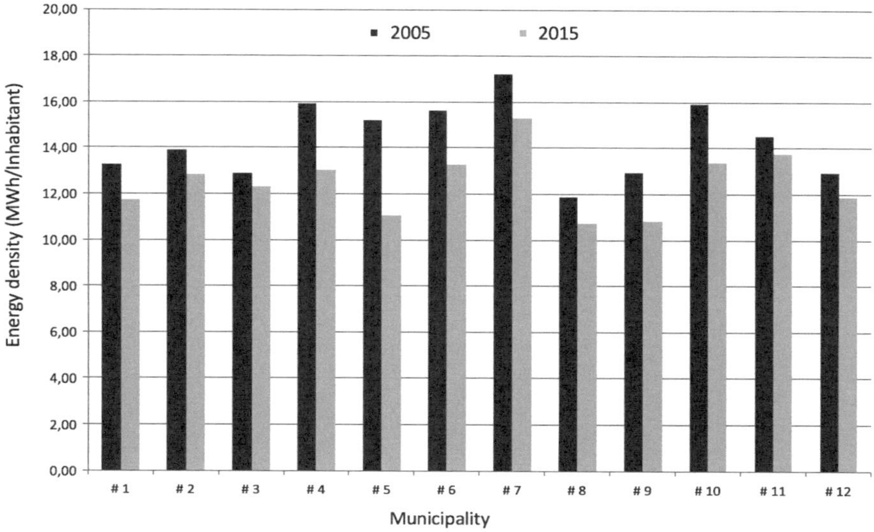

Fig. 7.5 Reduction of energy density between 2005 and 2015 (Dall'O' et al. 2018)

Considering the Italian context, the selected municipalities are of medium to small size. Regarding the winter climatic situation, responsible for energy consumption for winter air conditioning, the situation is homogeneous. Homogeneous is also the socio-economic situation of the chosen municipalities.

The reduction of energy density between 2005 and 2015 is shown in Fig. 7.5. The average reduction is around 10–15% except for the municipality of Cinisello Balsamo which marks a 27%: in this case, such a significant reduction could be the result of the construction of a new district heating system. For the municipalities of Melzo and Villa Cortese, however, the more modest reduction (around 5%) in consumption is due to a considerable expansion of the tertiary building sector (housing).

The calculation of energy consumption indicators of Fig. 7.5 refers to real consumption (end-users) based on the information officially provided by utilities which supply energy through networks (mainly electricity, natural gas, hot water supplied by district heating).

The data on the final consumption of the municipality intended as public administration (schools, public offices, public lighting, etc.) had to be added to the information contained in the energy bills for the reference year. According to the guideline for SEAP drawn up by the Joint Research Centre of Ispra (JRC), industry-related energy data has not been considered by all municipalities involved in our analysis.

The reduction in consumption between 2005 and 2015 depended on some objective factors that affected the 12 municipalities:

- implementation of energy planning policies (e.g. municipal energy plan);
- approval of a new building code that makes compulsory higher energy performance on higher buildings for both new buildings, renovations;

Table 7.4 General information of the municipalities under study (Dall'O' et al. 2018)

Energy vector	2005 (MWh/in.)	2010 (MWh/in.)	2015 (MWh/in.)	2005–2015 (%)
Electric energy	2.51	2.67	2.54	0.9
District heating	0.03	NA	0.22	538.4
Natural gas	8.24	7.20	6.64	−19.4
GPL	0.17	0.19	0.41	143.5
Heating oil	0.03	0.00	0.00	−99.9
Diesel	1.85	1.69	1.90	2.7
Gasoline	1.28	0.73	0.52	−59.5
Biofuel	0.02	0.06	0.14	672.5
Other biomass sources	0.22	0.20	0.16	−26.8
Thermal solar	0.00	0.00	0.02	11,611.8
Geothermal	0.00	0.01	0.01	70.2

- introduction of national economic benefits (e.g. financial support or tax deduction up to 65%) for citizens that invest in energy retrofit actions or invest in the use of renewable energy sources;
- diffusion of the energy labelling and promotion of energy audit of residential buildings;
- promotion of Energy Performance Contracts (EPC).

In the period 2005–2015, the energy sources used have evolved towards fuels with less environmental impact, as shown by the data in Table 7.4.

The economic benefits provided to citizens by the national government, mainly, greatly contributed to increase the market of energy retrofit and renewable energy sources in the residential sector (mainly substitution of windows, thermal insulation of the building envelope, thermal solar and PV solar, boiler replacement, installation of thermostatic valves on the radiators, etc.).

Research has focused heavily on this, and the summary of results is shown in Table 7.5 where the following evaluations can be obtained:

- identification of the actions included in the SEAPs of individual municipalities;
- identifying the most frequent actions in the different SEAPs;
- verifying the status of implementation of the actions (completed, ongoing, not yet started).

An examination of the table provides the following observations considering all the municipalities:

- 17 types of action have emerged, and none of these have been implemented on all municipalities;
- 30% of the actions are deemed to be completed, and 50% of the actions are ongoing;
- 20% of the actions are not yet started.

Table 7.5 State of implementation of planned actions in the municipalities considered in the study (Dall'O' et al. 2018)

	Public building energy audit	Design and implementation of energy-saving projects operated by ESCos	Design and implementation of energy-saving projects operated by ESCos	Acquisition of public lighting equipment and network	Public lighting equipment refurbishment	Interior lighting retrofitting
# 1	–	–	–	O	N	O
# 2	O	O	–	C	O	O
# 3	C	N	O	O	N	O
# 4	–	–	C	–	–	–
# 5	C	–	O	–	O	–
# 6	C	C	–	C	O	O
# 7	–	–	O	–	C	–
# 8	C	O	–	O	N	N
# 9	C	O	–	C	O	N
# 10	O	–	–	N	N	N
# 11	O	–	C	–	O	–
# 12	O	N	–	C	O	O

	PV systems on municipality buildings	Solar thermal plants on municipality buildings	Green public procurement	Gradual replacement of municipal fleet	Separated waste collection rationalisation	Pedestrian and cycle mobility development
# 1	O	N	O	O	C	O
# 2	O	C	N	O	O	O
# 3	–	–	–	O	C	O
# 4	O	O	C	O	–	C
# 5	–	–	O	–	–	O
# 6	O	O	N	O	O	O
# 7	O	–	C	O	–	O
# 8	P	O	–	O	O	O
# 9	O	O	N	C	O	O
# 10	C	N	–	N	C	O
# 11	C	–	O	O	–	O
# 12	N	–	–	O	O	O

(continued)

Table 7.5 (continued)

	Environmental zones, restricted traffic area	District heating realisation/renovation	Sustainable urban development—P.G.T. realisation and updating	Municipal Building Code (R.E.)	Urban Traffic Plan (P.G.T.U.)	Municipal Public Lighting Plan (P.R.I.C.)
# 1	O	C	N	–	N	–
# 2	O	–	N	O	–	N
# 3	O	–	N	C	O	–
# 4	–	–	–	C	–	–
# 5	C	C	C	C	–	–
# 6	O	–	O	O	O	N
# 7	O	O	O	O	–	–
# 8	N	–	C	C	O	N
# 9	C	C	O	C	N	N
# 10	N	–	N	C	N	C
# 11	–	–	–	C	–	–
# 12	N	–	C	C	N	–

	Energy front office for citizens	Web site and social network system	Zero miles to market	Events and training courses for energy stakeholder	Training courses for municipal technical staff	Environmental education for schools
# 1	C	O	C	O	C	N
# 2	C	O	–	O	C	C
# 3	C	–	O	O	O	O
# 4	O	C	–	O	C	–
# 5	C	–	C	O	–	–
# 6	C	O	N	O	O	O
# 7	C	–	–	N	–	N
# 8	C	N	N	O	O	O
# 9	C	O	–	O	C	O
# 10	–	–	C	N	N	N
# 11	O	C	–	–	C	–
# 12	C	–	–	–	O	O

C Completed; *O* Ongoing; *N* Not yet started

The scope of the COM was to accelerate the achievement of the target of reducing gas emissions by 2020. It is interesting to note that for as many as seven out of 12 of these municipalities, this target has been surpassed (some significantly) and the remainder is not far from this same target.

The overall assessment of the municipalities/local authorities of this study, taken as a group, is positive. The targets for the reduction of climatic gas emissions have been achieved by seven municipalities out of 12 with an advance of 2 years and the others are very close to this target.

Implementing and management of a SEAP have difficulties but the positive elements are many as follows:

- thanks to the SEAP, the municipalities are equipped with energy and environmental planning tools;
- thanks to the SEAP, municipalities implement and maintain an energy accounting function that involves the entire technical and administrative structure;
- the need to ensure a state of progress every 2 years (SEAP update and emission inventory update) makes this tool alive, which becomes a very useful tool for correcting the choices made.

This research has proven to be an element in comparison with what already exists: a comparison between sustainable planning policies is useful and synergistic, especially for small-sized communities such as those we have investigated.

7.4 Conclusions

This chapter has addressed the issue of energy planning on an urban and territorial scale, defined as Green Energy Planning. Considering the effects of energy on the environment, proper planning of energy resources contributes greatly to act against climate change.

In recent years, we have been living in a transition phase between a traditional energy model, based on the use of fossil energy sources and centralised power generation, and a sustainable model based essentially upon renewable energies, such as solar energy, biomass and wind energy and distributed power generation.

Imagining a new model of energy planning means referring to new paradigms, starting from the SDG 7, which aims to "ensure access to affordable, reliable, sustainable and modern energy for all", and in the first part of the chapter some arguments have been provided which should be duly reflected upon.

However, energy planning cannot remain in the realms of theoretical speculation but is something for a concrete application. For this very reason, a reference model has been presented which obviously can be adapted to specific realities but which in its structure is fully functional for all cases.

In Green Energy Planning fortunately, we do not start from zero but many positive experiences have been made and many good practices can be used as models to refer to when defining actions. The Sustainable Energy and Climate Action Plans

(SECAPS), implementation tools of the European Project Covenant of Mayors, have been considered important references in this chapter, and for this reason they have been treated and deepened; in addition to considering the issue of decarbonisation, the SECAPS also take into account the climate risk and vulnerability assessment; dealing with these two closely related elements with an integrated approach is strategic.

Considering our urban areas, with their inefficiencies and their complexities, the energy transition towards a sustainable model is a process that will last a long time, but above all it is a process that requires an awareness and participation on the part of citizens. This is nonetheless the only strategy which can affect the necessary changes.

References

Berghi S (2017) Energy use in urban transport sector within the sustainable energy action plans (SEAPs) of three big cities. Energy Procedia 126:414–420

Bertoldi P (2018) Guidebook 'How to develop a sustainable energy and climate action plan (SECAP)': Part 1, 2, 3. EU, JRC. Available at: https://ec.europa.eu/jrc/en/publication/eur-scientific-and-technical-research-reports/guidebook-how-develop-sustainable-energy-and-climate-action-plan-secap. Accessed 2 Sept 2019

Cajot S, Peter M, Bahu JM, Guignet F, Koch A, Maréchal F (2017) Obstacles in energy planning at the urban scale. Sustain Cities Soc 30:223–236

Chiesa G, Dall'O' G (2006) Gestione delle risorse energetiche nel territorio. C.E.A. Casa Editrice Ambrosiana, Milan

CoM (2018) How to prepare for floods, heatwaves and other climate change impacts. Available at: https://www.covenantofmayors.eu/support/adaptation-resources.html. Accessed 10 Sept 2019

Creutzig F, Baiocchi G, Bierkandt R, Pichler P-P, Seto KC (2015) Global typology of urban energy use and potentials for an urbanization mitigation wedge. Proc Natl Acad Sci USA 112(20):6283–6288

Croci E, Lucchitta B, Janssens-Maenhout G, Martelli S, Molteni T (2017) Urban CO_2 mitigation strategies under the Covenant of Mayors: an assessment of 124 European cities. J Clean Prod 169:161–177

Dall'O' G (2014) Smart city, la rivoluzione intelligente delle città. Società Editrice Il Mulino, Bologna

Dall'O' G, Sarto L, Pizzi A, Cristina Reggiani S (2018) Sustainable Energy Action Plans: monitoring the implementation of strategies for sustainability in Northern Italy. In: International conference seismic and energy renovation for sustainable cities, Catania (Italy)

De Santoli L (2016) Smart grid: strategia per la comunità dell'energia su scala urbana. AICARR, Delfino Editore

Delponte I, Pittaluga I, Schenone C (2017) Monitoring and evaluation of Sustainable Energy Action Plan: practice and perspective. Energy Policy J 100:9–19

Energy & Strategy Group (2014) Smart grid executive report. Available at: http://www.energystrategy.it/area-riservata/smart-grid-executive-report.html. Accessed 2 Sept 2019

European Commission (2016) Directive of the European Parliament and of the Council on the promotion of the use of energy from renewable sources (recast). COM(2016) 767 final, 2016/0382 (COD), Brussels, 30.11.2016

European Environmental Agency (2016) Towards clean and smart mobility: transport and environment in Europe. ISBN 978-92-9213-739-7, ISSN 2443-7662. https://doi.org/10.2800/090074

Heinrich Böll Foundation (2018) Energy Atlas, facts and figures about renewables in Europe. Available at: https://www.boell.de/sites/default/files/energyatlas2018_facts-and-figures-renewables-europe.pdf.pdf?dimension1=ds_energyatlas. Accessed 2 Sept 2019

International Energy Agency (2019) World energy balance—overview. Available at: https://webstore.iea.org/world-energy-balances-2018. Accessed 2 Sept 2019

International Organization for Standardization—ISO) (2018) EN ISO 50001:2018. Energy management systems—requirements with guidance for use

Lettner G, Auer H, Fleischhacker A, Schwabeneder D, Dallinger B, Moisl F, Roman E, Velte D, Ana H (2018) Existing and future PV prosumer concepts. Technische Universitaet Wien, Fundacion Tecnalia Research & Innovation, with the collaboration of PVP4Grid consortium. Available at: https://www.pvp4grid.eu/wp-content/uploads/2018/08/D2.1_Existing-future-prosumer-concepts_PVP4Grid_FV.pdf. Accessed 2 Sept 2019

Meneghetti D (2018) E-Mobility e sostenibilità urbana, integrazione delle infrastrutture di ricarica nello spazio urbano. Tesi di laurea, Facoltà di Architettura del Politecnico di Torino

Pablo-Romero M, Pozo-Barajas R, Sanchez-Braza A (2016) Analyzing the effects of Energy Action Plans on electricity consumption in Covenant of Mayors signatory municipalities in Andalusia. Energy Policy J 99:12–26

Piterà L (2016) Le tecnologie e la regolamentazione in Smart Grid: strategia per la comunità dell'energia su scala urbana. AICARR, Delfino Editore

SIMPLA Project (2018) Guidelines for the harmonization of energy and mobility planning. Available at: http://www.simpla-project.eu/en/guidelines/. Accessed 10 Sept 2019

SuM4All (2017) Global Mobility Report 2017, tracking sector performance. Available at: https://sustainabledevelopment.un.org/content/documents/2643Global_Mobility_Report_2017.pdf. Accessed 2 Sept 2019

United Nations (2018) Analysis of the voluntary national reviews relating to sustainable development goal 7. Available at: https://sustainabledevelopment.un.org/index.php?page=view&type=400&nr=2583&menu=35. Accessed 2 Sept 2019

Part II
Energy Efficiency and Sustainability in Urban Planning

Chapter 8
Methods and Tools for Urban Energy Planning

Giuliano Dall'O' and Simone Ferrari

Abstract Cities are responsible for around 70% of global energy demand and are considered as having a crucial role in effective abatement of global energy consumptions. The topic is largely discussed in available literature, which reveals the great diversity of applied approaches and the necessity to move towards the concept of smart energy systems, focussing on synergies among different energy sectors. However, considering the large share of responsibility of the building sector, this chapter focussed on related energy demand assessment. After an initial introduction, which spotlights the most complex elements, the chapter presents some methods of analysis to evaluate the energy performance of the existing building stocks. Subsequently an overview is presented of the methods and tools for determining the energy demand of buildings within urban energy planning, paying particular attention to those that rely on hourly profiles. The assessment of hourly energy demand of the existing building stock, as well as the prediction of its variation due to energy efficiency measures, are fundamental activities for planning strategies of distributed generation, district heating and/or cooling networks, renewables integration, energy storages, etc., all necessary in moving towards smart energy districts.

8.1 Methods for Estimating the Energy Performance of Existing Building Stock

In the sustainable planning of a city or a community, the energy retrofit of the existing building stock represents a strategic goal if important greenhouse gas reduction results are to be achieved. The energy analysis of all the buildings in a city or a neighbourhood, however, is by no means easy since the information available on

G. Dall'O' (✉) · S. Ferrari
Architecture, Built Environment and Construction Engineering, ABC Department, Politecnico di Milano, Via G. Ponzio, 31, 20133 Milan, Italy
e-mail: giuliano.dallo@polimi.it

S. Ferrari
e-mail: simone.ferrari@polimi.it

G. Dall'O' (ed.), *Green Planning for Cities and Communities*,
Research for Development, https://doi.org/10.1007/978-3-030-41072-8_8

the thermal-physical characteristics of the buildings and about the management and control of the associated plants and services installed are mostly unknown.

A classic energy audit methodology cannot be applied because it would be too expensive and time-consuming. On the other hand, knowledge of energy quality is essential because it allows the urban or the energy planner to define the current situation (baseline) and to plan strategies for energy retrofit actions evaluating the technical-economic aspects and the possible barriers to be overcome.

An understanding of the energy performance in buildings in a whole city or an entire district is important for sustainable energy planning strategies that accelerate the energy renovation process in existing buildings that are not energy efficient.

Researchers from the Department of Architecture, Built Environment and Construction Engineering (ABC) of the Milan Polytechnic worked on the subject of the energy assessment of existing building stocks. In the following sections the results of two research studies on this topic will be presented and discussed, becoming methodological approaches.

8.1.1 A Methodology for Energy Performance Classification of Buildings at Urban Scale

A methodology for energy performance classification of buildings on an urban scale, largely based on information which is already available on building stock (i.e., cartographic documentation, thematic maps, geometric data and others) was proposed in the study of Dall'O' et al. (2011). Data regarding the energy performance of buildings were collected using energy audits on sample buildings, which were selected using a statistical approach. Using the tools in a Geographic Information System (GIS) platform, the integration of two data sources allowed for a low cost, comprehensive framework of the energy performance of buildings in Carugate, a small town located in the Lombardy region of northern Italy. The methodology is outlined in the flowchart in Fig. 8.1.

This documentation retrieval phase includes the collection of all information that may be useful for analysis, for example, site maps, aerial photogrammetric surveys and building cadastral data. Internet is a very useful source that contains a large amount of information, if the maps are sufficiently updated, tools as Google Maps® very useful, particularly for detecting the status of the buildings' roofs.

The management of the energy characteristics data of buildings is based on a GIS platform. Many municipal technical departments already have a GIS platform, and in these situations, the work is easier because it is then sufficient to integrate additional information in the same platform.

In some cases, the basic digital map must be corrected, for example, when completing the boundary lines or closing the polylines of each building. GIS implementation involves the creation of a database. The elementary object is the building, which is

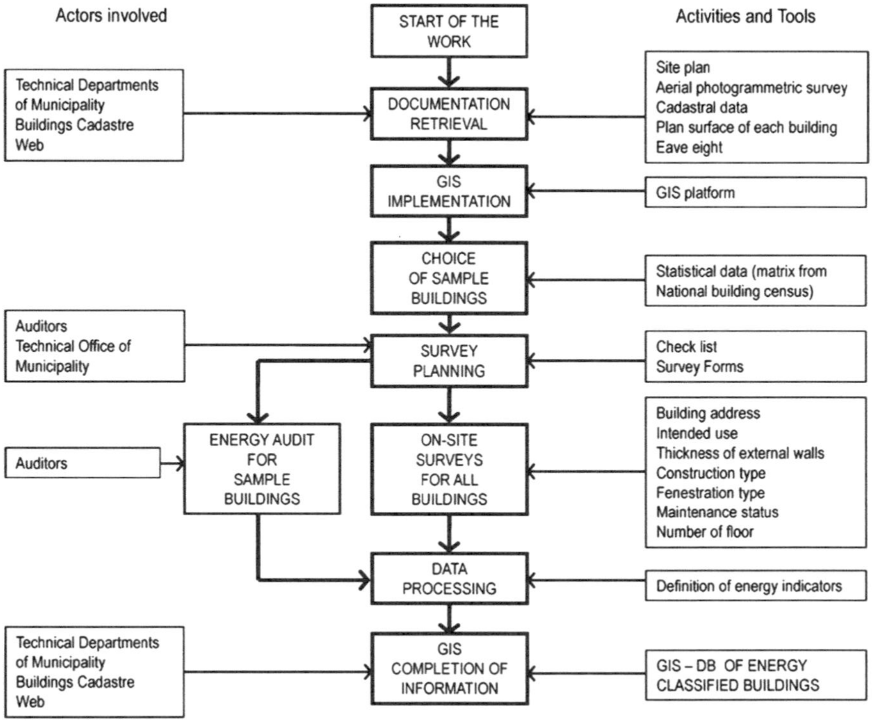

Fig. 8.1 Flowchart of the methodology for the energy performance classification of the buildings (Dall'O' et al. 2011)

associated with a record. The fields of each record define the attributes of a building's characteristics (e.g., area, volume, the condition of the facilities and energy performance). Each building is identified by a code that must be defined at this stage.

From the statistical data of the population census, it is possible to obtain information on existing building stock to define a matrix of the "number of apartments per building/construction period". From this matrix, it is possible to define the characteristics of the buildings to be identified within the building stock (sample buildings representative of each construction period) and to perform a detailed investigation with a specific energy audit.

The purpose of survey planning is to efficiently and effectively plan the next phase of an on-site survey. Checklists and survey forms must be prepared to facilitate the survey. All the paths of the auditors must be planned to limit unnecessary footwork on the streets. Proper planning reduces the time required while maintaining high quality results. In many cases the mapping of buildings from the outside can be facilitated using the Google Maps® "Street-view" function. The purpose of this activity is to obtain all of the information that is needed to complete the framework of the status of the building stock from an energy efficiency and conservation perspective.

The next phase involves the selection of some representative buildings on which to perform a detailed energy audit. The scope of these audits is to calculate the primary energy needs for winter heating (EP_H) expressed in kWh/m^2 per year: calculations must be performed using the calculation method that has been adopted locally for energy certification. The choice of the number of samples for the energy audits of the buildings must be made on the basis of the existing building stock. To this end the national census is a useful source that provides data on buildings and houses. In this case study, considering that in Carugate there are 1320 buildings, the total number of sample buildings, 93, is approximately 7% of the entire building stock. Energy audits are not necessary if the municipality has a large number of energy certifications in hand.

The EP_H data for the sample buildings are initially divided up into groups of construction periods: in the case under study seven construction periods (before 1919, 1919–1945, 1946–1960, 1961–1980, 1981–2002, 2003–2006, 2007–2010) were considered.

The correlation between the shape coefficient, or S/V ratio and EP_H must be calculated to obtain a linear function between the shape coefficient of the building S/V and EP_H. In the calculation phase, it is important to verify that the coefficient of determination R^2 of each line is acceptable (values are considered acceptable if R^2 is greater than 0.75). Figure 8.2 shows the regression lines that relate the EP_H to the S/V ratio for different construction periods.

The equations of the regression lines were the used to complete the information in the database of the GIS platform. Data on the energy performance (EP_H) of each building were obtained automatically by the software using the calculation function

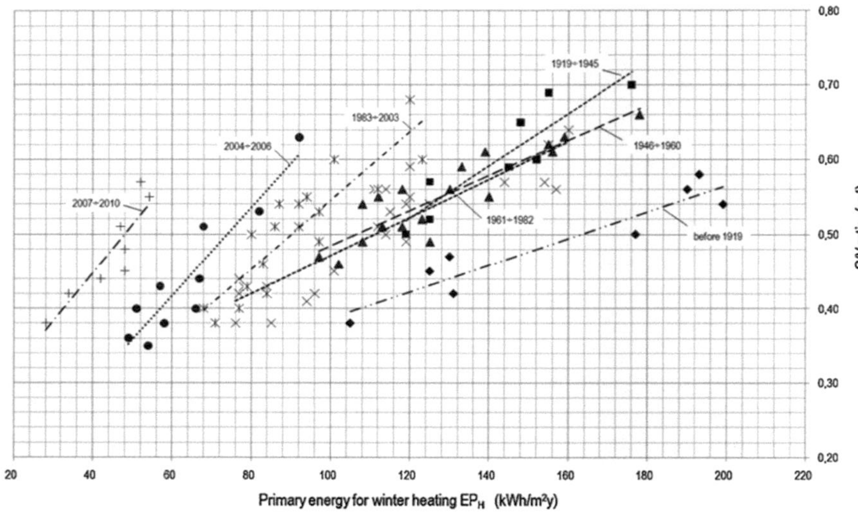

Fig. 8.2 Regression lines that relate the EP_H to S/V ratio for different construction periods (Dall'O' et al. 2011)

tools which employ the equations derived on the basis of the energy audits of a sample of the buildings, functions of the construction period and the S/V ratio (construction periods and S/V ratio were information already available in the database).

Once all of the data are included in the GIS platform, the technical department of the municipality/local authority uses a tool that is capable of performing a large number of territorial energy management and planning functions.

Some examples of its potential uses are listed hereunder:

- estimation of the energy class of existing buildings that can be updated with the official implementation of energy certificates;
- monitoring of estimated energy consumption and verification of any reductions in energy consumption and consequentially of CO_2 emissions due to improvements in energy efficiency and/or use of renewable energy sources (solar thermal, solar PV);
- real-time verification of the results obtained from the implementation strategies to improve the energy performance of buildings or to promote renewable energy (the platform is able to man-age the different scenarios in a selective way by considering, for example, interventions beginning with the most inefficient buildings or from buildings belonging to a certain class of construction period);
- informing the general public of the local area through a web interface about the energy quality of their buildings and encouraging them to perform energy audits in some cases.

A GIS platform greatly improves the user interface through the display of graphical tools. In Fig. 8.3, for example, buildings are associated with a colour depending on their estimated energy class.

By browsing on the digital map, it is possible to query the database selectively, for example, for the assessment the energy performance of buildings for a certain construction period or for a selected area of the territory. It is also possible acquire information about a single building.

The methodology was validated by comparing the consumption data from all the residential sectors, which was calculated by the software, with those derived indirectly from the conventional energy (natural gas in this study) which was actually consumed. Data on natural gas usage were provided by the local gas utility.

8.1.2 Indicators for Energy Planning from Energy Certification Database

Energy certification of buildings, mandatory under the European Directive EPBD (EC 2002) provides interesting data on the thermo-physical properties and geometry of existing buildings. Although the energy certificate is intended to provide the characteristics of individual buildings, thereby stimulating the real estate market towards ever better energy performance, data management of the certificates issued over time,

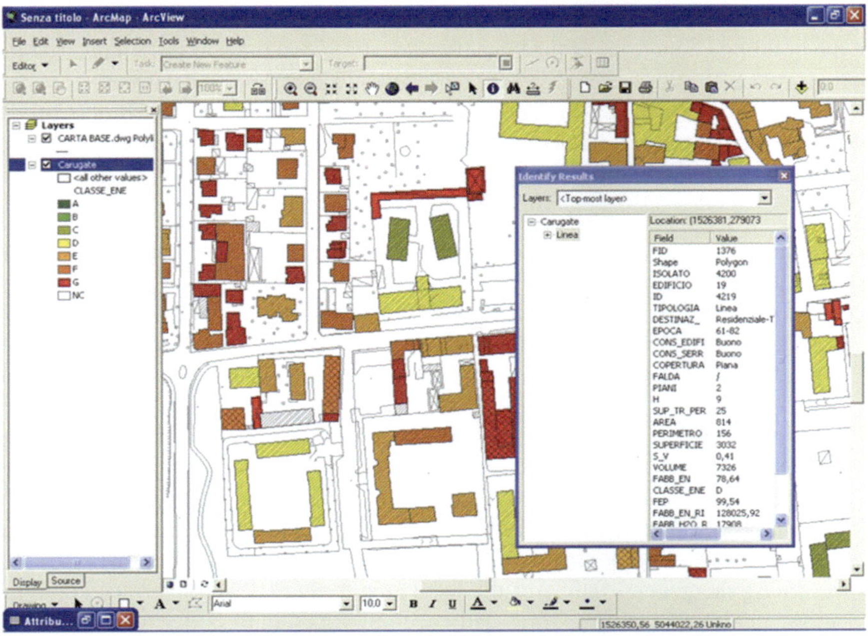

Fig. 8.3 User interface of the GIS platform when the database was complete (Dall'O' et al. 2011)

using a national or regional energy cadastre, makes available a data base which is useful for energy planning in the building sector.

The use of the data provided by the energy certification of buildings to obtain energy performance indicators related to the existing building stock was the objective of a study carried out by Dall'O' et al. (2015).

This document is based on the results of a benchmarking study of the Lombardy Region, northern Italy. By integrating data from the energy cadastre (175,778 energy certificates) with the statistical data obtained from the national census, indicators were obtained on the energy performance of existing buildings.

The energy indicators obtained, characterised by building type and construction period, normalized as a function of Degree-Days, became an effective tool for energy planning at local and regional scales. In the specific case, the energy indicators have been used to estimate the potential for energy retrofitting of existing buildings in the Lombardy Region. The same indicators could be used by municipalities/local authorities for energy planning at the municipal or district level.

In Lombardy Region the energy certification procedure requires that the energy assessor, registered in the regional professional register of energy assessors, must transfer the entire set of the data contained in a single Energy Performance Certificate (EPC) into the regional cadastre, automatically: in this way the energy certificate database is updated in real time.

In order to promote transparency of the activities of all regional public administration, the Lombardy Region, as of 2013, has released the data on the EPCs. Relevant data of each EPC can be read on-line by any citizen. Alternatively, the entire database of energy certificates can be downloaded in order to allow off-line processing.

Table 8.1 shows the available data: these refer to the main characteristics of the energy certificates.

The published database does not contain all the characteristics of the buildings but only the main ones; however these are sufficient for the purposes of the study.

Table 8.1 Structure of the database of the EPC cadastre in Lombardy Region (Dall'O' et al. 2015)

Symbol	Field description	Units
EPC #	Energy Performance Certificate Code	–
A_{ID}	Energy assessor ID number	–
Loc	Location (municipality)	–
DD_W	Winter standard Degree-Days	°C
BT	Building type (e.g. residential, school, office, etc.)	–
CY	Construction year	–
A_g	Gross floor area	m^2
A_n	Net floor area	m^2
V_g	Gross volume	m^3
V_n	Net volume	m^3
S	Dispersant surface	m^2
GO_R	Glass surface/opaque surface ratio	–
U_e	Average U-value of the building envelope	$W/m^2\ K$
U_r	Average U-value of the roofing	$W/m^2\ K$
U_w	Average U-value of the windows	$W/m^2\ K$
U_b	Average U-value of the basement	$W/m^2\ K$
EC	Energy class (ranging from A+ to G)	
ET_H	Heat demand indicator	$kWh/m^2\ y$
EP_H	Primary energy indicator	$kWh/m^2\ y$
Vent	Ventilation type (e.g. natural, mechanical, etc.)	–
AC	Air changes	1/h
P	Boiler capacity	W
HT	Heating system (e.g. boiler, heat pump, CHP, etc.)	–
Ft	Fuel type (e.g. gas, oil, etc.)	–
SPV	Solar PV system type (e.g. monocrystalline silicon, amorphous silicon, etc.)	–
PV	Solar PV surface	m^2
STH	Solar thermal type (e.g. flat plate, vacuum, etc.)	–
TH	Solar thermal surface	m^2

Table 8.2 List of the parameters used to filter the data contained in the energy cadastre in Lombardy Region (Dall'O' et al. 2015)

Parameter description	Unreliability of the parameter
Net floor surface (A_n)	$A_n < 50$ m^2
Net volume (V_n)	$A_n < 120$ m^3
Ratio of dispersant surface and heated volume (S/V_g)	$S/V_g < 0.2$ or $S/V_g > 1.5$
Calculated average height (V_n/A_n)	<2.4 m or >6 m
U-values for opaque surfaces	$U < 0.15$ W/m^2 K or $U >$ 2.60 W/m^2 K
U-values for transparent surfaces	$U < 0.8$ W/m^2 K or $U >$ 6.00 W/m^2 K
Heat demand indicator/primary energy indicator (ET_H/EP_H)	<0.5 or >1.5

EPCs of buildings allowed the Lombardy Region to acquire a considerable amount of information about the housing stock: at the date of the study (September 15, 2013) the total number of registered EPCs, referred to the residential sector, amounted to 1.154610.[1]

A first analysis of the database of the energy cadastre made it possible to delete records corresponding to the buildings in which the data fields were incomplete, after this data-cleaning process, the number of EPCs was reduced to 850,970.

The EPCs in the database refer to certifications that may relate to entire buildings or individual flats. The objective of the study, however, was to define the characteristics of entire buildings. In order to achieve this goal, only records in which all surfaces of the building envelope (roofs, walls, basement) had a value greater than 0 were selected. After this processing it was found that of the entire archive the certifications that refer to complete buildings are 233,212.

The platform of the energy cadastre of the Lombardy Region does not verify the consistency of the data contained in the downloaded EPCs. For this reason it was necessary to clean up the database by removing all records that contain obvious errors. The EPCs were then filtered, based on the criteria defined in Table 8.2.

The EPCs which proved compliant, and suitable for the purpose, were reduced to 175,778, a value which nonetheless represents a large amount of available data.

The next step was the definition of a reference matrix.

The Italian national census classifies existing buildings on the basis of two parameters of interest: construction period and number of flats per building.

[1]The According to ISTAT, the Italian national census of population and housing (accessed on September 2019), in Lombardy there are about 4,827,000 apartments; considering that in the energy cadastre of Lombardy Region (accessed on September 2019) there are about 2,067,000 EPCs, 43% of the existing residential building stock is certificated.

Table 8.3 Matrix used for the classification of buildings

Construction period	Number of flats per building					
	1	2	3–8	9–15	16–30	>31
<1930						
1930–1945						
1946–1960						
1961–1976						
1977–1992						
1993–2006						
After 2006						

The construction period can be reasonably related to the construction technology involved, while the number of apartments per building determines the size of the building and the S/V ratio.

Concerning the first parameter, the buildings were divided into 7 classes: the construction period is one of the items of information contained in the EPCs database and therefore readily available. The clustering according to the six classes, starting from the EPCs of the energy cadastre, was possible because the field "number of flats per building" was available for each EPC.

Table 8.3 shows the matrix used for the building classification in accordance with the chosen parameters. The matrix that has emerged consists of $7 \times 6 = 42$ cells. The choice of this matrix is important because it is compatible with the data available via the national census and, as shall be seen later, allows the creation of interesting synergies between the energy cadastre and the database of the national census.

Starting from the data of EPCs one can define, for each of the cells of the matrix, three types of representative buildings:

- buildings which preserve their original energy performance;
- buildings which have been partially retrofitted to improve their energetics (replacement of windows).
- buildings which have undergone a complete energy retrofit (replacement of windows and building envelope insulation).

The possibility of obtaining three matrices is very useful since it allows one to test the potential for improving energy performance.

Figure 8.4 shows the flow diagram of the procedure used to define the representative buildings. The definition of the representative buildings within each cell was made, in each step, using a recursive procedure with analysis of frequencies on the basis of indicators in order to select the really significant samples, eliminating those characterised by the parameter values which are least recurrent. The analysis was performed for each of the 42 cells of the matrix and was therefore quite complex.

The first filter of the selection concerns the indicator of the quality of the building envelope, expressed by an average U-value. The second selection concerns the energy performance of the transparent surfaces of the building envelope. At this stage three

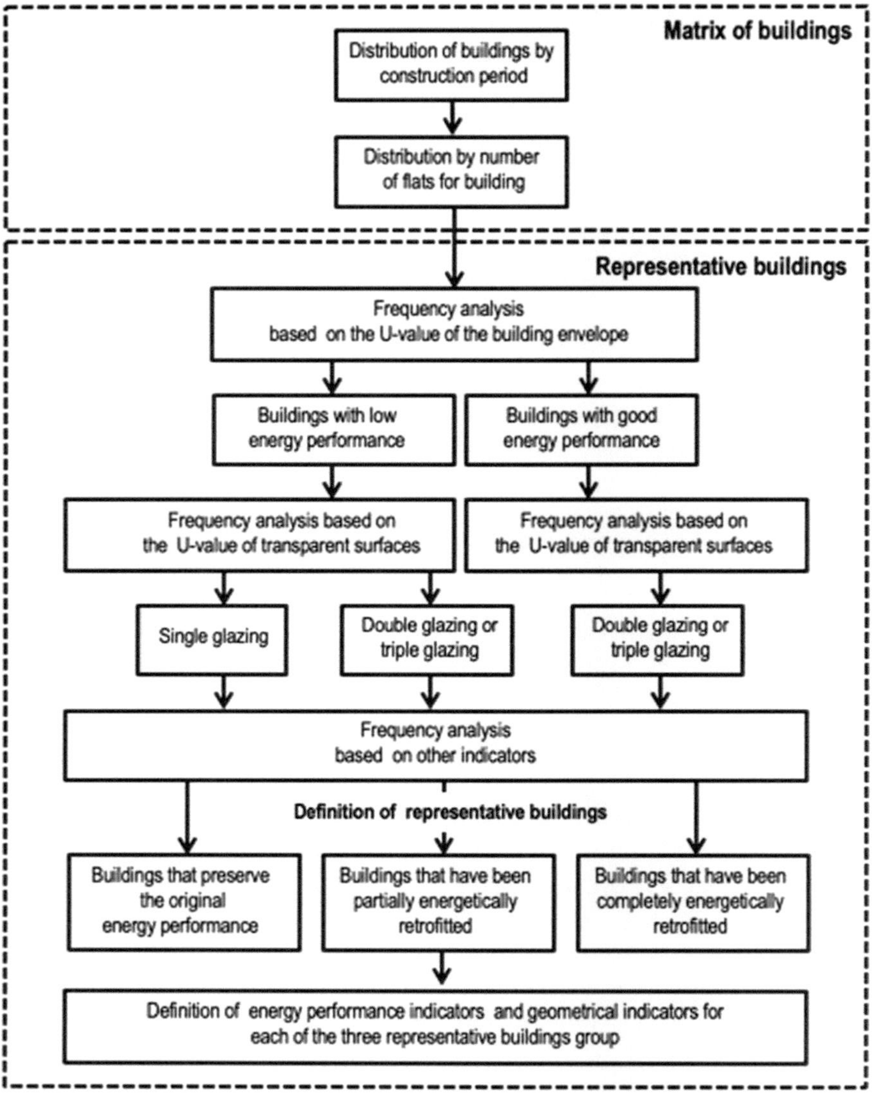

Fig. 8.4 Diagram of the procedure for the definition of the representative buildings within each cell of the matrix of Table 8.3 (Dall'O' et al. 2015)

classes of buildings are highlighted: buildings which preserve their original energy performance, buildings that have been partially retrofitted to improve their energetics (window replacement) and buildings that have undergone a complete energy retrofit (replacement of windows and building envelope insulation).

However the definition of the representative buildings which then constitute the final matrices do require additional steps, through filtering based upon other indicators e.g. the average *U*-value of the roof, the average *U*-value of the basement, ET_H (specific heat required by the building for space heating during the winter season), EP_H, the ratio between the heating capacity and the net floor surface, the overall average efficiency of the heating system (expressed as ET_H/EP_H) and the S/V ratio.

The identification of the representative buildings within the single cells of the matrix used allow the extrapolation of a series of useful items of information to characterise, from the energy point of view, the residential buildings of the existing building stock of the Lombardy Region.

The final result of this statistical analysis is represented by sets of three matrices: the values reported in the matrixes of Table 8.4, for example, define the representative statistical values of EP_H, expressed in kWh/m^2 year, for each cell of the used matrix.

Analogous matrices define other parameters, both energy and geometric performance of representative buildings, including:

Matrix $M1_{1,2,3}$ Average *U*-value (W/m^2 K) of the opaque vertical envelope;
Matrix $M2_{1,2,3}$ Average *U*-value (W/m^2 K) of the transparent surfaces;
Matrix $M3_{1,2,3}$ Average *U*-value (W/m^2 K) of the roofs;
Matrix $M4_{1,2,3}$ Ratio between the opaque surfaces of the vertical building envelopes and the net floor area;
Matrix $M5_{1,2,3}$ Ratio between the transparent surfaces of the vertical building envelopes and the net floor area;
Matrix $M6_{1,2,3}$ Ratio between the roofs surfaces and the net floor area,

where the subscripts of matrices 1, 2 and 3 indicate the three situations examined: (1) buildings which preserve their original energy performance, (2) buildings that have been partially retrofitted to improve energetics and (3) buildings that have undergone a complete energy retrofit.

The indicators derived from the analysis of the database of the energy cadastre of the EPCs in Lombardy Region can be used in the form of matrix, in the context of energy planning, both at regional and local level (town/city level).

The starting point for further processing is the availability of a matrix (construction period – number of flats per building) in which each cell is given the value of the total net area of the existing building stock: in the case of Italy, this matrix can be derived from the data of the national census. The structure of the matrix, and consequently the number of cells over which the existing building stock is divided, depends upon the availability of data and can be adapted accordingly.

Having available the matrix that defines the net floor areas, which can be denominated the "M0" matrix, and the matrices with the indicators described above, the calculations are simple and can be performed using a spreadsheet, via normal operations of addition, subtraction, multiplication and division between cells.

When the Lombardy Region updated its Regional Energy and Environmental Plan, for the evaluation of the potential for energy savings in the residential sector, it used this method, based on statistical data of the existing building stock.

Table 8.4 Average value of EP_H (kWh/m^2 y) for the representative buildings, broken down by period of construction, number of flats per building and conservation status (Dall'O' et al. 2015)

Construction period	Number of flats per building					
	1	2	3–8	9–15	16–30	>31
Buildings that preserve their original energy performance						
<1930	342	358	236	206	227	202
1930–1945	382	380	287	246	221	267
1946–1960	386	364	282	236	212	198
1961–1976	381	340	263	235	202	192
1977–1992	302	298	270	215	199	206
1993–2006	142	138	124	123	100	111
After 2006	57	52	44	40	35	34
Buildings that have been partially retrofitted to improve energetics						
<1930	311	289	224	201	169	177
1930–1945	332	324	224	187	162	160
1946–1960	330	282	253	218	161	162
1961–1976	300	309	222	197	184	158
1977–1992	237	229	217	165	164	163
1993–2006	–	–	–	–	–	–
After 2006	–	–	–	–	–	–
Buildings that have undergone complete energy retrofit						
<1930	90	68	54	56	54	94
1930–1945	76	70	62	46	53	91
1946–1960	98	93	63	52	85	110
1961–1976	95	95	89	93	79	98
1977–1992	90	88	83	96	109	95
1993–2006	–	–	–	–	–	–
After 2006	–	–	–	–	–	–

A limitation of the proposed method lies in the fact that all the energy assessments are based on a theoretical calculation: in fact the energy performances calculated for the preparation of EPCs refer to the technical standards applicable and take account of a standard use of the building. In other words, the real energy consumption of buildings can be different from the theoretical one and consequently even the potential for savings may be different.

In order to bring the assessments of potential savings calculated on the basis of the needs arising from standardized algorithms, related to energy certification, into line with actual consumption measured on buildings of the residential sector, it is necessary to define the relationship between the two variables involved. The Lombardy Region has estimated that the theoretical consumption for space heating in the residential sector is 27% higher than that actually observed. Therefore in its analysis it has taken account of this corrective index.

The innovative element of this study lies in the fact that the characterisation of the building stock does not require field investigations across the territory, measurements nor energy audits but is rather based upon existing information which is already available; this being that of the energy certificates registered in the databases.

8.1.3 Application of Neural Network for Evaluating Energy Performance of Buildings

A further piece of research directed by Khayatian et al. (2016) evaluates the possibility of calculating the energy performance of existing buildings using the database of the energy register containing the EPCs with a different approach to that described in the previous paragraph.

The objective of the study was to fit an estimator to the Calculation software used for estimating the energy performance of building for EPC purposes, by benefiting from the large number of energy certificates that are available online.

Due to the complexity of the calculation procedure, a black-box approach using Artificial Neural Networks (ANN) was adopted for replicating the relation between the input parameters and the respective ET_H value. Several combinations of inputs were tested for defining the most suitable implementation of features. It was observed that using 12 features (Table 8.5) was sufficient for successfully predicting the ET_H value (specific heat required by the building for space heating during the winter season) with an acceptable margin of error. Meanwhile, small abnormalities were observed in the overall performance of the network. Initial investigations on the predicted values revealed that the observed abnormalities were related to defective input data.

Further explorations were also conducted in order to outline the most important factors which affect the model's performance. It was observed that the model consisted of sufficient number of samples, yet additional features were required for obtaining a greater accuracy. A sequence of 100 models were trained due to the stochastic weight initialisation in training ANNs. In order to evaluate each entry, a Gaussian distribution of 100 predictions were created based on the trained models. This approach provided a probability of occurrence for each certificate rather than

Table 8.5 Average classification of utilized features (Khayatian et al. 2016)

Category	Features
Direct	Degree days, net volume, net floor area, dispersant surface, opaque to glazed ratio, year of construction, thermal conductivity (walls, windows, roof, basement)
Derived	Average floor height, opaque surface area, glazed surface area, construction period, non-linear features

a definitive statement. As a consequence, the procedure of monitoring each entry is prioritised based on its level of reliability i.e. probability of occurrence. Such approach may be replicated for similar EPC datasets with comparable data. Final results indicated that $95 \pm 3\%$ of entries fall within the limits, so the proposed tool is reliable.

The application of machine learning in building energy estimation provides the opportunity to extract predictions in a rapid manner. Therefore, the methodology can also be used as an easy tool to support energy planning on a territorial scale.

8.2 Methods for Estimating Buildings Energy Demand at District Level

In the framework of distributed energy planning, evaluating reliable energy profiles of different sectors has a prominent role. At the same time, it is a quite challenging task, since the availability of actual energy profiles is not common.

For guiding stakeholders involved in local smart energy planning to choose the proper method for assessing the buildings energy demand, Ferrari et al. (2019a) surveyed 70 studies, adopting different methods including some case-study applications and identifying those adopting hourly energy profiles.

From the methodological perspective, a set of criteria for classifying the selected contributions was defined and, as final results, tables summarizing the characteristics of the principal methods and a selection of studies providing directly usable energy profiles were reported. Broadly speaking, this piece of research demonstrates that the potential replicability of analysed methods is constrained to the availability of datasets and, more particularly, underlines the necessity of defining reliable hourly energy profiles in order to develop accurate energy scenarios. The principal findings of the research are reported in the following.

In the context of review articles on methods for buildings energy demand estimation, the study of Swan and Ugursal (2009) is one of the most exhaustive. They made a first broad distinction between:

- Top-down methods, whose output data are estimated based on aggregated input data through econometric or technological correlations;
- Bottom-up methods, based on use of subordinated input data and highly detailed and reliable calculations.

Starting from the methodological classification reported in the aforementioned article (Swan and Ugursal 2009), investigated studies have been classified hereunder, consistently with the main scope of this research, first according to the time resolution of used energy demand data and then, to the adopted data source/estimation method, as follows (Fig. 8.5):

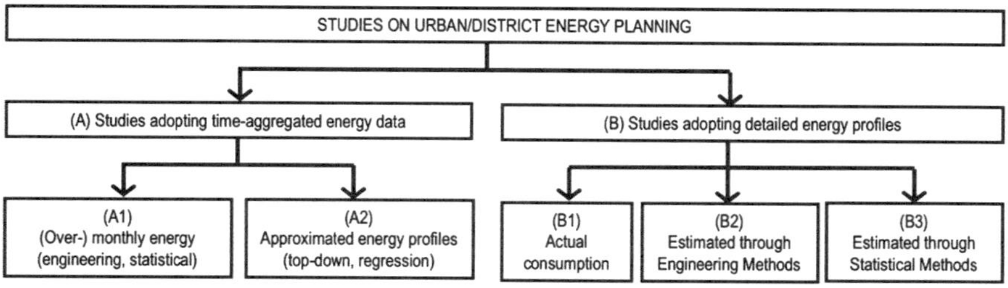

Fig. 8.5 Scheme of adopted classification. Adapted from Ferrari et al. (2019a)

- Studies adopting time-aggregated energy data, as:
  - Over-monthly energy data;
  - Approximated energy profiles;
- Studies adopting detailed energy profiles, as:
  - Actual consumption profiles;
  - Energy profiles estimated through engineering methods
  - Energy profiles estimated through statistical methods.

Following, all identified methodological categories are outlined, discussed and compared.

8.2.1 Studies Adopting Time-Aggregated Energy Data (A)

Studies adopting (over-)monthly energy data (A1). In some studies, the annual or monthly energy demand was calculated, either based on engineering or statistical methods, for assessing the balance of the potential energy supply with the energy loads. Although such approach is typical of truly early stage first analysis, it clearly returns highly approximated outcomes which can be substantially far away from more detailed results.

Documented experience referred to engineering methods revealed assessments of energy consumptions for buildings according to local monthly based calculation procedure, or by means of commercial building energy simulations, assuming single thermal zones and retrieving properties from direct survey, statistical and commercial data and software libraries, or via a simplified steady-state calculation procedure.

Documented experience based on statistical methods revealed energy consumptions estimated thanks to statistical data on population and/or buildings characteristics, or through correlations with weather data (e.g. energy signature (CEN 2008).

Studies adopting approximated energy profiles (A2). In order to provide more accurate energy assessments while coping with serious lack of data, some studies adopted procedures for deriving approximated profiles from top-level data, thanks to correlations with weather data, building use category or typology consistency.

Top-down approaches were used to proportionally derive hourly profiles at local level from spatially aggregated hourly profiles. In one case the energy demand hourly profiles for a town have been derived by scaling-down the previously determined national hourly profiles to the consumptions reported in the municipal energy balance, and then adjusted to take into account the local climatic conditions and a future scenario. Similarly, in another case the measured national profile was scaled-down to the annual energy consumption of the town. For two case studies the lower scale profiles were defined from the available overall urban hourly energy data, based on the distance from the urban centre (i.e. urban texture) and on the different end-use energy sector.

Finer data can be derived also from temporally aggregated data, by means of regression. Starting data to be regressed can be obtained by means of statistical data or simplified calculation procedures (e.g. steady state or quasi-steady state procedures, calculation based on total installed power of electrical appliances, etc.).

Regarding the time resolution of approximated energy profiles, regressions were carried out in some case from annual/monthly data into daily, over-hourly or hourly ones while in other cases from daily data to hourly and from regional hourly to local hourly.

8.2.2 Studies Adopting Detailed Energy Profiles (B)

Studies adopting actual energy consumption profiles (B1). For the accurate estimation of energy fluxes, the availability of actual hourly based energy consumptions for the investigated case study is of course the optimal condition. Hourly data can be generally retrieved by monitoring campaigns in delimited areas, such as a University campus, or provided by electricity distribution utilities or operators of district heating networks.

Studies adopting energy profiles estimated through Engineering Methods (B2). The adoption of engineering methods to assess the energy demand of buildings enables one to have very accurate estimations, when based on dynamic energy simulation. Despite this, detailed building energy simulations on urban scale could not always be justifiable in terms of the time necessary, owing to the required level of detail and quantity of information. Hence simplifications in the calculation methods or in the number and complexity of the modelled buildings are often adopted.

This can be achieved by simulating "Representative buildings" (Ferrari and Zanotto 2016a), Building Archetypes or samples, representative of the considered building stock, adopting simplified calculation procedures for deriving buildings thermal energy profiles such as the EN ISO 13790 simplified hourly method (ISO 2008).

More accurate estimations of thermal energy profiles can be carried out by means of detailed dynamic building energy simulation tools, such as TRNSYS (Klein et al.

2014), EnergyPlus (DOE 2018), ESP-r (http://www.esru.strath.ac.uk/Programs/ESP-r.htm), etc.

Studies adopting hourly energy profiles estimated through Statistical Methods (B3). Statistical methods include a wide range of techniques.

Space heating and cooling, domestic hot water (DHW) and electricity consumption profiles can be determined based upon statistical analysis of data from direct surveys and measurements on buildings representative of different typologies and/or uses in order to determine, through probability distribution function, regression analysis or Machine Learning methods, hourly energy profiles of the stock being considered.

The latter methods are increasingly adopted for forecasting energy loads due to the ability of learning from past behavior so returning in most robust outcomes.

8.2.3 Discussion and Conclusion

For accomplishing reliable distributed energy planning, the adoption of detailed energy demand profiles is essential. However there is a general unavailability of detailed consumption profiles at local level.

From the research article (Ferrari et al. 2019a) the Authors saw that those studies relying on time aggregated energy data (A) often lead to a very much approximated estimation.

In fact, the adoption of seasonal energy data (A1) is responsible for the neglect of energy fluctuations which are in fact relevant when planning smart and RES (Renewable energy sources) integrated energy systems, including assessment of storage systems. Moreover, approximated energy profiles (A2) derived from spatially aggregated data through top-down approach, by placing on the same level a small community and a Country, or from temporally aggregated data derived from regressions based on the outdoor temperatures, neglecting other weather parameters, such as the solar radiation (Ferrari and Zanotto 2016b), or building usage patterns (which are usually responsible for a more variable profile), can be also critical.

Regarding the studies adopting detailed energy profiles (B), apart from those cases where actual consumption profiles are available (B1), robust methods for the estimation of reliable profiles, being either engineering-based (B2) or statistical-based (B3), must be chosen.

For reducing the computational effort of engineering modelling, the clustering of a building stock in a series of representative buildings is a well-established approach that allows, if combined with accurate dynamic energy simulations, an acceptable approximation of the energy evaluations to be maintained (Ferrari and Zagarella 2016). However, a critical point of these studies could be the assumption of standard values of the building usage patterns instead of an accurate assessment (Ferrari and Zanotto 2016c). Studies adopting distributions, which are based on a statistical approach but as detailed as those from engineering methods, could be a way to overcome this issue.

Among the statistical methods, those which are based on machine learning enable one to make reliable estimates from uncertain data and even reduce computational effort, overcoming the limitations of the basic statistical correlations. However the adoption of statistical methods relies on measured/simulated data which are not everywhere accessible.

8.3 Tools for Urban Energy Planning

For low carbon district/urban energy planning, several computational tools have been developed although guidelines for a more diffuse use in common planning, outside the academic ambit are needed. Ferrari et al. (2019b) assessed 17 selected easy-access and well documented tools based on: analysis type, operation spatial scale, output time scale, use and licence.

General information, functionalities, structure, graphic user interface, required input data and outputs were described. A plenty of information was provided on the data quality and level of details and the related constraints to implement an energy system analysis. The possible final user of the above mentioned research is guided in choosing the tool based on the related outputs and potentialities, but also on available data for a given context and the required skills.

Based on the above mentioned research, this sub-chapter focuses on the tools liable to be used outside the academic environment (i.e. by public administrations, urban energy planners). Hence, nine tools having user-friendly interface have been examined, as classified in Fig. 8.6, according to: time resolution of outputs (i.e. over-hourly based or sub/hourly based outputs), and kind of licence (i.e. subject to payment or free of charge). As a result, *coverage is given to* 3 tools that report over-hourly output data of energy demand-supply, and so can be adopted at very early stage of urban energy planning, and 6 tools that enable energy calculations on at least an hourly base, which is essential for modelling the detailed interactions among variable energy fluxes at urban level.

The general information on the nine user-friendly tools are reported in the following sections: the descriptions of the six hourly-based tools have been treated in greater detail.

8.3.1 User-Friendly Tools with Over-Hourly Based Outputs

District-ECA was developed by the Fraunhofer Institute for Building Physics with the IEA-ECBCS Annex 51 partners for supporting those operators who are active in the first-stage of planning energy-efficient districts, for quickly evaluating the effects on both the supply and the demand sides strategies. The software is freely downloadable after making a preliminary registration and comprises a set of tools with

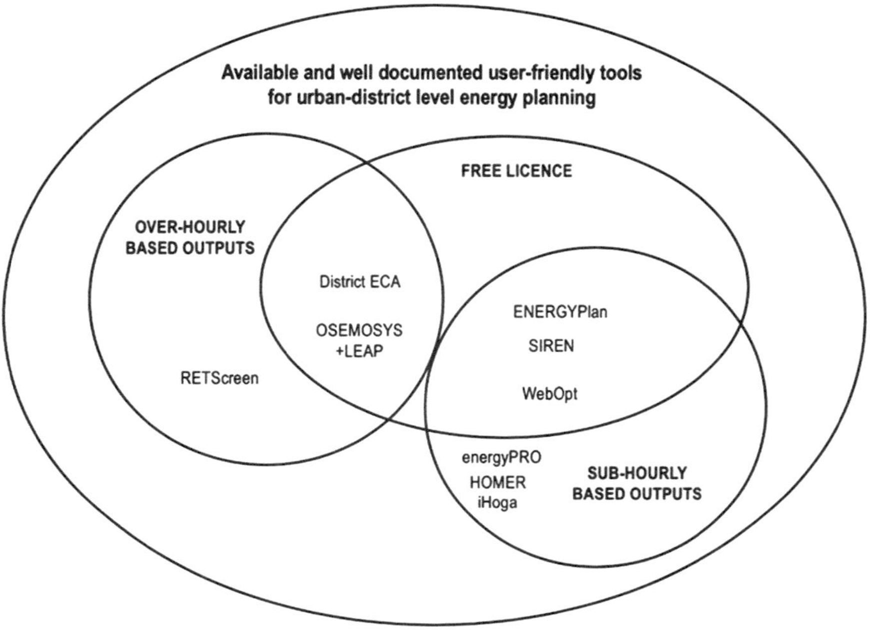

Fig. 8.6 Scheme of the method adopted for the classification of the tools. Adapted from Ferrari et al. (2019b)

miscellaneous scopes[2] (Fraunhofer Institute of Buildings 2013). The main core is an easy-to-use tool (i.e. Energy Assessment of Districts) for accomplishing a monthly based assessment of energy balance and related emissions of a district. A single building modelling of related envelope and systems should be done, the technologies provided mainly refer to space heating, weather data and typical solutions are constrained to IEA-ECBCS Annex 51 participating Countries.

OSeMOSYS (Open Source Energy MOdeling SYStem) is an open source modelling tool for long-term energy assessment and planning[3] (Howells et al. 2011). Its most common graphic user interface (GUI) is the commercial software LEAP maintained at the Stockholm Environment Institute[4] and referred as a decision support software, enabling users in extensive data management and reporting, assessing medium/long-term scenarios including changes in energy use, related emissions, resources deployment, for any economic sector. Most of its calculations occur on an annual time-step, while the time horizon is unlimited.

Since it is a very general-purpose tool, required input data broadly depend upon the case study (e.g. top-down, bottom-up, economic, social data, etc.).

[2]https://www.district-eca.com/index.php?lang=en. Accessed 27 September, 2019.

[3]http://www.osemosys.org/. Accessed 27 September, 2019.

[4]https://www.energycommunity.org/default.asp?action=home. Accessed 27 September, 2019.

So far as energy related data are concerned: annual values for demand and production, eventually by sector and fuel, are required; additionally, costs of systems and energy, system efficiencies and capacities, energy profiles, emission factors, interest and inflation rates can be indicated.

RETScreen[5] (Minister of Natural Resources of Canada 1997) is a commercial tool developed at the Natural Resources Canada department of the Government of Canada as a clean energy management software system. Currently, it consists of an Excel-based software package and a Windows-based one. It is possible to model systems of any scale, developing different scenarios (buildings envelope and systems energy efficiency measures, various electricity and thermal energy supply plants, integration of RES, etc.).

Required information includes: general data and settings on calculation, climatic location, type of project, sector, system features such as capacity, delivered energy, efficiency. In addition, CO_2 emissions per each fuel and financial data (e.g. inflation rate, lifetime, debt ratio range, initial costs, incentive and grants, annual Operation and Maintenance (O&M) and fuel costs) can be inserted.

8.3.2 User-Friendly Tools with Hourly or Sub-hourly Based Outputs

EnergyPRO, is a commercial tool developed by the Danish company EMD International A/S for accomplishing techno-economic analysis and optimisation of energy projects with a combined supply of electricity and thermal energy from a multiplicity of different energy producing units[6] (EMD 2017).

It is typically used for analysing scenarios including district heating (DH), CHP, combined cooling, heat and power (CCHP), but can be also used for projects including geothermal energy, solar collectors, PV, wind farms, hydro pumping stations, storage systems.

The user should select the type of assessment among: Design, with emphasis on energy conversion and operation payments for one-year calculations, Finance for multi-years investment analysis, Account, for a more detailed economic assessment, Operation, for optimizing the operation on a daily period. The user can provide information on external conditions by adding data, i.e. loads and energy prices, as time series. It is possible to assess regional level scenarios by defining more interconnected distributed production sites.

When describing the energy exchange among the possible sites, required information includes: the start and end site, the transmission direction, the energy service (heat, process heat, cooling), and the transmission capacity and loss which can be defined by functions or time series. The calorific value of fuels with related units of

---

[5]http://www.nrcan.gc.ca/energy/software-tools/7465. Accessed 27 September, 2019.

[6]http://www.emd.dk. Accessed 27 September, 2019.

measurement, eventually the maximum usable amount of the fuel storage, the offered fuel to production units per each month can also be indicated.

An unlimited number of demands per each service (heat, process heat, electricity and cooling) can be defined. In the tool library, arbitrary profiles and, specifically for the German context, profiles from national standards of electric and thermal energy are included. Otherwise, it is possible for the user to specify his/her own data both as annual or time-series (i.e. hourly or sub-hourly based) amounts, both as fixed values or dependent upon weather data (i.e. average ambient temperatures) and with year-to-year variations by adding a defined fixed or variable index for specifying the time series development over the years in question. So far as energy supply is concerned, the included units are: recovery of rejected heat, CHP, electric boiler, absorption/electric chiller, heat pump, flat plate/evacuated tubes solar collector, PV, wind farms and storage systems. Usually required data, depending on the single technology, are: capacity, production, fuel, power curve, inlet and outlet temperatures, if the operation is dependent on another unit, charging/discharging power; height difference of hydro-pumping storage reservoirs, dimensions and technical features, etc. The operation strategy, except for a user-defined one, is basically driven by a minimisation of net present cost (NPC) through assigning levels of priority to production units. Emissions of CO_2, NOx and SO_2 per unit of mass can be defined. It is possible to define the electricity market, choosing between fixed tariff and spot market.

HOMER is a commercial tool developed at the U.S. National Renewable Energy Laboratory to assist in the design of micropower systems and to facilitate the comparison of power generation technologies across a wide range of applications (HOMER Energy 2016).

The software can perform three assessments: simulation, to model hourly the micropower system behaviour and determine the techno-economic feasibility over its operating lifetime; optimisation, to simulate different system configurations in search of the one that satisfies the technical constraints at the minimum NPC; sensitivity analysis, to evaluate the impact on the optimisations results of changes in the input parameters even in a long-term perspective.

First, the climatic zone of the context under study must be specified, in order to determine the RES availability. Thereafter four types of energy load are considered: the primary load, i.e. electricity demand, the deferrable load, i.e. a demand for electricity that can be met at any chosen time within a certain time interval (e.g. water pumps, ice makers, and battery-charging stations), the thermal load, the hydrogen load. Primary, thermal and hydrogen loads must be defined by indicating average daily values of energy and power, a peak power and a daily hourly schedule.

The profiles can be modified either by introducing some weekly/monthly variations or adding a percentage of disturbance which will randomly change the curve. The tool includes some default arbitrary hourly profiles, which are the same for each energy service but differ per end-use sector.

Otherwise it is suggested to use external datasets mainly referring to US context or, if possible, by importing one's own hourly data.

There is a wide library of supply/conversion/storage units, but others can be created by the user. Generally speaking, the required information includes: system capacity, the plant lifetime, the unit costs of investment, of replacement and operation and management with related frequency, and whether the electric current is alternating or direct. Fuels can be defined based on the library or by inserting the type, unit prices, calorific value, GHG emissions for each substance.

In case of modelling RES-based systems, other data, regarding the efficiencies and technical characteristics of supply systems as well as average monthly availability of resources, are required. Once the energy system has been modelled, the user must define the settings of the system assessment and optimisation. The required economic parameters are: nominal discount rate, expected inflation rate, project lifetime, system fixed capital and operation and maintenance (O&M) costs, capacity shortage penalty, and currency.

For the definition of optimisation constraints, maximum annual capacity shortage, minimum RES fraction and operative reserve as a percentage of load or RES output, are required. Possible economic penalties and annual limits of GHG emissions production can be set. The principal outputs, both in tabular and graphic form, are the total and annually cash flows by component and either per NPC or annualised costs, details about the electricity production and consumption, per component and load types, annual emissions of pollutants, possible ranking of developed scenarios according to NPC and/or related to sensitivity analysis.

iHoga (Improved Hybrid Optimization by Genetic Algorithm), developed at the University of Zaragoza, is a commercial software for the simulation and optimisation of hybrid stand-alone systems of electric power generation based on RES[7] (Dufo-López 2017). An educational version can be freely downloaded, although the lack of several functionalities and some computer-related requirements (i.e. internet connection and Windows version up to 8) widely limit its use, so it has not been taken into account for this review.

The expected load consumption of alternating/direct electric current, hydrogen and/or water has to be defined, by either adopting monthly average values or selecting default profiles to which one eventually applies a percentage of random variation. Included profiles come from literature and case studies of stand-alone systems. Alternatively the user's own time-step based profiles can be adopted by importing files with rows of data referring to hourly power, mass flow rate for hydrogen and volume flow rate for the water. There is the facility to indicate data on solar irradiation, wind and hydrogen either as average monthly values from NASA or as own time-step-based files.

The simulation time step, between a minute and an hour, and time horizon must be chosen and in the case of the lifetime of the whole system, the user has to choose between a mono-objective (based on NPV minimisation) and a multi-objective optimisation (based on minimisation of NPV, equivalent CO_2 emissions or unmet energy load) otherwise more advanced alternatives can be defined.

[7] https://ihoga.software.informer.com. Accessed 27 September, 2019.

Regarding financial data, the nominal interest rate, the expected annual inflation rate, the system lifetime/assessment period, the currency, the installation costs and variable initial cost, as both the fixed cost and its percentage to initial cost must be entered. Total cost of the different solutions and CO_2 emissions are represented graphically.

EnergyPLAN is a freeware deterministic software developed at Department of Development and Planning at Aalborg University, for assisting the design of national energy planning strategies on the basis of technical and economic analyses. It is downloadable by means of preliminary registration[8] (Lund 2017).

It has been used in several scientific studies, both academic and not (Connolly et al. 2015). Two alternative analysis can be performed: technical simulation, for balancing either thermal or electric energy fluxes, and market-economic simulation, for assessing the system feasibility on the basis of annual costs. Electricity, heating, cooling, industrial and fuel, transport, water consumptions can be modelled; for all of them, the necessary input data are the total annual energy requirements and related hourly distributions.

The hourly distribution is the energy consumption profile for a leap year, where each value is normalized to the maximum hourly value. In the library, energy demand distributions are included at town level, which usually come from utility companies for some contexts which have been investigated, and at national level, coming from energy balance or STRATEGO project (Connolly et al. 2015). Additionally, the efficiencies of the heating and cooling production units must be specified; the contribution from solar thermal energy, by entering the capacity of the heat storage, given in days of average heat demand, and the share of consumers having installed solar thermal collectors, with the related total annual production and hourly distribution, can be modelled as well as, when existing, a flexible share for the electricity consumption.

The supply side is separately modelled with reference to the energy service (i.e. heat and electricity, electricity only, heat only, thermal plant fuel distribution, waste, liquid and gaseous fuels) for whose plants the annual production and/or capacity, the operation efficiencies and the production hourly distribution are required; in the particular case of more than one DH demand, the data must be reported for each defined group.

Moreover, the carbon dioxide content by kilogram of fuel can be inserted. For the performance of a market-economic analysis the user is required to insert the interest rate, and, for each plant unit, the investment cost, the lifetime and the percentage shares of fixed operation and maintenance costs (in order to permit the software to calculate the annual amount).

In the additional subsection, as suggested in the manual, other costs can be indicated, such as those referred to the retrofitting of buildings. Hourly energy balances, annual/monthly fuel consumptions, CO_2 emissions and costs of the defined system are provided as outputs, which can be exported in an excel sheet while charts of hourly fluxes can be also plotted.

---

[8]http://www.energyplan.eu/. Accessed 27 September, 2019.

SIREN (SEN Integrated Renewable Energy Network) Modelling Toolkit is a freeware tool developed at the Australian not-for-profit organization called Sustainable Energy Now team and is currently freely distributed as a beta version.[9] The software's goal is used to calculate energy generation for renewable energy power stations and developing electricity long-term scenarios. In detail, it aims to determine the optimal locations to access RES, minimise grid connection costs and meet the varying demand on the grid, while achieving the best in terms of efficiency, cost effectiveness and energy security.

The whole software is made of a geo-referenced tool (SIREN), an energy calculation tool (SAM Power Models), provided by the National Renewable Energy Laboratory, and a set of Excel worksheets for the energy system optimisation (Power Balance tool).

The project area can be defined based on coordinates and the related imported satellite image. The satellite maps are used for visualizing, locating and modelling both the existing electricity network and the new RES plants. Apart for Australia territory for which default data are available, the hourly weather and electricity loads data and, possibly, the existing geo-referenced electric network layout must be imported. Possible supply technologies are: biomass, PV (fixed, tracking or rooftop), geothermal, solar thermal, wind, hydro and wave.

Required data are: location, power capacity, area, capital, operation and maintenance costs, hourly power production and, in the specific cases of wind farms and PV, respectively the type, the rotor diameter and the number of turbines and the orientation of the panels.

Connecting lines must be specified with regard to type, cost and maximum carrying capacity. By means of the SAM Power Models, hourly electricity balance is reported and exported to the Power Balance tool, which quantifies and cost-optimises the various amounts of storage and generation technologies for completely balancing the system; finally, CO_2 emissions can be calculated as outputs.

WebOpt stands for Distributed Energy Resources Web Optimization Service and is the freeware academic online version, accessible through preliminary registration[10] (Stadler et al. 2013), of the Distributed Energy Resources Customer Adoption Model (DER-CAM) software, developed at the Berkeley Lab. Even if, compared to DER-CAM software, the browser-based version is more limited and has also been less frequently adopted in existing studies it is a quite complete tool, able to assess an integrated energy system. WebOpt permits the modelling and optimisation of energy scenarios as a function of the criteria of economic costs and/or carbon dioxide emissions.

Three alternative optimisation strategies can be achieved: a cost minimisation; a CO_2 emissions minimisation or a multi-objective analysis (i.e. CO_2 minimization with cost constraint, for which it is required to fix a maximum allowable energy cost for purchasing electricity from the grid). Hourly energy loads for electricity,

---

[9] http://www.sen.asn.au/modelling_overview. Accessed 27 September, 2019.

[10] https://building-microgrid.lbl.gov/projects/distributed-energy-resources-web. Accessed 27 September, 2019.

electrically powered cooling, electrically powered refrigeration, space heating, water heating and/or natural gas should be imported as normalized profiles to 1 GWh.

They can be inserted using one building default profile, suggested for first analysis, or copying and pasting in each cell the user-defined data or selecting one profile from the ASHRAE U.S. Climate Regions. While, if the user-defined load data are entered, there is a requirement for a monthly average hourly value, the typical hourly based weekly profile, differing if possible between working and weekend days, and the peak. Specifically, the cooling and the refrigeration loads are always expressed in the electricity consumed by an electric chiller with a coefficient of performance of 4.5.

There is the possibility of modelling in detail the energy costs of electricity and natural gas, with options on indicating: differences of cost on a seasonal/monthly base, costs fixed and variable part, different time of use. System investment and O&M costs, lifetime and maintenance periods, efficiencies and capacity must be defined. The percentage of schedulable load, the solar radiation daily profile and the carbon dioxide emissions per energy unit are additional items that can be defined. In Information on total annual costs, payback time, installed plants capacities, the detailed hourly energy profile are provided as results of the simulation.

8.3.3 Discussion and Conclusion

For supporting the decision-makers in urban distributed energy planning, several tools are available today. From the work of Ferrari et al. (2019b), nine user-friendly tools for the development of RES integrated energy scenarios in urban contexts have been selected and their main features have been reported in this chapter.

Previous studies have shown that several differences characterise the available tools, so it is not possible to identify an optimal one valid for any situation, but the choice strictly depends upon the specific goals of the project.

In particular, for modelling the interactions among variable energy fluxes and for evaluating the effects of several energy supply units, it is essential to adopt tools with at least hourly-based outputs.

In this regard, concerning the 6 sub/hourly-based tools described in this chapter, the following commentary can be added: iHoga and energyPRO can run with time-steps until 1 min. In addition, very easy employment of Homer, EnergyPLAN and WebOpt has been observed but the online use of the last mentioned has the drawbacks of slowness and limited basic functions (e.g. copy and paste, lack of shortcuts).

In general terms their use is facilitated by their excellent support platforms, including video-tutorials, lists of related publications, user manuals and sometimes guides on required data resources; among these tools, energyPRO, Homer and Energy-PLAN benefit from the most complete training material. Of additional value is the integration with spatial representation, i.e. georeferenced satellite maps in SIREN or network graphical representation in energyPRO and iHOGA, which allow one to spatially analyse the energy fluxes, optimally locate RES plants, estimate distributed network losses, etc.

There is the possibility of modelling and comparing more than one scenarios at time but only when using Homer, SIREN and WebOpt. Regarding default loads, all tools foresee the possibility of importing one's own timeseries for specific case studies, which is also the option most encouraged. Otherwise, default profiles are either of an arbitrary form or constrained to specific contexts.

Some tools are more appropriate for assessing the electrical energy sector, i.e. HOMER, iHoga, SIREN. Accordingly, in HOMER, modelling thermal energy depends on preliminary modelling of the electrical energy involved; however, it is able to consider up to two thermal loads, with some heating supply components, while no plant for space cooling is specifically included. In contrast, other tools are able to model more sectors, thereby enabling the user to assess integration and flexibility in urban energy systems.

References

Connolly D, Hansen K, Drysdale D (2015) STRATEGO creating national energy models for 2010 and 2050 work package 2—background report 1

Dall'O' G, Galante A, Torri M (2011) A methodology for the energy performance classification of residential building stock on an urban scale. Energy Build 48:211–219

Dall'O' G, Sarto L, Sanna N, Tonetti V, Ventura M (2015) On the use of an energy certification database to create indicators for energy planning purposes: application in northern Italy. Energy Policy 85:207–217

Dufo-López R (2017) iHOGA version 2.4 user's manual. University of Zaragoza, Spain

EC (2002) Directive 2002/91/EC of the European Parliament and of the Council of 16 December 2002 on the energy performance of buildings

EMD International A/S (2017) User's guide—energyPRO

European Committee for Standardizing (2008) CEN 15603, energy performance of buildings. Overall energy use and definition of energy ratings

Ferrari S, Zagarella F (2016) Assessing buildings hourly energy needs for urban energy planning in southern European context. Procedia Eng 161:783–791

Ferrari S, Zanotto V (2016a) Defining representative building energy models. Building energy performance assessment in southern Europe. SpringerBriefs in applied sciences and technology. Springer, Cham, pp 61–77

Ferrari S, Zanotto V (2016b) Climate-related assessment of building energy needs. Building energy performance assessment in southern Europe. SpringerBriefs in applied sciences and technology. Springer, Cham, pp 99–118

Ferrari S, Zanotto V (2016c) Implications of the assumptions in assessing building thermal balance. Building energy performance assessment in southern Europe. SpringerBriefs in applied sciences and technology. Springer, Cham, pp 35–45

Ferrari S, Zagarella F, Caputo P, D'Amico A (2019a) Results of a literature review on methods for estimating buildings energy demand at district level. Energy 175:1130–1137

Ferrari S, Zagarella F, Caputo P, Bonomolo M (2019b) Assessment of tools for urban energy planning. Energy 176:544–551

Fraunhofer Institute for Building Physics IBP (2013) Manual district energy concept adviser

HOMER Energy (2016) HOMERPro version 3.7—user manual. Boulder, USA

Howells M, Rogner H, Strachan N, Heaps C, Huntington H, Kypreos S, Hughes A, Silveira S (2011) OSeMOSYS: The Open Source Energy Modeling System. An introduction to its ethos, structure and development. Energy Policy 39:5850–5870

International Organization for Standardization (2008) ISO 13790, energy performance of buildings—calculation of energy use for space heating and cooling

Khayatian F, Sarto L, Dall'O' G (2016) Application of neural networks for evaluating energy performance certificates of residential buildings. Energy Build. 125:45–54

Klein SA, Beckman WA, Mitchell JW et al (2014) TRNSYS—a transient system simulation program user manual. The Solar Energy Laboratory, University of Wisconsin, Madison

Lund H (2017) EnergyPLAN. Advanced energy systems analysis computer model. Documentation version 13

Minister of Natural Resources Canada (1997) RETScreen software online user manual, 1997–2005. Available at: http://publications.gc.ca/site/eng/search/search.html?st=1&ssti=&e=0&f=0&adoof=0&ast=user+manual+&cnst=&hpp=10&psi=1&rq.ssp=-5. Accessed 27 Sept 2019

Stadler M, Marnay C, Cardoso G, Donadee J, Siddiqui A, Wang S, Feng W, Lai J, Bhattacharya P, Mégel O, Groissböck M, Yu-Ko C (2013) WebOpt user manual. DER Web Optimization Service (WebOpt), Lawrence Berkeley National Laboratory (LBNL), pp 1–32

Swan LG, Ugursal VI (2009) Modeling of end-use energy consumption in the residential sector: a review of modeling techniques. Renew Sustain Energy Rev 13(8):1819–1835

U.S. Department of Energy (2018) EnergyPlus version 8.9.0 documentation

Chapter 9
Energy Retrofit Strategies in the Building Sector

Giuliano Dall'O'

Abstract In the energy-environmental balance of a city, the building sector uses on average 40% of the energy taken from fossil fuel sources, thereby significantly contributing to air pollution and the emission of greenhouse gases. Green planning strategies must therefore carefully consider this sector, promoting energy retrofit actions that can contain environmental impacts. The energy and environmental redevelopment of buildings must not only be considered an emergency but also an interesting opportunity for growth in the construction market; however, some critical issues, not only of a technical but also economic and social nature, must be assessed. This chapter, focusing on the issue of energy and environmental redevelopment of buildings in cities and communities, on the territorial scale, aims to provide green planners with useful tools to approach their strategies in a pragmatic manner.

9.1 Buildings in the Cities, a Complex Framework

Cities and communities are the social, economic and cultural points of aggregation of human activities and buildings, with urban infrastructures, are pieces of a mosaic that have grown over time. Buildings are the elements that most characterize a city: their shape, their aesthetic appearance, their composition within the urban fabric contribute to making a city attractive or neutral, inclusive or inhospitable.

The energy consumption of buildings depends above all on the function they perform in creating a protected environment for humans, from the climatic point of view in winter, through the heating system, but often also in summer through the air conditioning system. Summer and winter air conditioning is not the only cause of energy consumption.

Energy is also consumed for artificial lighting, for installations and for all the equipment needed to make what Le Corbusier defined "the living machine".

G. Dall'O' (✉)
Architecture, Built Environment and Construction Engineering, ABC Department, Politecnico di Milano, Via G. Ponzio, 31, 20133 Milan, Italy
e-mail: giuliano.dallo@polimi.it

G. Dall'O' (ed.), *Green Planning for Cities and Communities*,
Research for Development, https://doi.org/10.1007/978-3-030-41072-8_9

The environmental impact of buildings is considerable. According to European Union statistics,[1] buildings are responsible for approximately 40% of energy consumption and 36% of CO_2 emissions in the EU, making them the single largest energy consumer in Europe. At present, about 35% of the EU's buildings are over 50 years old, and almost 75% of the building stock is energy inefficient. At the same time, only 0.4–1.2% (depending on the country) of the building stock is renovated each year.

Renovation of existing buildings could therefore lead to significant energy savings and play a key role in the clean energy transition since it has the potential to reduce the EU's total energy consumption by 5–6% and lower CO_2 emissions by about 5%.

The awareness of how necessary it is to intervene in the building sector to reduce energy consumption from fossil fuels is well established at a political level. SDG 7 (affordable and clean energy) and SDG 11 (sustainable cities and communities) confirm governance strategies that had already been adopted in many countries.

A significant commitment in the direction of promoting energy efficiency in the building sector has been made by the EU. The EU's legislative framework for energy-efficient buildings was strengthened in 2002 with the adoption of the Energy Performance of Buildings Directive or EPBD (EC 2002) which included four main aspects:

- Establishment of a calculation methodology: Member States were to implement a methodology for the calculation of the energy performance of buildings, taking account of all factors that influence energy use.
- Minimum energy performance requirements: regulations would need to set minimum energy performance requirements for new buildings and for large (>1000 m^2) existing buildings when they were refurbished.
- Energy performance certificates: an energy performance certificate would need to be made available whenever buildings were constructed, sold or rented out.
- Inspections of boilers and air conditioning: regulations would be needed, requiring inspections of boilers and heating systems (with the possibility of alternative approaches such as providing advice), as well as inspection of air conditioning systems.

Over the years, the evolving regulatory framework has mobilized significant resources and shown itself to have a positive impact on Europe's energy efficiency. The above-mentioned EPBD, transposed by the Member States, was updated by Directive 2010/31/EU (EU 2010) on energy performance of buildings (EPBD recast) which widened the scope of the EPBD and raised the bar. It introduced the concept of Nearly Zero-Energy Buildings (NZEB) and a target of 2018/2020 for their introduction.

The European Commission has not simply restricted itself to enacting laws but has also carried out an effective monitoring action to measure its effectiveness.

[1] https://rezbuildproject.eu/2018/12/03/sustainable-buildings-fight-climate-change/ (Accessed 20 September 2019).

The "EPBD evaluation report" (EC 2016), presented by the European Commission in November 2016, addresses the effectiveness, efficiency, relevance, coherence and EU added value of the Directive. The evaluation report identified many positive effects and only limited regulatory failures; at the same time, it identified the need for simplification and streamlining of measures, enhancement of compliance, better linking with financial support, adaptation to technological developments and a strong need to increase building renovation.

The 2018/844 Directive (EU 2018) marks a further change with respect to the previous legislative framework and does so by starting to assign to buildings the role that is theirs within the transformation of modern society in a sustainable sense. The Directive does not enter into the merits of how this should be achieved, leaving this decision to the Member States. Each of these, in fact, must identify its own long-term redevelopment strategy in order to have a decarbonized and highly energy-efficient building stock by 2050, facilitating, in terms of costs, the effective transformation of existing buildings in NZEB. This strategy can be used to address the issue of safety in the event of fires, as well as to deal with the risks associated with intense seismic activity.

The redevelopment strategy must also include measurable progress indicators in order to verify the trend at 2030, 2040 and 2050 and include the details relating to the implementation of the policies and actions envisaged.

The Directive suggests, as a measure to achieve targeted and cost-effective restructuring, the introduction of an optional "Building Renovation Passport" system, a personalized document for each building that details how and when to implement a deep (in-depth) redevelopment in stages, or transform an existing building in a NZEB thanks to partial intervention works distributed over a period of 15–20 years.

The energy renovation of buildings is a process that has started, and the first results, as highlighted by the EPBD evaluation report, are tangible. The guidelines defined by the European directives, and substantially accepted by the market, show that the energy performance improvement targets can be obtained through a mix of strategies:

- defining high performance energy standards for new buildings (ideally tending to zero-energy buildings);
- improving the energy performance of the existing building stock both through regulatory legislative instruments and through financing policies (Fig. 9.1);
- taking advantage of renewable energy sources to speed up the decarbonization; strengthening the role of energy certification of buildings which, by informing citizens about the energy quality of the building, can guide market choices (energy efficiency as an added value in real estate).

The energy retrofit of buildings is a complex issue: intervening in an urban context that has developed over the years, or in some cases over centuries, through a stratification of building interventions made of ever-changing forms and technologies is by no means easy. The problem is not the single building in itself but the context in which the building is inserted, which often places constraints on redevelopment (e.g the impossibility of exploiting solar energy or the difficulty in modifying shapes).

Fig. 9.1 New façade sampling for an energy retrofit project in Dresden (D). *Credits* The Author 2012

For buildings designed in periods in which energy was cheap and environmental sustainability was not an issue, it is sometimes difficult to increase energy performance to high levels (e.g. NZEB or ZEB).

Energy inefficiency is sometimes not the only critical aspect of the building: the inadequacy of protection against seismic activity of the location, for example, can make structural interventions the priority. The overall costs of restructuring, in many cases, can be higher than those of demolition and rebuilding.

There is a question that a green planner must always ask him/herself: is it necessary to redevelop all existing buildings or is it appropriate to make a selection and only redevelop those buildings that have intrinsic value within the city?

The energy retrofit of buildings with an "urbanistic" approach, i.e. an approach on a larger scale that considers the surrounding neighbourhood and the city, allows for more effective environmental, economic and social choices: the energy and environmental improvement of buildings should, above all, become an energy and environmental enhancement of the city, also considering the social aspects.

Urban regeneration must be used as an essential opportunity to make urban spaces more liveable and stimulating for citizens. The exterior aspect of the renovated building is therefore important because it assists in making the city more attractive (Fig. 9.2).

Fig. 9.2 Energetically renovated facade of an old social housing building in Dresden (D). *Credits* The Author 2012

The energy crisis first and the environmental crisis subsequently have catalysed the energy issue by making the energy performance of buildings apparently the unique objective.

Energy efficiency is certainly important; however, the environmental sustainability of the building must not be neglected, considering aspects that cannot be "measured" with the metre of energy performance, for example, the quality of building materials, the impacts of the building on the environment and the relationship of the building with the context. A zero-energy building can be sustainable from the energy point of view but unsustainable for the community if it is located outside the urban context, in an area badly served by public transport, which means the necessity of using private vehicles to reach it. The sustainability of living is more important than the sustainability of the building.

Environmental certification protocols such as Leadership in Energy and Environmental Design (LEED®) promoted by USGBC (US) or Building Research Establishment Environmental Assessment Method (BREEAM®) promoted by BRE (UK) enhance these aspects and therefore should be more widely employed.

9.2 Energy Balance of Buildings

Retrofit measures on the building envelope (or shell for US technicians) have the purpose of making more sustainable the relationship between the indoor space (or confined environment) in which it is necessary to ensure optimal comfort conditions and the external environment.

Heat fluxes through the opaque and/or transparent walls of the envelope must be controlled considering that the strategies in summer and in winter may change according to the different external climatic conditions. The general concepts of the strategies to be adopted can be summarized as follows:

- improvement of the thermal quality of the opaque and transparent envelope by mean of the increasing of the thermal resistance (reduction of the U value);
- control of solar radiation with the aim to making the most of the free heat contributions in winter while ensuring protection from the sun in summer;
- control of the lighting component of the solar radiation, favouring the technical solutions that use natural lighting but at the same time, protecting the interior spaces from the effects of glare;
- reduction of uncontrolled air infiltration.

To better understand how the retrofit strategies may be defined, it is important to refer to the energy balance on the building as illustrated in a simple schematic way, Fig. 9.3 fulfils that purpose for a winter energy balance (Dall'O' 2013).

The objective of an HVAC[2] system is to maintain within the indoor spaces the design operating conditions, such as a constant the air temperature of 20 °C in the winter season. The building interacts with the external environment, where the climatic conditions may vary: when the outside temperatures are lower, a heat flux to the outside is generated through the walls that define the envelope.

The thermal losses from the envelope increase as the outside temperature decreases, with a dependency upon the values of the thermal resistance of the individual components (opaque vertical walls, windows, roofs, basements, etc.). Better insulation, increasing the thermal resistance to the passage of heat, therefore helps to reduce heat losses.

To ensure the healthy environment of the internal spaces, it is necessary to maintain suitable internal air quality (IAQ). A correct air ventilation of internal spaces is therefore necessary, but on the other hand, it does contribute significantly to increasing heat losses since a part of the inside air must be continuously replaced with external fresh air which must be heated. The global energy losses for heating are the sum of the losses due to thermal dispersions through the envelope and the losses due to ventilation of spaces.

In the energy balance of a building, however, not only the heat losses but also the heat gains should be considered. Internal free gains are provided not only by people and equipment within the building (e.g. lighting systems, appliances, etc.) but also as

---

[2]HVAC is short for heating, ventilation and air conditioning. The system is used to provide heating and cooling services to buildings.

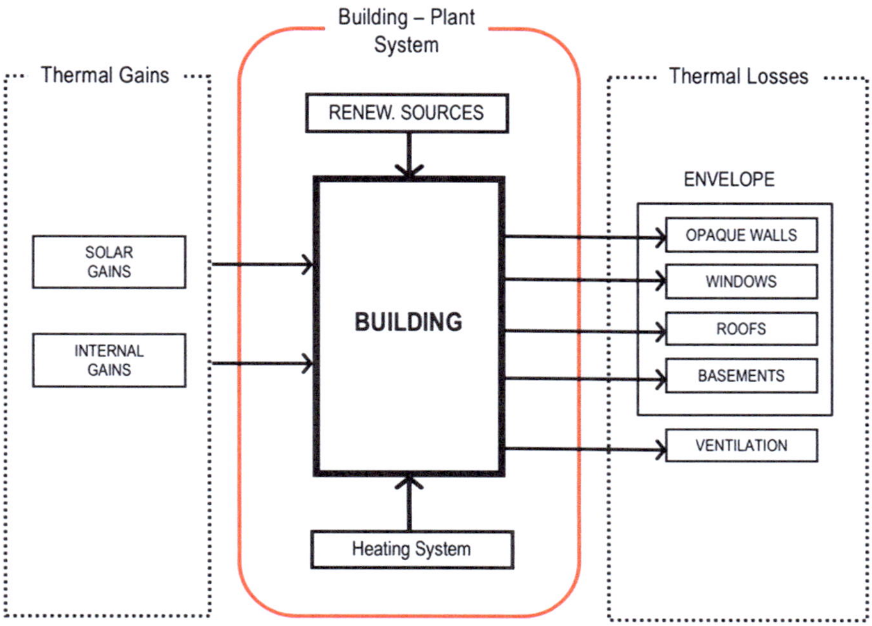

Fig. 9.3 Simplified energy balance for a building during heating season (Dall'O' 2013)

natural heat coming from the contributions of solar radiation through the transparent and opaque components of the building envelope.

If the objective is to maintain a constant internal comfort temperature, the difference between energy losses and energy gains must be compensated by the installation of an HVAC system: its contribution to the energy balance could be dramatically reduced by employing renewable energy sources (e.g. solar thermal, solar PV, biomass, etc.) to power it.

The energy balance shown in Fig. 9.3 considers only winter heating. In the overall energy assessment for the winter season, one should consider also other types of energy uses, such as those related to the production of domestic hot water, just as for the summer energy uses related to air conditioning must be taken into consideration (Dall'O' 2013).

The energy balance for summer cooling is more complex, as shown in the simplified diagram of Fig. 9.4.

The summer cooling load is greatly influenced by the construction components and materials of the building envelope and in particular by the transparent surfaces that interface with the solar radiation: the shielding and shading devices and the optical characteristics of glazing (in particular the characteristics of reflection) are very important to minimize the heat intake due to solar radiation.

During the winter heating, all free heat contributions, whether internal or external gains, are useful in the energy balance; during summer cooling, the opposite applies

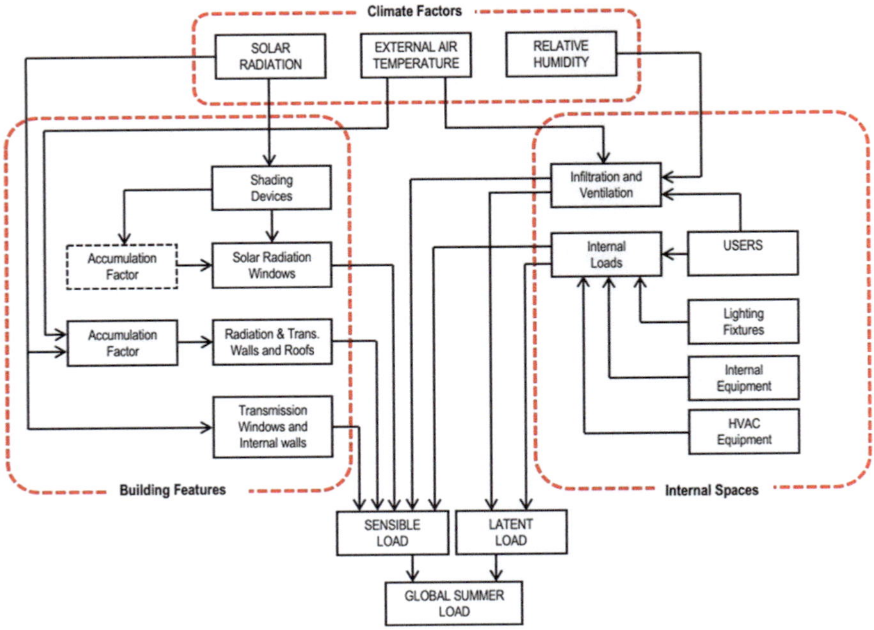

Fig. 9.4 Simplified calculation diagram for the summer cooling energy balance on a building (Dall'O' 2013)

since all the incoming heat (solar radiation) and generated heat (people, lighting, equipment, etc.) must be removed by the HVAC system. For this reason, the retrofit measures that aim to reduce these loads are very useful in saving energy besides their effects in increasing thermal comfort.

9.3 Green Energy Audit of Buildings

Reducing the energy consumption of an existing building and the environmental impact is often possible and convenient. However, initiating the process of implementing the changes is not simple if the customer–owner has little expertize and little or no awareness of the energy status of his/her building.

The technical instrument to effect this reduction is the energy audit, defined by the ISO 50002:214 Standard[3] as "a systematic analysis of energy use and energy consumption within a defined energy audit scope, in order to identify, quantify and report on the opportunities for improved energy". The main elements of an audit process are

[3]ISO 50002:2014 "Energy audits—Requirements with guidance for use".

- An analysis of building and utility data, including study of the installed equipment and analysis of energy bills;
- A survey of the real operating conditions;
- An understanding of the building behaviour and of the interactions with weather, occupancy and operating schedules;
- The selection and the evaluation of energy conservation measures;
- An estimation of energy-saving potential;
- The identification of customer concerns and needs.

A complete audit procedure has been proposed in the frame of the AUDITAC Project[4] and HARMONAC projects[5] to help in the implementation of the energy performance of buildings (EPB) directive in Europe and to fit to the current European market.

- *Benchmarking stage*: While normalization is required to allow comparison between data recorded on the studied installation and reference values deduced from case studies or statistics, the use of simulation models, to perform a code-compliant simulation of the installation under study, allows one to assess directly the installation, without any normalization being necessary. Indeed, applying a simulation-based benchmarking tool allows an individual normalization and allows avoiding size and climate normalization (Bertagnolio and Lebrun 2008).
- *Preliminary audit stage*: Global monthly consumptions are generally insufficient to allow an accurate understanding of the building's behaviour. Even if the analysis of the energy bills does not allow accurate identification of the different energy consumers present in the facility, the consumption records can be used to calibrate building and system simulation models. To assess the existing system and to simulate correctly the building's thermal behaviour, the simulation model has to be calibrated on the installation under study. The iterations needed to perform the calibration of the model can also be fully integrated in the audit process and help in identifying required measurements and critical issues (Andre and Bertagnolio 2010).
- *Detailed audit stage*: At this stage, on-site measurements, sub-metering and monitoring data are used to refine the calibration of the building energy simulation (BES) tool. Extensive attention is given to understanding not only the operating characteristics of all energy-consuming systems but also situations which cause variations in load profile on short- and longer-term bases (e.g. daily, weekly, monthly, annual). When the calibration criteria have been satisfied, the savings related to the selected energy conservation measures (ECOs) can be quantified (Bertagnolio et al. 2010).
- Investment-grade audit stage: At this stage, the results provided by the calibrated BES tool can be used to assess the selected ECOs and orientate the detailed engineering study.

---

[4]AUDITAC—Field Benchmarking and Market development for Audit methods in Air Conditioning.

[5]HARMONAC—Harmonizing air conditioning inspection and audit procedures in the tertiary building sector.

The problem is not only to evaluate strategies for improving performance but also to define a correct path regarding how to implement them in a rational way, considering not only the technical constraints but also the economic and legal aspects.

The decision to start any project stems from motivation; the stronger the motivation, the easier it will be to put it into effect.

The Green Energy Audit procedure developed by Dall'O' (2013) is not limited to reduce energy consumption; rather, it is aimed at a far more important goal: improving the overall sustainability of the building.

At the methodological level, the word "green" implies a series of choices:

- the definition of measures that lead to a reduction in the consumption of resources; conservation of energy then becomes conservation of resources;
- criteria of choice for interventions can be addressed from the outset with these indicators; the auditor then must have two objectives (or a proportioned mix of the two): to maximize energy performance and to maximize environmental quality;
- measures that use renewable energy are preferred (e.g. solar thermal, solar photovoltaic and biomass);
- when defining measures, the auditor should consider all natural solutions that can help control the climate and light in the building, such as green roofs, green facades, natural shading systems, passive solar and daylighting systems;
- evaluation of sustainability targets according to the recognized protocols (e.g. LEED®, BREEAM®, etc.).

The aim of a Green Energy Audit is to identify measures that can improve the energy performance and sustainability of the building in question.

The cost/effectiveness analysis allows the identification of the most advantageous scenario, while the intersection with the LEED® credits suggests the most sustainable measures and in general terms, the solutions and technologies with environmental benefits. For example, the assessment of measures aimed at improving thermal insulation of the building should consider the benefits of reducing demand for virgin materials and reducing waste: it may be possible to use salvaged, refurbished or reused materials or use materials with a recycled content.

In the construction phase, a construction waste management plan should be developed and implemented with the aim of recycling and salvaging non-hazardous construction materials and demolition debris.

Numerous other environmental aspects could be considered in the assessment of measures and their application: restoring vegetation on the site after construction, reducing the heat island effect using vegetation or materials with a high reflectance, employment of alternative surfaces and techniques to increase on-site infiltration and to reduce pollutants from stormwater runoff.

The main difference between the energy audit and Green Energy Audit approach lies in the way in which the retrofit strategies are defined. In a conventional energy audit, retrofit actions are normally defined on the basis of cost/effectiveness parameters. Some measures considered in the Green Energy Audit are not cost-effective at all because they have no affect on energy saving (e.g measures which optimize the air quality).

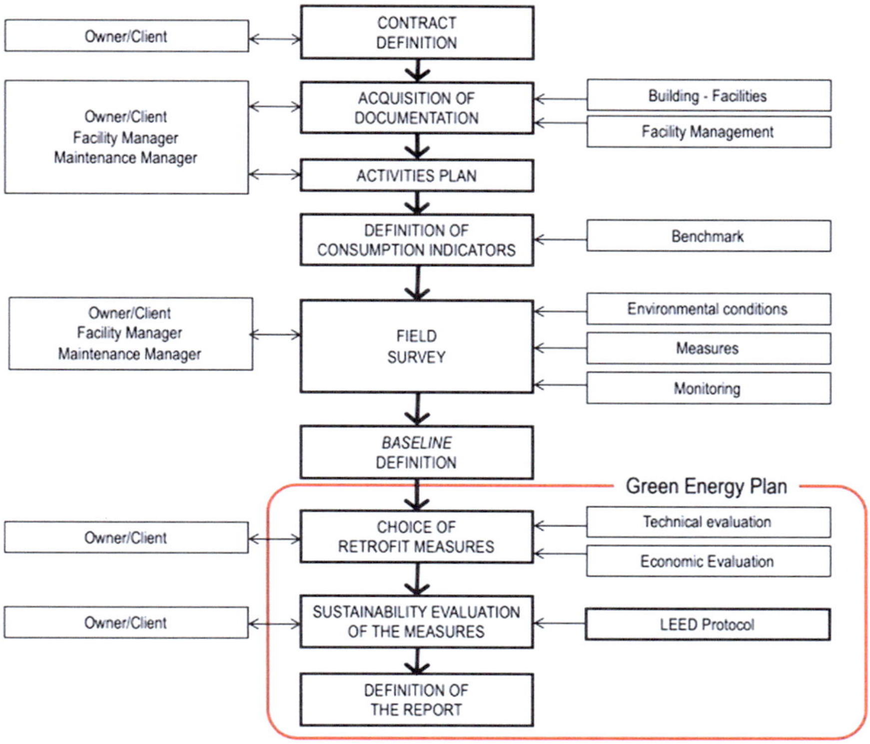

Fig. 9.5 Flux diagram of the Green Energy Audit process (Dall'O' 2013)

A Green Energy Audit is aimed at improving the overall sustainability of the building, so it is necessary to consider environmental aspects which compensate the reduction of cost effectiveness.

For this very reason, in the procedure presented, a prior evaluation is made of the LEED® classification that can be potentially achieved. The procedure of Green Energy Audit is shown in Fig. 9.5 (Dall'O' et al. 2012a; Dall'O' 2013).

9.4 Planning of Energy Retrofit Actions

9.4.1 Towards an Integrated Strategy

Energy retrofit actions, or ECOs, can significantly reduce the energy consumption of existing buildings. With reference to the diagrams in Figs. 9.3 and 9.4, it is possible to optimize the energy performance of the building envelope (external walls, roofs, basements) by reducing the energy flows that pass through it.

Some energy retrofit actions may concern HVAC systems: by increasing the energy performance of plant components, even by replacing a part of them, it is possible to reduce inefficiencies by better exploiting the potential of given fuels or energy vectors.

Renewable energy sources (e.g. solar thermal, photovoltaic solar, biomass) can make an effective contribution to the energy and environmental balance of the building. Their "weight" in the energy balance is greater if the building as a whole is more efficient: their use is therefore recommended as long as all the energy retrofit actions on the building envelope and on the HVAC systems have already been planned.

Energy retrofit actions should cover all energy uses: winter heating, summer cooling, domestic hot water production, lighting and other electrical uses.

In an environmentally sustainable project, actions should also be envisaged to reduce the use of resources, such as the reduction in wastage of drinking water, as well as in the choice of construction materials, the preference necessarily being for those with lower environmental impact.

Consistent with the diagram in Fig. 9.5, the choice of energy retrofit strategies must be verified not only from the technical point of view (i.e. the feasibility of the action) but also from the economic point of view. The actions that concern the building envelope, more expensive and with longer payback times, should be applied in cases where the building requires significant maintenance.

9.4.2 Thermal Insulation of the Roofs

The energy retrofit actions on roofs depend on their morphology. In the case of pitched roofs, thermal insulation could be applied from the inside or from the outside. External thermal insulation is convenient when the roof requires rebuilding. Thermal insulation of the building's roof allows one to obtain energy benefits not only in winter but also in summer since the roof is a part more subject to solar radiation.

The application of these measures is convenient if the space below the pitched roof is normally used/occupied, otherwise it is more convenient to thermally insulate the floor slab which borders with the space below.

An interesting technique also from the environmental point of view is that of the "green roofs system". A green roof is a roof partially or completely covered with vegetation and a growth medium, planted over a waterproofing membrane. It may also include additional layers such as a root barrier and drainage and irrigation systems. This technique can be applied both on pitched roofs and on flat roofs, but the latter is decidedly more common thanks to the significant associated advantages that can be obtained.

The green roof can be applied, without distinction to any building regardless of the intended use: there are many examples of application of green roofs on residential constructions or office buildings, industrial warehouses, shopping malls, schools and recreational buildings (Dall'O' 2013).

Green roofs permit many advantages to be obtained for a building, such as absorbing rainwater (water is stored by the substrate and then taken up by the plants, from where it is returned to the atmosphere through transpiration and evaporation), providing thermal insulation (depending on the used material and the thickness of the layer), reducing summer cooling needs (by evaporative cooling), helping to lower urban air temperatures and mitigate the heat island effect (green roofs can contribute to obtain credits on LEED® protocol), creating a habitat for wildlife.

9.4.3 Thermal Insulation of the Façades

So far as the facades of existing buildings are concerned, the most efficient retrofit technique is that of insulation from the outside. The External Thermal Insulation Composite System (ETICS) is an optimal solution for the energy retrofit of the external walls of existing buildings (this technology is also commonplace as solution for new buildings).

ETICS is a system applied from the outside of the wall, usually including an adhesive, a levelling mortar, an insulation panel, an alkali-resistant reinforcement grid, a primer and a finishing coat, as well as sealants and accessory materials for the installation.

The main advantage of this measure derives from the continuity of the insulation, which eliminates thermal bridges and the risk of surface and interstitial condensation. The measure increases the thermal inertia and the sound insulation of walls, thus improving both the thermal and the acoustic comfort for users. The operation is carried out relatively quickly and completely from the outside, with very limited discomfort for users from the works.

If the thermal insulation of the walls from outside is not possible (e.g. for architectonic constraints), the thermal resistance of the walls can be increased through internal insulation using a counter-wall.

Amongst the green technologies for the built environment, green façades or green walls are interesting solutions which can be used for energy and/or environmental improvement retrofit of existing buildings.

The greening systems for vertical walls can be divided into two categories: green façades and living walls (vertical gardens). Green façades are made of green climbing plants that grow directly on the façade or in front of the façade if supported by structures such as grids and cables. Living wall systems (LWS) are constituted by pre-vegetated panels or integrated systems fixed to a structural wall or to a frame, with hydroponics for the growth of vegetation.

The installation of a green wall provides several advantages:

- *in summer*, there is a reduction of the solar load on the protected wall surface, estimated at around 30–40%. Shading generated by the vegetation reduces the maximum surface temperature by about 10 °C on days with higher radiation (the phase shift of the peak of heat is around 8 h);

- *in winter*, the presence of the vegetation on the façade reduces convective phenomena, increasing the overall thermal resistance of the wall.

One of the additional benefits of the vegetation regards the capture and retention of fine dust particles on the surface of the leaves. Particles smaller than 10 μm, present in particular in the urban areas, may cause damage to health because they can be inhaled into the human respiratory system.

Green façades may have a positive impact on both physical and mental health and well-being. Green views and access to green spaces in cities help to relieve the everyday pressures of crowds and noise (Dall'O' 2013).

In the energy balance of a building, windows are critical elements. In fact, within this single component, many needs must be satisfied: thermal insulation, solar control, environmental comfort, visual comfort, acoustic comfort and ventilation.

Where energy performance or structural performance cannot be guaranteed, one should verify the state of the windows and possibly plan a global replacement. In order to reduce the U value of the glazing, two strategies can be adopted: a surface coating treatment of the glass panes (e.g. low-emissivity, selective coatings, etc.) and the filling of the interspaces between the panes with gases with low thermal conductibility (e.g. Argon, Xenon, Krypton) instead of air.

A window must guarantee the energy performances not only in winter but also in summer, whereas in the winter, the effects of solar radiation are positive (solar gains), and in the summer, solar radiation is the main source responsible for the thermal loads. On designing the new windows, the control of solar radiation (e.g through the installation of external or internal solar shading systems) must be considered in order to reduce the cooling demand of HVAC systems and increase the thermal comfort of the users.

The use of vegetation for sun protection of façades or windows is a natural solution that can contribute to an increase in the sustainability of the building. The objective is to provide an independent green structure that could be installed adjacent to the walls to be protected or positioned at a certain distance from it.

The possibility of using vegetation with deciduous leaves allows the use of the energy benefits of solar radiation in winter (solar gains) but at the same time protects the walls or the windows in summer.

9.4.4 Improvement in the Efficiency of HVAC Systems

For HVAC systems, retrofit actions must be aimed at achieving the following objectives:

- guaranteeing thermal comfort and IAQ both in summer and in winter;
- generate heat, or cooling, with maximum energy efficiency;
- allow local control of environmental conditions.

As regards heat generation, short-term interventions could consist in replacing the existing boiler. The replacement of an existing boiler with a new one of higher performance (e.g. condensing boiler) helps to increase the heat generation efficiency of the system, hence reducing operating costs for winter heating and DHW.

Amongst the combustion heating technologies, the one that offers the best energy performance is condensing technology. In condensing boilers, high efficiency is obtained (typically up to 108% if referred to lower heating value) by recovering the latent condensing heat of the water vaporcontained in the fumes.

Before replacing the existing boiler, it is advisable to perform a global calculation of the energy needs of the building so that the new boiler is chosen with a corrected heating capacity. Indeed many existing buildings were erected and equipped before laws or rules on energy efficiency were introduced. Since the size of the boiler was not well-defined, boilers were normally oversized with a heat capacity in most cases double that of the required capacity.

Condensing boilers, even if efficient, use fossil fuels (methane). For environmental problems, the current tendency is to not use more traditional fuels but to refer to electricity. In a "green" vision of cities, the generation of heat and cooling will be increasingly orientated towards the use of heat pumps.

In reversible models with a single appliance, it is possible to produce heat for winter heating and cold for summer air conditioning. Heat pumps, by consuming electricity, integrate well with photovoltaic solar systems. NZEB buildings have long used this configuration (solar PV and reversible heat pump).

In existing buildings, much of the energy wastage, which can be estimated at up to 30%, are due to the inefficiencies of the control system. Therefore, great attention must be paid to this aspect. The efficient regulation systems, using a local control, allow the comfort conditions to be managed in a timely manner and allow the use of free external energy supplies (e.g solar radiation and for lighting systems, natural light) or internal (heat developed by people and equipment in winter).

In this sector, technological innovation has evolved considerably, and the market, even via Internet resources, is marketing efficient, simple to use and low-cost home and building automation control systems.

9.4.5 Renewable Energy Sources

Renewable energy sources, in particular solar thermal, solar photovoltaic and biomass, will play a fundamental role in the development of green cities, and in part, they are already doing so.

The theme is obviously complex and for the use of renewable energy sources in green planning, a chapter of this book (Chap. 10, the role of renewable energy sources in green planning of cities and communities) has been dedicated to the topic. This sub-section is limited to an expression of some general considerations.

The drastic reduction of solar PV costs makes it convenient for its installation in existing buildings in "green connected" plant configurations. Solar PV directly produces electrical energy, and this is a great advantage because this energy can be used for electrical uses but also for thermal uses (e.g winter heating and hot water production) and in an efficient manner by employing a heat pump.

A feature of this technology is its application flexibility: the solar modules can be installed as an integral part of the building's roof but can also be integrated into the façade. The technology also lends itself to insertion and integration on an urban scale, for example, in vehicle parking shelters.

Solar thermal technology only produces heat that can be used for uses such as environmental heating, hot water production or processes. The strong competition of solar PV makes this technology applicable to situations in which the only objective is to produce hot water without special needs. It should be recalled that solar PV, thanks to the connection with the public electricity grid, can exploit all the energy captured, while the solar thermal can only operate in combination with thermal storage and is therefore configured as a stand-alone plant. Architectural integration of solar thermal is more difficult, and maintenance costs are greater.

9.5 Deep Renovation of Buildings

The activity of retrofitting is now growing, thanks to the financial incentive policies (in Italy, for example, with tax deductions, it is possible to cover up to 75%, of the costs related to an energy retrofit). Other elements that play in favour are the greater awareness on the part of citizens/users on environmental issues and the real estate marked which is beginning to recognize the value of efficient and sustainable buildings.

However, the CO_2 reduction targets for 2030 and 2050 are very ambitious and require greater efforts.

All new buildings, which are built respecting the high-energy-efficiency standards imposed by national legislation (in the EU at least NZEB), contribute to reducing CO_2 emissions in the building sector; however, it is unthinkable to envisage a complete replacement of the existing building stock in so few years. The sector of existing buildings, built in periods in which energy efficiency was not a problem and hence very inefficient, is the area on which to act decisively.

An energy requalification that acts on buildings in a partial way, for example, through the replacement of the boiler or the replacement of doors and windows, is simply not enough and more needs to be done.

The Energy-Efficiency Directive (EED) (Directive 2012/27/EU) (UE 2012) and the revised Energy Performance Building Directive (EPBD) (Directive 2018/844) (EU 2018) both contain provisions for increasing Europe's renovation rate. The greatest challenge to reducing energy use in buildings lies in increasing the rate, quality and effectiveness of building renovation since the current rate is merely 1.2% per year. Effective renovation approaches need to be widely demonstrated and then replicated to ensure that the renovation rate rises to between 2 and 3% each year.

The US Department Of Energy (DOE) promotes energy technology and innovation in the USA. The DOE has set a "Better Buildings Initiative" challenge that pushes for a 20% reduction of energy across a portfolio of commercial and industrial buildings by 2020.

The DOE has designed "Advanced Energy Retrofit Guides"[6] for existing buildings in order to assist building stakeholders in the selection of energy-efficiency improvements. Within these guides, they claim "deep retrofits can reduce a building's energy use by over 50%. Deep retrofit projects combine many O&M and standard retrofit measures in an integrated whole-building design approach" (Shnapp et al. 2013).

The winning strategy to accelerate the process of renewing existing building stock is "deep renovation". But what is exactly its meaning? An answer to this question is given by Shnapp et al. (2013) that in their research work identify some definitions.

- *Deep Renovation* or *Deep Energy Renovation* is a term for a renovation that captures the full economic energy-efficiency potential of improvement works, with a main focus on the building shell, of existing buildings that leads to a very high-energy performance. The renovated buildings energy reductions are 75% or more compared to the status of the existing building/s before the renovation. The primary energy consumption after renovation, which includes, inter alia, energy used for heating, cooling, ventilation, hot water and lighting after the deep renovation of an existing building, is less than 60 $kWh/m^2/yr$ (definition often used in Europe).
- *Deep Retrofit* or *Deep Energy Retrofit* implies replacing existing systems in a building with similar ones that are of higher quality and performance, which leads to a better energy performance of an existing building. The primary energy consumption includes energy used for heating, cooling, ventilation, hot water, lighting, installed equipment and appliances. After the deep retrofit, the buildings energy reduction is 50% or more compared to the status of the existing building/s the retrofit (definition mainly used in the USA).

Deep Reduction or *Deep Energy Reduction* is a term used in the USA for a deep renovation or a deep refurbishment, which aims at more than 75% reduction in energy use in comparison with that prior to the improvement[7]:

- *Zero-Carbon-Renovation*: A deep renovation with large-energy consumption reductions, where the energy needed to supply the resisting need is carbon neutral;
- *Zero-Energy-Renovation*: A deep renovation with large-energy consumption reductions, where the energy needed to supply the resisting need is supplied as renewable energy on site.

An interesting contribution, from an economic point of view, to the topic of deep (in-depth) renovation was commissioned by EURIMA, the European Insulation Manufacturers Association (Hermelink and Müller 2010). Scope of the study provides a notion of economic viability for single and multi-family homes across the EU, taking a case study analysis as basis. The study discussed the economic feasibility of

[6] Available on: https://www.energy.gov/eere/buildings/advanced-energy-retrofit-guides (Accessed 20 September 2019).

[7] This is a definition commonly used by the Thousand Home Challenge, find more information at: http://thousandhomechallenge.com/.

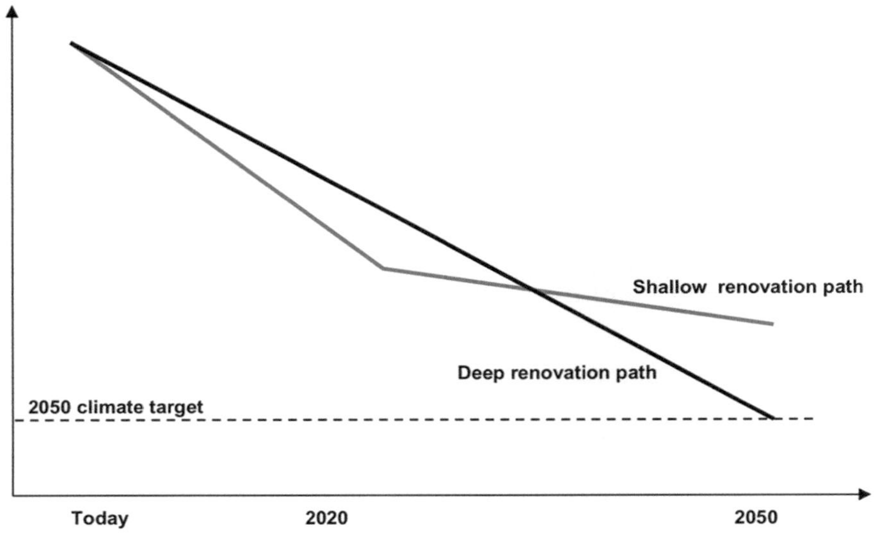

Fig. 9.6 Speed trap of shallow renovation (Hermelink and Müller 2010)

deep renovations for single and multi-family homes using the German "Low Energy Building Stock" programme as a case study.

According to the study, dynamic simulations for future building stock scenarios have clearly shown that "deep renovation at reasonable speed" is a more promising strategy to reach the 2050 climate targets than "shallow renovation at high speed" (shallow meaning that current national requirements for renovation are just met).

It is important to do projections till 2050 as projections till only 2020 may be misleading and suggest an inadequate strategy. As the schematic paths in Fig. 9.6 show, at the interim milestone 2020, a "shallow renovation" strategy might look better than a "deep renovation" strategy while surprisingly failing to achieve the crucial 2050 target.

The usually performed short-term projections contain a severe risk of becoming irreversibly caught in this "speed trap of shallow renovation": usually renovations with the best investment/saving ratio are done at the start of a nation's renovation programme. These are the buildings having the highest absolute saving potential. When these buildings undergo a "shallow" renovation, huge absolute savings remain untapped.

As regards the US market, Cluett and Amann (2015) published a report regarding residential deep energy retrofit. The report explored energy-efficiency programmes that target in-depth energy savings through substantial improvements to existing residential buildings.

The authors have stated that "deep energy retrofit case studies show that existing technology and practices can result in energy savings of 50% or more. However, there are still many barriers to adoption and delivery on a large scale". The barriers highlighted by the study are the following:

- *Workforce*: one of the barriers to implementing in-depth energy retrofit measures was access to a workforce with the right skills.
- *Market interest and acceptance*: encouraging homeowners to undertake in-depth energy retrofits at the time of major home renovations is a promising opportunity.
- *Financial barriers*: it is difficult to complete an in-depth energy retrofit that saves more than 50% of the energy used in a home without significant financial investment. Many homeowners simply cannot afford the upfront capital investment necessary for in-depth energy improvements.
- *Cost effective*: cost effectiveness is a challenge for utility programmes involving in-depth energy retrofits. More data from completed projects are needed in order to assess the cost effectiveness of comprehensive energy retrofits as well as individual measures.

9.6 A Case Study of In-Deep Renovation: Energiesprong

One of the most important in-depth renovation case studies is the Energiesprong project. The idea of Energiesprong originated in the Netherlands, where it began as a government-funded innovation programme to drive an improved energy-efficient standard in the Dutch market.

Today in the Netherlands, over 5000 homes have been retrofitted to desirable, net zero-energy houses at no extra costs for the residents. The mission of the Energiesprong Foundation is to scale this approach to more markets, to create an industry which is able to design, produce and deliver whole house retrofits with excellence across millions of houses.

Currently, Energiesprong teams are active in the Netherlands, France, the UK, Germany and Northern Italy. In New York State (RetrofitNY) and California, initiatives inspired by Energiesprong are working on a solution for the USA.

After an Energiesprong retrofit, a home is net zero energy, meaning it generates the total amount of energy required for its heating, hot water and electrical appliances. It also provides superior indoor comfort.

This can be achieved by using new technologies such as prefabricated facades, insulated rooftops with solar panels, smart heating and ventilation and cooling installations. A refurbishment comes with a long-term performance warranty on both the indoor climate and the energy performance for up to 40 years. A complete home makeover can be completed in less than 10 days, and some have been done in as little as a day.[8]

An energy redevelopment is Energiesprong when it meets the following four conditions:

[8]Source: https://energiesprong.org/about/.

Fig. 9.7 Installation phase of the prefabricated elements of the building envelope in a residential building in UK, within the Energiesprong project. *Credits* The Renewable Energy Hub, https://www.renewableenergyhub.co.uk/blog/what-are-energiesprong-homes/

- *Accessible*: to finance the redevelopment of the building through a drastic reduction in consumption, intervention and maintenance costs.
- *Quality*: the performance of the building is provided and guaranteed in the long-term by the same enterprise who carries out the intervention. This constitutes an intrinsic guarantee of quality.
- *Non-intrusive*: the redevelopment is carried out in two weeks and allows tenants to stay in the building for most of the time. This is made possible thanks to the engineering and optimization of the entire process.

In general, an Energiesprong renovation or new build is financed by future energy cost savings plus the budget for planned maintenance and repairs over the 30 years following the renovation works. Figure 9.7 shows an example of deep renovation process using prefabricated elements.

This allows residents to maintain the same cost of living. In the case of housing associations, tenants pay the housing association an energy service plan which is the equivalent of their previous energy supplier's bill.

9.7 Energy Retrofit Potentials in a Case Study

The topic of energy renovation of buildings is the subject of numerous studies and much research by professors and researchers of the Department of Architecture, Built environment and Construction engineering (ABC) of the Politecnico di Milano.

This case study describes an innovative approach for the analysis of the potential energy savings of retrofitting existing building stocks: it considers in particular the actual technological and economic constraints of the implementation of feasible energy-efficiency measures (Dall'O' et al. 2012b).

The analysis was applied to five municipalities in the province of Milan that have signed the Covenant of Mayors, committing to meet and exceed the 20% CO_2 reduction objective of the EU by 2020. Because the scale of the analysis is municipal, the resolution of the requested data is high. In order to achieve realistic and achievable energy savings, the authors created an energy cadastre and conducted a large in-field survey for each municipality.

By determining the characteristics of the building stock, one then knows which energy retrofit actions are feasible from a technical, legal and economic point of view. The result is a tool that does not overestimate the potential energy savings, helping the public administration/local authorities to define their energy-saving policies.

The methodology for the analysis of the potential energy savings of energy retrofits of existing building assets is outlined in the flow chart in Fig. 9.8.

The different phases of the process are shown in the central portion of the flow chart; the actors (operators and technicians) involved are listed on the left, and the activities and tools are described on the right.

In this study, the authors wanted to emphasize the methodology and the analysis of the potential surfaces. Thus, only improvements to the building envelope, such as the thermal insulation of external walls and roofs and window replacement, were considered.

A comprehensive methodology also includes the analysis of the potential energy savings to be derived from increasing the efficiency of the heating system and household appliances and of installing photovoltaic and thermal panels on the roof.

As identified by Nemry et al. (2008) in order to improve energy efficiency of existing buildings, the most significant options are

- replacement of windows;
- additional facade insulation;
- additional roof insulation.

The first phase consists of collecting all the information that may be useful for analysis on an urban scale, for example, site maps, aerial photogrammetric surveys and orthophotos of the area. Maps available in digital formats are helpful, especially on the Internet, for example, in the digital archives of Provinces and Regions (thematic maps). Google Earth® is also very useful if the maps are sufficiently updated, particularly to detect building roofs.

A cadastre was created in order to classify each building by typology and age of construction.

On-site surveys were useful to determine the state of conservation of the buildings upon which the possible retrofitting depends.

Each building is identified by the block typology and a unique number (Fig. 9.9). This classification was implemented in a database to combine several characteristics and form an energy cadastre.

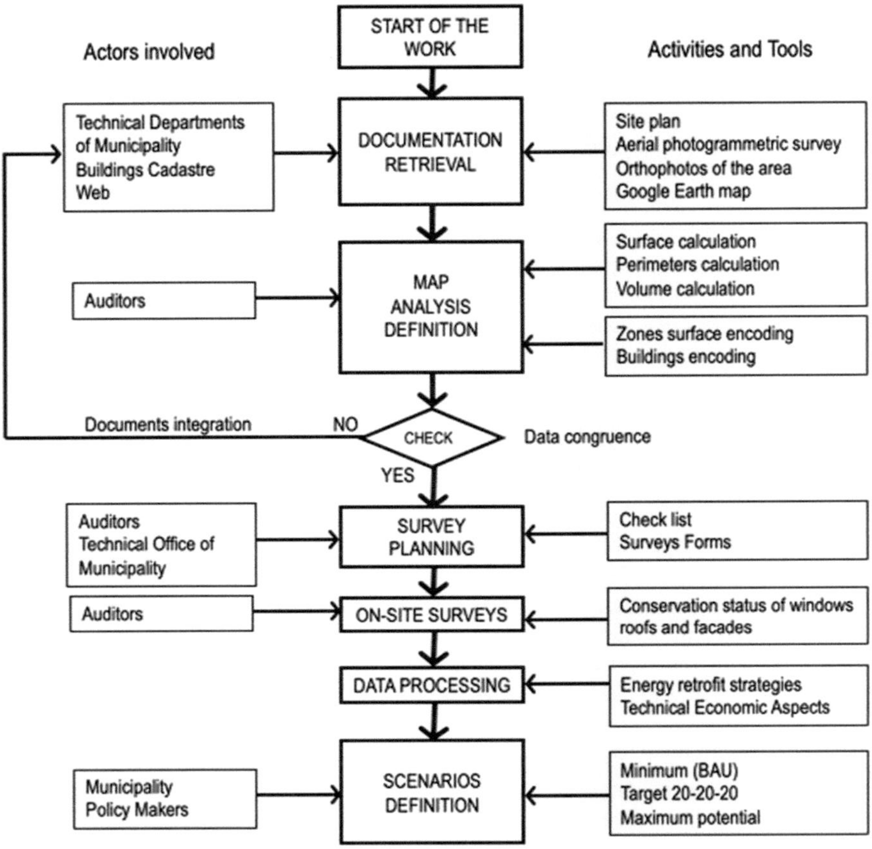

Fig. 9.8 Flow chart of the methodology (Dall'O' et al. 2012b)

Survey results for the state of conservation of the buildings are visible in the graph of Fig. 9.10. The percent of buildings in poor condition is lower for recently constructed buildings.

In the energy cadastre, the net area, transparent surface, opaque surface and roof area for each building were all calculated. Owing to the characteristics surveyed, it was possible to distinguish the potential retrofitting surfaces, i.e. to distinguish between those surfaces where energy retrofitting was possible and those surfaces for which no retrofitting could be envisaged.

Examining the walls of the building envelope, the areas were divided into three categories:

- ETICSable buildings on which it is possible to retrofit using an ETIC; i.e. buildings with plastered walls in poor condition and without constraints;

Fig. 9.9 Extract of the cadastre map with encoding (Dall'O' et al. 2012b)

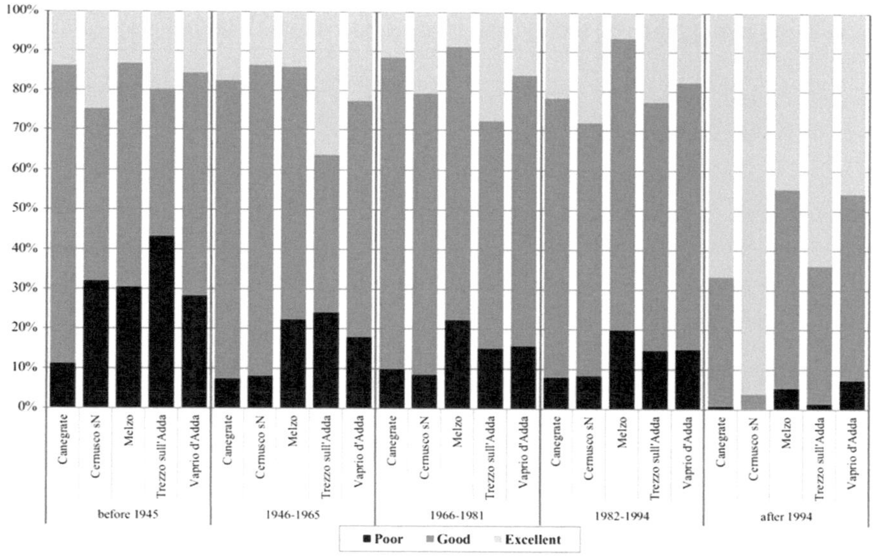

Fig. 9.10 State of conservation of building stock (Dall'O' et al. 2012b)

- Partially ETICSable buildings on which it is partially possible to retrofit using ETICS; i.e. buildings with plastered walls in good condition and without constraints;
- Non-ETICSable buildings on which it is not possible to retrofit using an ETICS; i.e. buildings with non-plaster walls, in excellent conditions or with constraints (e.g historic buildings).

The second scenario assumes that the required efforts, in terms of regulation implementation and related investments, are made to reach the goal of 20% reduction of CO_2 emissions (where 2005 is assumed as baseline).

The third scenario defines the maximum potential of the improvement of the building envelope efficiency considering the real constraints. This last scenario is very important because it provides clear and objective information about the achievable limit if the true goal is to maintain the building stock (Fig. 9.11).

The results of this study, which is based on a real model (five municipalities), provide two elements for reflection: the first is that in intervening in existing buildings sometimes there are objective constraints that do not allow retrofit intervention on the envelope (e.g architectural constraints), the second is that a scenario based on market rules (BAU) is not sufficient to achieve the greenhouse gas reduction targets: a robust action plan is needed, and economic incentives are needed. The Energiesprong model seen in the previous section can be considered to be a winning scheme.

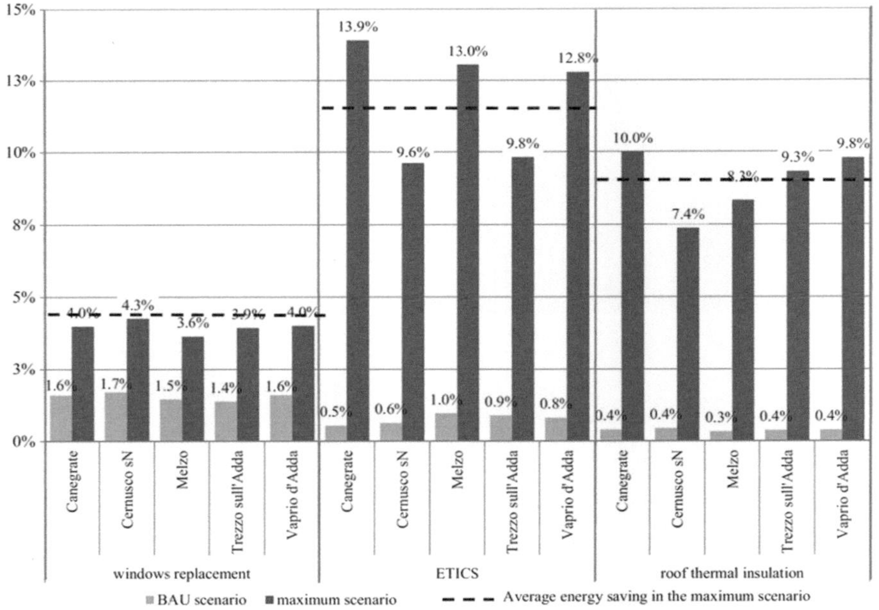

Fig. 9.11 Potential energy savings in the BAU and maximum saving scenario (Dall'O' et al. 2012b)

9.8 Conclusions

The 2015 Paris Agreement on climate change following the 21st Conference of the Parties to the United Nations Framework Convention on Climate Change (COP 21) boosts the Union's efforts to decarbonize its building stock. Taking into account that almost 50% of Union's final energy consumption is used for heating and cooling, of which 80% is used in buildings, the achievement of the Union's energy and climate goals is linked to the Union's efforts to renovate its building stock by giving priority to energy efficiency, i.e. making use of the "energy efficiency first" principle as well as considering deployment of renewables.

The European Union is committed to developing a sustainable, competitive, secure and decarbonized energy system. The Energy Union and the Energy and Climate Policy Framework for 2030 establish ambitious EU commitments to further reduce greenhouse gas emissions by at least 40% by 2030 as compared with 1990, to increase the proportion of renewable energy consumed, to make energy savings in accordance with Union level ambitions and to improve Europe's energy security, competitiveness and sustainability (EU 2018).

These ambitious goals are faced with an existing building stock which on the one hand constitutes an interesting energy retrofit market but on the other shows a series of critical aspects which must be overcome.

If the approach to the energy redevelopment of buildings considered the city as an object of interest and was not concentrated on the single building, it could promote energy redevelopment policies that also become policies for requalification and renewal of the city as a whole.

An urban strategy, through green planning, should make choices in the interest of the community by promoting an energy requalification that does not a priori exclude the demolition of existing buildings and the replacement of these with new and energy-efficient buildings (NZEB or ZEB).

For the energy recovery of existing buildings that deserve to be maintained due to their peculiarities (architectural, historical, environmental, functional, social and cultural values), rather than waiting for a market development according to a BAU scheme, it is useful to promote strong actions: the Energiesprong model, discussed in this chapter, has proved its effectiveness.

An urban strategy is correct if it concerns itself with energy efficiency, but other aspects must also be considered. The large-scale application of rating systems based on environmental certification protocols for buildings should be facilitated in order to enhance the relationship between buildings and the urban environment in which they are increasingly integrated.

References

Andre P, Bertagnolio S (2010) Development of an evidence-based calibration methodology dedicated to energy audit of office buildings. Methodology and modeling. In: 10th REHVA World Congress—Clima 2010

Bertagnolio S, Lebrun J (2008) Simulation of a building and its HVAC system with an equation solver. Application to benchmarking. Build Simul 1(3):234–250. https://doi.org/10.1007/s12273-008-8219-4

Bertagnolio S, Andre P, Lemort V (2010) Simulation of a building and its HVAC system: application to audit. Build Simul 3(2):139–152. https://doi.org/10.1007/s12273-010-0204-z.hdl:2268/9501?locale=en

Cluett R, Amann J (2015) Multiple benefits of multifamily energy efficiency for cost-effectiveness screening. Report A1502, American Council for an Energy-Efficient Economy. Available at: http://www.ourenergypolicy.org/wp-content/uploads/2015/06/a1502.pdf. Accessed 20 Sept 2019

Dall'O' G (2013) Green Energy Audit of Buildings, a guide for a sustainable energy audit of buildings. Springer, London

Dall'O' G, Speccher A, Bruni E (2012a) The Green Energy Audit, a new procedure for the sustainable auditing of existing buildings integrated with the LEED protocols. Sustain Cities Soc 3:54–65

Dall'O' G, Galante A, Pasetti G (2012b) A methodology for evaluating the potential energy savings of retrofitting residential building stocks. Sustain Cities Soc 4:12–21

EC (2002) Directive 2002/91/EC of the European Parliament and of the Council of 16 December 2002 on the energy performance of buildings

EC (2016) EVALUATION of Directive 2010/31/EU on the energy performance of buildings. Available at: https://ec.europa.eu/energy/sites/ener/files/documents/2_en_autre_document_travail_service_part1_v2_0.pdf. Accessed 13 Sept 2019

EU (2010) Directive 2010/31/EU of the European Parliament and of the Council of 19 May 2010 on the energy performance of buildings

EU (2018) Directive (EU) 2018/844 of the European Parliament and of the Council of 30 May 2018 amending Directive 2010/31/EU on the energy performance of buildings and Directive 2012/27/EU on energy efficiency (text with EEA relevance)

Hermelink AH, Müller A (2010) Economics of deep renovation: implications of a set of case studies. Report commissioned by EURIMA—European Insulation Manufacturers Association. Available at: https://www.eurima.org/uploads/ModuleXtender/Publications/51/Economics_of_Deep_Renovation_Ecofys_IX_Study_Design_FINAL_01_02_2011_Web_VERSION.pdf. Accessed 13 Sept 2019

Nemry F, Uihlein A, Makishi Colodel C, Wittstock B, Braune A, Wetzel C et al (2008) Environmental improvement potentials of residential buildings (IMPRO-building). Office for Official Publications of the European Communities (JRC Scientific and Technical Research Series), Luxembourg

Shnapp S, Sitjà R, Laustsen J (2013) What is a deep renovation definition? Technical report, Global Buildings Performance Network. Available at: http://www.gbpn.org/sites/default/files/08.DR_TechRep.low_.pdf. Accessed 13 Sept 2019

UE (2012) Directive 2012/27/EU of the European Parliament and of the Council of 25 October 2012 on energy efficiency, amending Directives 2009/125/EC and 2010/30/EU and repealing Directives 2004/8/EC and 2006/32/EC Text with EEA relevance. Available at: https://eur-lex.europa.eu/legal-content/EN/TXT/?uri=celex%3A32012L0027. Accessed 12 Sept 2019

Chapter 10
The Role of Renewable Energy Sources in Green Planning of Cities and Communities

Paola Caputo

Abstract Recent estimates suggest that cities account for about 70% of the global energy demand and thereby can be identified as key players to decarbonise the energy generation sector. Among the possibilities to face this challenge, the integration of renewable energy sources into the energy supply mix has a prominent role. Currently, many cities have drastically reduced their energy demand in all final uses and are even boosting a process of 100% energy transition to renewable energy sources. This goal is clearly ambitious not only from an economic point of view, but also even from the technical standpoint, meaning that this integration of renewables is not straightforward. Hence, an effort from academic researchers in order to develop proper methods and tools to support energy planning at urban scale as well as efforts from public services administrations to acquire a proactive role is desired. In this chapter, in order to provide insights into heterogeneous expertise involved in the sector, a framework on the renewables integration at city scale is outlined, highlighting the current main issues and challenges that are encountered, describing some selected experiences and investigating the possible technological solutions.

10.1 Introduction

Among the most relevant causes of global warming is the notable increase observed over the last century in the anthropogenic greenhouses gas (GHG) emissions. This increase, during the past twenty years, is mainly due to the combustion of fossil fuels. In detail, the Intergovernmental Panel on Climate Change (IPCC) reported that CO_2 emissions from fossil fuel combustion and industrial processes contributed to the global GHG emission increase to the tune of 78%, passing from about 15 $GtCO_2$ in 1970 to 32 $GtCO_2$ in 2010 (IPCC 2014).

Furthermore, it is largely proven that there is no possibility for a further deployment of fossil resources, which are physically and economically limited, since the

P. Caputo (✉)
Department of Architecture, Built Environment and Construction Engineering (ABC), Politecnico di Milano, Via G. Ponzio, 31, 20133 Milan, Italy
e-mail: paola.caputo@polimi.it

G. Dall'O' (ed.), *Green Planning for Cities and Communities*,
Research for Development, https://doi.org/10.1007/978-3-030-41072-8_10

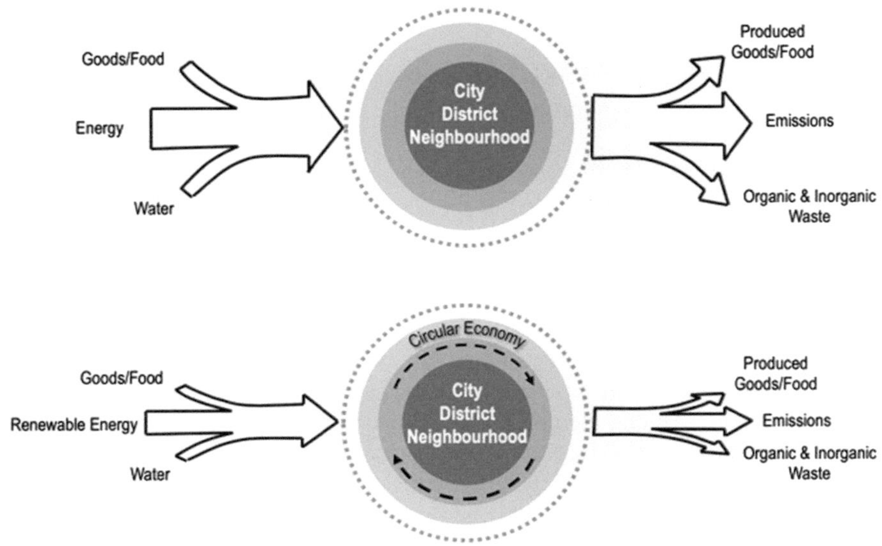

Fig. 10.1 Role of RES in circular urban metabolism (UN-Habitat 2018)

time needed for the accumulation of fossil fuels is drastically longer than that necessary for their extraction. In detail, looking at global level, at the end of 2018, the reserves accounted for 1730 billion barrels of oil, 196.9 trillion cubic feet of gas, and 1055 billion tons of coal. Based on the current global usage of fossil fuels and under the current techno-economic conditions, this corresponds to about 50, 51, and 132 years of availability, respectively (BP 2019).

The issue of climate change and depletion of resources can be tackled in the framework of urban metabolism,[1] because cities need energy as input for their operation and produce GHG as output. Therefore, optimising urban metabolism permits a reduction of these two global critical factors. This optimisation can be achieved by reducing the inputs needed, increasing the recycle and reuse processes in urban areas and their surroundings, and decreasing the outputs emitted (waste energy and matter; pollutants). Hence, the optimisation of urban metabolism can be considered also as a shift from a linear to a circular urban paradigm (Fig. 10.1).

Based on this approach, once energy demands are reduced, in order to avoid a further exploitation of fossil fuels, a strong integration of renewable energy sources (RES) for energy supply allows a tendency towards self-sufficiency and decarbonisation. Moreover, initiatives for decarbonising urban areas can also offer the opportunity to renovate degraded areas and control the urban micro-climates.

---

[1]One approach used to understand and transform urban areas into more efficient systems is the metabolic one, which started with the metaphor of a city as a living organism. The first concept of a city's metabolic requirements as "all the materials and commodities needed to sustain the city's inhabitants at home, at work and at play" (Wolman 1965).

Renewable and non-fossil sources are namely hydropower (the most aged and prominent for power generation), wind, solar (thermal, photovoltaic, and hybrid), geothermal, biomass, and biofuels, but also ambient energy, wave and other ocean energy.

Large RES integration is one of the targets set by the Agenda 2030 (UN 2015b) and notably can have several positive impacts. It does not only allow a reduction in environmental impacts, but can also contribute to reduce the energy imports, to increase the energy system capability, and even to create job opportunities. However, RES integration is not an easy task for both technical and economic reasons.

For instance, some of the renewables are programmable, while others are not programmable (such as energy from solar, wind, wave, and ocean), meaning that their availability is affected by variable natural conditions. In any case, the most important barrier to increase the renewable quota[2] is the trend in energy demand, which, despite the recent policies, continues to be ever-increasing.

As confirmation that cities and urban districts represent a suitable scale for promoting RES, recently, considering the rising trend of auto-consumption, the so-called Renewable Energy Directive (RED) has introduced the role of renewable **energy communities**. These are groups of end-user customers, particularly households, entitled to produce, consume, store, and sell renewable energy, including doing so via renewables power purchase agreements, as well as sharing the energy within the community (EU 2018a). Generally speaking, thanks to this regulation, organized local communities (from building tenants up to whole towns) will promote self-consumption from RES, an active role of consumers, a more efficient use of energy, a widespread access to energy, and at the same time a reduction in energy prices. This regulation could therefore support important initiatives towards the development of low-carbon settlements.

10.2 The European Energy Context

International statistics demonstrate that in the last 20 years investments in RES have increased all over the world. For example, China tripled its RES installed power between 2010 and 2016, increasing from 230 to 685 GW. In the same period, RES power increased from 321 to 534 GW in Europe and from 28 to 54 GW in Italy, which therefore represented more than 1/10 of Europe as renewable power. However, this almost incredible upward trend cannot yet be considered as signal of an actual path towards a global decarbonisation.

Without a doubt, the EU has assumed a fundamental role in the energy policies. Once set the targets for 2020, among which a 20% of gross final energy consumptions[3] covered by RES (EU 2008), it has driven a rapid transition passing from a

[2]i.e. the ratio (final energies satisfied by RES)/(total final energies).

[3]*Gross final consumption of energy* means the energy commodities delivered for energy purposes to industry, transport, households, services including public services, agriculture, forestry and fisheries,

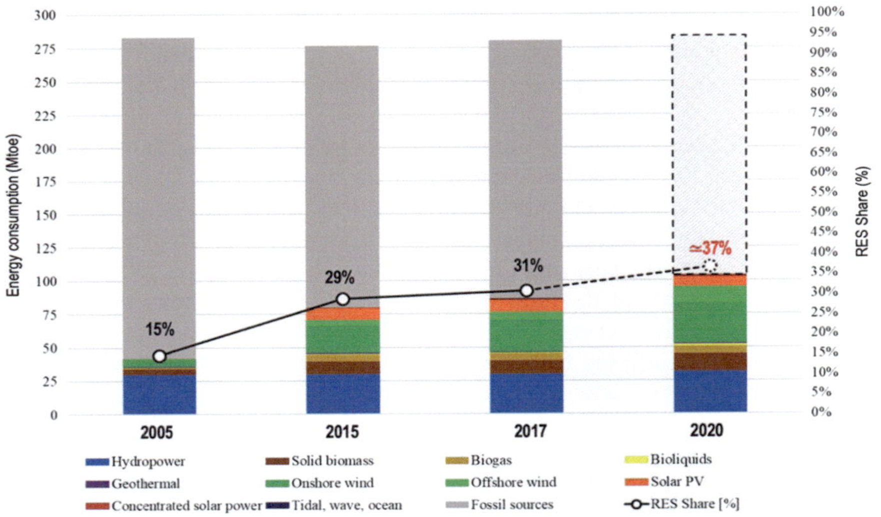

Fig. 10.2 RES contribution to gross final energy consumptions of electricity in EU-28 (figure elaborated based on EEA 2018)

share equal to 10.6% in 2007 to 16.7% in 2015,[4] thus obtaining a cut of the primary fossil fuel consumption of about 12% from 2005 to 2016 (EEA 2018).

Looking at the energy sectors (Figs. 10.2 and 10.3[5]) in Europe, RES mainly contribute to electricity (31% in 2017), while lower shares are reported in thermal uses (19%) and transportation (7%); similar shares are reported for Italy, i.e., 34%, 20%, 6%, respectively[6] (EEA 2018). On a Country level, the current RES share widely varies throughout EU, ranging from over 30% of gross final energy consumptions in Austria, Denmark, Finland, Latvia, and Sweden to below 9% in Belgium, Luxembourg, Malta, and the Netherlands (EEA 2018).

Italy has overtaken in advance the fixed target (i.e. a RES share of 17% by 2020), achieving a share of 18.3% in 2017, also thanks to several campaigns of incentives that have fostered the diffusion of photovoltaics (PVs) and wind turbines installation (MiSE 2019), even if hydro remains the most exploited source for power generation after wind. At a lower scale, the Decree on burden sharing (MiSE 2012) had fixed the RES shares, split between each Italian Region in order to comply with the European targets for 2020. Also, in this case, the trend is variable among Regions, including a large number having already exceeded the fixed targets, notably 5 of which for more than 50% (Calabria, Friuli Venezia Giulia, Provincia Autonoma di Bolzano, Umbria,

the consumption of electricity and heat by the energy branch for electricity, heat and transport fuel production, and losses of electricity and heat in distribution and transmission (EU 2018a).

[4]Eurostat. https://ec.europa.eu/eurostat/data/database (accessed 23/10/2019).

[5]In Figs. 10.2 and 10.3, fossil fuels consumption and RES share on the total final consumptions in 2020 have been estimated proportionally to the previous years.

[6]Eurostat. https://ec.europa.eu/eurostat/data/database (accessed 23/10/2019).

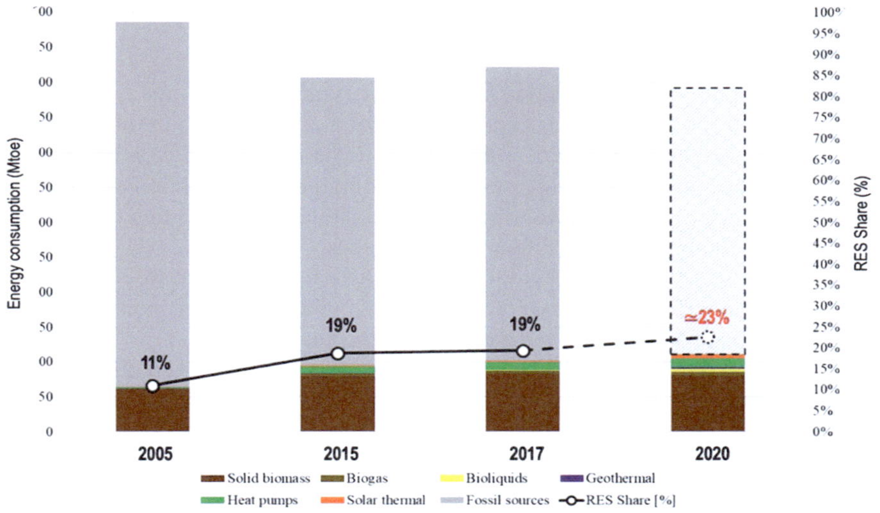

Fig. 10.3 RES contribution to gross final energy consumptions of heating and cooling in EU-28 (figure elaborated based on EEA 2018)

Valle d'Aosta, Veneto) and a few cases that have not yet reached the required outcome (Campania, Lazio, Liguria, Sicily) (GSE 2019). Figures 10.2 and 10.3 clearly show that, despite fossil fuels still represent the major contributor in the energy supply mixes, the achievements can be considered to be encouraging.

Looking at a midterm perspective, the EU has enacted the Directive 2001/2018 (EU 2018a), mandatorily requiring the increment of RES penetration in the EU gross final energy consumption to at least 32% by 2030. This scenario should be able to reduce the GHG emissions by 45% for 2030 and 60% by 2050. On the other hand, the Italian government has approved the National Energy Strategy (SEN) for 2030 (MiSE 2017), aspiring to an increase in the global RES share up to 28%, mainly in the electricity sector (RES share coverage of 55%).

In a long-term perspective, to limit the temperature increase to 1.5 °C as required at the Conference of Parties in Paris (UN 2015a), it is actually imperative to drive the most rapid transition path which provides for the halving of the fossil fuel consumptions by 2050, pushing towards a zero-carbon scenario at global level.

Already in 2009, the EU had set its long-term strategy for reducing GHG emissions by 80–95% in 2050, by means of a set of scenarios that call for disruptive and radical measures in all sectors. In fact, starting from measures aimed at strongly increasing the energy efficiency in industry and buildings, the largest integration of RES, mainly wind and solar, (from at least 40% up to 100%) was outlined. However, an electrification of the energy sector is going to become predominant: this could also have some serious implications, in terms of high-quality energy consumptions and grid stability.

10.3 The Role of Energy Planning in Decarbonizing Cities

A large part of the global population lives in cities, which in fact accounts for 70% of the global energy demand and is expected to dramatically increase in future, in step with the urban population growth (UNEP 2015). Urban planning affects behavior of cities in the long term, and proper energy planning allows the optimisation of the global efficiency of cities considering urban morphology and building density, use of buildings and trends of the different energy needs.

Cities are open and complex systems where heterogeneous variables have to be taken into account in order to explore the multiple relationship between people, the built environment, mobility, and the quality of life (Fig. 10.4).

Focusing on the role of energy use in buildings, once a significant reduction in the amount of energy required for ensuring thermal and visual comfort has been achieved, by means of both passive measures to improve the envelope performance and choice of mechanical systems with high efficiency, the residual energy demand for space heating, hot water, and electric appliances must be produced with renewable sources in order to avoid further fossil fuel consumptions and related emissions (Butera 2013). The Roadmap 2050 (European Climate Foundation 2010) and the European Directive 2018/844/EU (EU 2018b) are in line with this approach, facilitating the cost-effective transformation of existing buildings into nearly zero-energy buildings (NZEBs). Hence, integration of RES can become an opportunity to renovate buildings with an integrated approach, where architects, mechanical engineers and energy or comfort experts work together in an interactive and iterative manner. Existing buildings, responsible for the largest built volumes and greater energy consumptions,

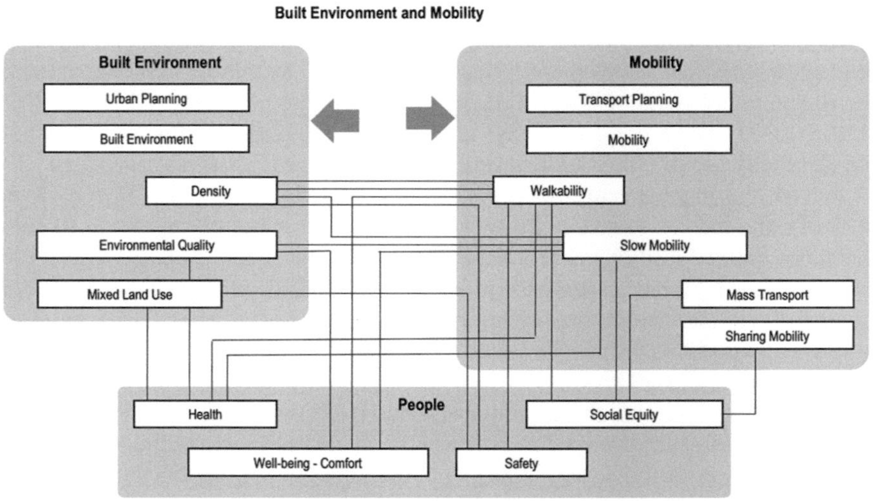

Fig. 10.4 Scheme of the complex system of relations that involve the built environment (UN Habitat 2018)

are the focus of this process, but their energy renovation is not always feasible owing to technical barriers (e.g. historical protected buildings, shadowed surfaces, etc.) and the necessary economic effort.

In this regard, a shift from the conventional intervention and works at the single building scale to the community level is more advantageous, thanks to the implicit economy of scale and greater cost-optimality, enabling more relevant energy savings for the same investments. To advance practically, it is fundamental to perform overviews of the national building stocks and to assess the most cost-effective measures by building types and climatic zones as well as the energy savings, which are achievable.

As an additional feature, buildings are becoming active and flexible players of a smart energy system. *Energy flexibility* means that buildings will be equipped with smart technologies (e.g. smart metres, devices for building energy management and demand response, etc.) which render them capable of managing their energy demand and generation in accordance with local climate conditions, user needs and grid requirements (Vigna et al. 2018). In this context, energy demand of buildings can be reshaped based upon the RES supply availability, and not the inverse (i.e. the supply follows the demand), while the excess on-site production can be exchanged with other consumers nearby. In this way, buildings have become smart actors integrated into, often complex, smart grids and the EU has in fact required the definition of a smart readiness indicator (EU 2018b).

All these aspects have to be included in urban energy planning, where, dependent upon local climatic and morphological conditions, on urban density and building stock features, on different programmable and un-programmable local RES availability, different measures can be defined systematically and effectively, benefitting from district energy systems and decentralised energy production and interacting also with other resource cycles (e.g. water).

Moreover, sustainable mobility can also be integrated, because, together with measures for influencing the behavior of people (e.g. car/bike sharing) and improving vehicle efficiency and emissions, there is also the possibility of integrating the sector with RES (i.e. biofuels) and electrifying it, with a potential for vehicle to grid (V2G), i.e. exchange of energy with the grid and making a contribution to shaping loads.

As illustrated in Fig. 10.5, innovative energy planning can be considered as a tool to create urban environments with 100% renewable energy sources: once energy consumption has been drastically reduced, these can be supported by local renewable energy sources also thanks to the storage and management of the systems).

10.4 Technologies for RES Integration in Urban Contexts

The implementation of renewable technologies within an urban context certainly needs to be accurately addressed, taking into consideration also architectural and

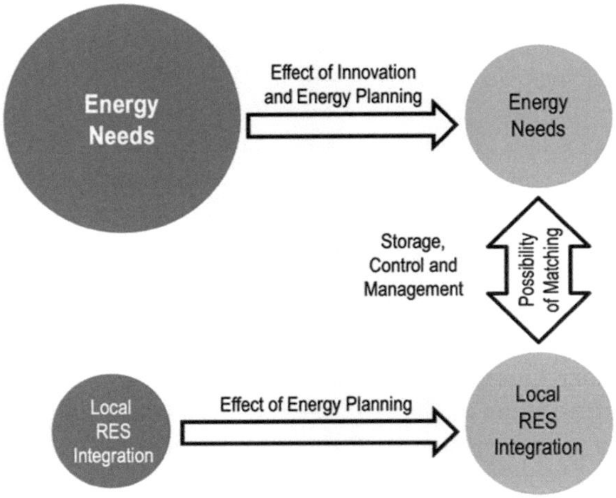

Fig. 10.5 Innovative energy planning at urban scale through integration between energy consumption reduction, local RES and storage and control management

technological integration. As mentioned, two scales can be distinguished: building and district, even if the interaction between them is possible and indeed recommended.

10.4.1 Solar Energy and Related Technologies

Solar energy technologies can be divided into thermal, photovoltaic (PV) and hybrid. PV devices convert solar energy into electricity. The productivity of a PV system is highly dependent upon the climatic context, i.e. to the availability of solar radiation. It is therefore extremely important to define the correct inclination and orientation of the modules, in order to maximise solar uptake.

Architectural integration represents a privileged sector for PV systems, with very promising prospects for growth, even in economic terms. In fact, building-integrated photovoltaics (BIPV) provides a variety of opportunities, such as the use of the land surface already occupied by buildings, the savings on support structures, the replacement of materials and components such as traditional roof elements, the possibility of using the energy produced on site according to the logic of distributed generation. In this framework, urban BIPV retrofit of façades has been envisaged as a possible twofold solution to reduce the energy demand of existing buildings and generate energy by using multi-functional BIPV façade systems (Fig. 10.6). In a recent piece of research (Saretta et al. 2019), an integrated GIS-based methodology has been developed providing a tool that allows the designer to estimate the theoretical solar potential in existing urban contexts, a precious item of information within urban energy planning.

Solar thermal systems convert solar radiation into heat for heating a fluid at different levels of temperature, depending upon the local conditions and on the selected

Fig. 10.6 Example of BIPV in energy retrofit in Zurich; PV is integrated in the façade. *Credits* Viridén + Partners AG

components. The two main types of collectors are: flat plate and evacuated tube. The decreasing cost of photovoltaic (PV) panels and of heat pumps for domestic hot water production is progressively reducing the cost gap between the solar thermal water heaters and the combination of PV with heat pumps; in most cases, the latter systems are already more cost effective.

Solar photovoltaic-thermal (PVT) collectors simultaneously produce electricity and heat, while increasing the PVT efficiency through cooling. These systems could be a competitive alternative or supplement to heat pumps, especially in areas with high energy density.

10.4.2 Hydropower

Hydropower refers to electricity derived from flowing water by turbines (Fig. 10.7) and is an ancient technology, and two basic configurations are available: with dams and reservoirs, or without. Hydropower dams with a large reservoir can store water over short or long periods to meet peak demand. The facilities can also be divided into smaller dams for different purposes, such as night or day use, seasonal storage, or pumped-storage reversible plants, for both pumping and electricity generation. Hydropower without dams and reservoirs means producing at a smaller scale, typically from a facility designed to operate in a river without interfering in its flow.

In countries like Norway, 99% of electricity comes from hydropower. While the world's largest hydropower plant is in China, with 22.5 GW of installed power and 80 $\div$ 100 TWh per year of electricity generation. In Italy, hydropower is widespread, but its application in urban contexts is quite limited.

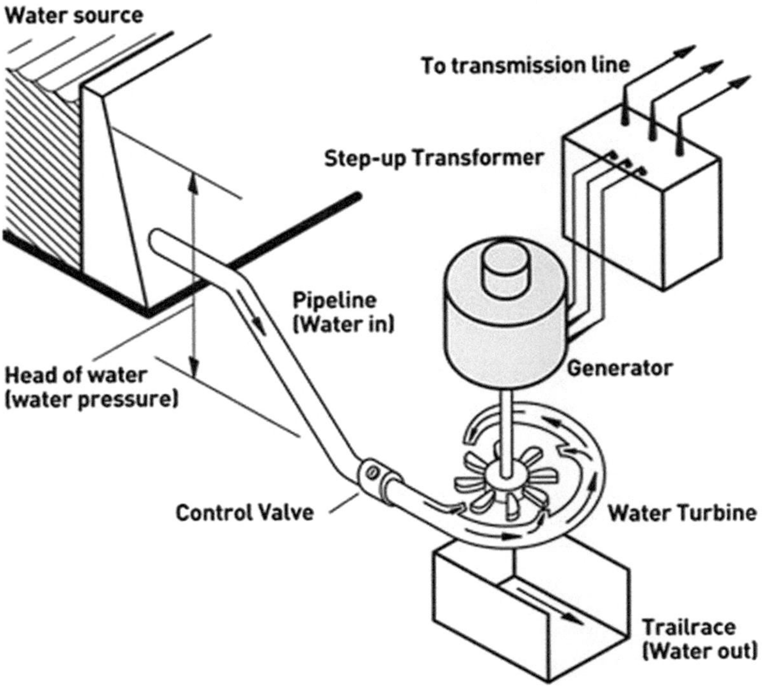

Fig. 10.7 Scheme of a micro-hydropower plant (20–500 kW). *Source* http://www.veoliawater2energy.com/en/references/micro-hydro-power-plants/ (accessed on 23/10/2019)

Small-scale micro-hydropower projects are more feasible and generally permit a reduction of the environmental impacts. These sizes are also very suitable for communities in remote locations.

10.4.3 Wind

Wind turbines are a well-established technology and represent a cost-effective system. Two categories are available: macro wind turbines that are installed for large-scale energy generation such as wind farms and micro wind turbines, designed with a vertical or horizontal axis, used for local electricity production. The main components of a wind turbine include blades, rotor, gearbox, and generator.

Wind farms can be also part of the landscape of peri-urban areas, especially in Northern Europe. The feasibility of a wind turbine installation, both at building and district scale, is dependent upon the availability of wind and on environmental issues. Therefore, with the exception of mountainous, offshore or highly windy areas, the adoption of wind turbines in cities is quite limited.

Fig. 10.8 Example of application of a wind power system on the roof of a building. *Source* https://www.climatetechwiki.org/technology/building-integrated-wind-turbines (accessed on 23/10/2019)

However, even if the cost-effectiveness of small wind generators is less than that of the large turbines, they are a valuable option and could make a considerable contribution to the energy balance of an urban area, in case of installation in the built environment and, in rare cases, on the roofs of the buildings (Fig. 10.8).

10.4.4 Biomass: A Focus on Urban Greenery

Biomass represents a valuable resource and an alternative to fossil fuels, for many reasons: availability, different typologies, programmability and storage, and technological maturity. Many technologies are available depending on the type of biomass to be processed, on the final use and on the economic conditions. Biomass available at neighbourhood scale derives from urban agriculture, maintenance of forests, parks, and gardens, the organic fraction of solid waste and sewage sludge. Wood biomass can be used directly (as direct combustion of pellets or wood-chips) or after gasification. The organic fraction of solid waste and sewage sludge can also be used to produce biogas. In addition, biofuels include a wide range of fuels that can be adopted for different final uses.

In order to be considered fully sustainable, the production and use of bio-energies have to be planned carefully, individualising a sufficient resource supply potential in an adequate range.

Biomass can be adopted for different final uses (electric or thermal), with various sizes and performances. The technical literature (Caputo et al. 2019) underlines as the most sustainable the use of wood biomass in mountain areas for district heating (DH) purposes, often equipped also with small CHP (combined heat and power) systems. As example, Fig. 10.9 shows the thermal station of the biomass DH plant in Obereggen (Bolzano, Italy).

Significant biomass potentials can be found not only in forest and rural areas, but also in urban contexts by means of pruning by-products from the urban greenery maintenance (Springer 2012), even more in combination with innovative projects aimed at urban reforestation. Urban green spaces are local ecosystems, which contribute substantially to the quality of urban life. They provide a valuable opportunity for recreational and cultural activities that improve the physical and psychological well-being of citizens. They also have a positive effect on ecological factors such as air quality and noise levels.

Besides the ecosystem services and benefits, the strengthening of the bond between natural and urban environment can improve the urban metabolism. Currently, municipalities do not use pruning by-products; on the contrary, they pay high costs for disposal in appropriate facilities. Neglecting bureaucratic and regulation issues, a significant potential is available by the optimisation of the management of urban pruning by-products, taking good account of the potential environmental, energy and economic benefits. To that end, many factors that differ from context to context have to be considered:

- Peculiar characteristics of each tree's species and the growing climate context,
- Growing location of the trees within the urban context (streets, parks, etc.),
- Management practices,
- Pruning techniques.

In this framework, in a recent research (Ferla et al. 2019), a GIS-based method was developed, applicable wherever information in terms of Urban Green Inventory and GIS are available. The GIS tool could lead to the implementation of circular economy practices, development of new skills, and the optimisation of municipal policies, in particular if urban pruning by-products can fuel local district heating systems.

10.4.5 RES Integration at Building Scale

Among RES applicable at building scale, the most commercially available and commonly used are:

- Heat Pumps (HP)[7];
- Solar Photovoltaic (PV);

[7]HP requires electricity, that can be also produced by RES, for its operation and permits the exploitation of renewable heat such as geothermal, groundwater, air or solar.

Fig. 10.9 Thermal station of the biomass DH plant in Obereggen (Bolzano, Italy). *Source* SEV, https://www.sev.bz.it/en/south-tyrol-energy-association/1-0.html (accessed on 23/10/2019)

- Solar Thermal (ST);
- Biomass Boilers.[8]

Their potential can be exploited by different technologies; each having specific characteristics, which determine its best use and suitability. It is therefore important to examine how a given technology would work for a particular building. According to Day et al. (2013), several issues should be tackled in defining energy supply by RES at building scale:

- Building type, function, and form;
- Building's energy demand;
- Building occupancy and usage patterns;
- Site access, plant space, and storage options;
- Local site and climate conditions;
- Synergies and conflicts between RES;
- Regulation, planning, and other constraints and requirements;
- Shading from neighbouring buildings, structures, and trees, etc.;
- Capital and maintenance costs;
- Economic incentives.

[8]Here neglected because their application is controversial in urban areas due to air pollution issues.

10.4.6 RES Integration at District Scale: A Focus on District Thermal Systems

RES can be applied at district scale, depending on the local availability and integrated by networks originating from a district energy system (Fig. 10.10).

District energy systems include networks devoted to electricity (i.e. mini-grids, smart grids) and thermal energy. For the latter, it is necessary to install a heating/cooling supply station and a grid of underground insulated pipes for distributing hot or cold water to multiple buildings in a district, neighbourhood or city (from a few buildings to thousands of buildings and homes across a city).

District heating and cooling systems (DHCS) are gaining ever greater relevance in the framework of climate change, debates over future energy scenarios and decarbonisation challenges. Often DHCS operate in cogeneration/CHP mode, thus acquiring also a great relevance in the path towards the future power paradigm based on distributed generation, due to the clear economic, technical, and environmental benefits achievable.

One advantage of DHCS is related to the possibility of integrating different RES as biomass, solar, geothermal, groundwater, or waste heat from industrial processes. The potential varies between cities though, not only because of the climatic conditions and of the difference in the availability of renewable resources, but also because generally a minimum level of heating density is required (IRENA 2016).

In the last century, DH has undergone a vast evolution and technological maturation, involving:

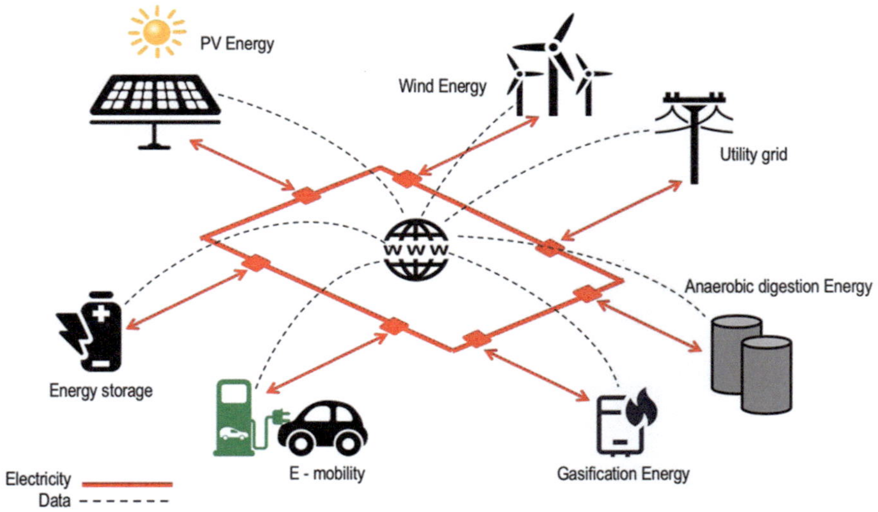

Fig. 10.10 Concept of integration of different RES in a mini-grid. *Adapted* From https://www.climatetechwiki.org/technology/building-integrated-wind-turbines (accessed on 23/10/2019)

- a decreasing supply temperature for heating and increasing supply temperature for cooling;
- a decreasing size and amount of material used for its components;
- an increased share of standardization and prefabricated components (Lund et al. 2014).

Approximately 6000 DH systems can be found all over Europe today, providing a coverage of 10% of the heating market, of which a part is integrated by renewable energies and waste heat recovery; their grids have an overall length of almost 200,000 km, 1/3 of the total length all over the world (Werner 2017).

DHS can also have the heat source fed by biomass, especially if wood derived from the maintenance of forests and from timber industry is available. Referring to the Italian context, the principal advantages are related to the possibility of exploiting an almost renewable source, programmable and suitable for thermal purposes, with positive effects in terms of environmental protection, climate change reduction, and saving of fossil primary energy, according to the European and national targets towards de-carbonisation. In addition, benefits on the local economy are evident, also due to the creation of local enterprises. The main disadvantages lie in the economic and normative instability (variation of price of fuels and changes or drastic reduction of the supporting mechanisms) and to the new regulatory policies aimed at the standardization of the management and monitoring conditions. Other unfavorable factors are the high investment costs, especially for the construction of the grid, and to other non-technical barriers, including eventual preconceptions against biomass combustion for fear of the release of air pollutants. From the energy point of view, the reduction of heat lost between heat generation and end-users appears to be the most important lever to improve the efficiency of these systems (Caputo et al. 2019).

As evolution, in (Lund et al. 2014), the concept of 4th Generation District Heating (4GDH) is clarified, including the relations to district cooling and the concepts of smart energy and smart thermal grids. The development of 4GDH involves meeting the challenge of more energy-efficient buildings as well as being an integrated part of the operation of smart energy systems able to include smart electricity and adoption of different RES and waste energy locally available. Finally, in (Buffa et al. 2019), the evolution of DH also towards 5th Generation District Heating and Cooling Networks is described, underlining the peculiarities of these networks for a given district.

10.5 Experiences of RES-Integrated Urban Areas in Europe

There are a diversity of interesting cases of study in Europe demonstrating the possibility of a path towards the 100% RES-based paradigm.

Table 10.1 Comparison of European case studies features

City	Vaxjo	Brescia	Galanta	Cernier	Lyon
EU Programme	Ready	Pitagoras	Geocom	Solution	Smart together
Retrofitted area (m^2)	54,192	265,000	18,183	12,744	35,000
Investment (M€)	12	12	1.08	2.29	7
Primary energy savings (GWh/year)	8.02	39.06	2.59	2.54	5.48
CO_2 emissions savings (tCO_2/year)	1778	7220	812	398	3534

https://smartcities-infosystem.eu/scis-projects/demo-sites/ready-site-v%C3%A4xj%C3%B6 (accessed on 23/10/2019)

For example, the city of **Güssing** (Austria)[9] implemented an economic model that included supplying 100% of its energy needs with local resources. The town is a net renewable energy exporter, producing about 10 times more energy than it needs and approximately 40 times more electricity than it can use. Besides this very positive energy situation, about 1000 jobs were created, more than 100 directly in the energy sector, and the town also became a tourist attraction and inspiration for the entire region.

Another example is the Beddington zero-energy district **(BedZED)**,[10] located in Hackbridge—London (England) and designed for adopting exclusively on-site sustainable energy sources. Here, 108 kW of BIPV cover close to 20% of BedZED's power needs and the average BedZED household uses 45% less electricity and 81% less heat than people in the surrounding communities. In addition, by its layout and carpool service, BedZED promotes walking, biking, and community life. These measures have resulted in a 65% drop in fossil fuel miles driven by BedZED residents. Another energy saving, in addition to water savings, was achieved by innovations such as low flow toilets, water recycling, and rainwater capture. Due to its sensitivity to environmental issues, BedZED has been the object of numerous awards and has a visitors centre.

Other experiences are available in Europe and some have been selected as cases of smart energy system transition that have been originated in urban areas, thanks to EU programmes. Related features are summarised in Table 10.1.

The Swedish city of **Växjö** has started on its pathway towards 100% RES by 2030 achieving a RES share of 60% in 2014. The city has 3 primarily wood fired CCHP plants, 4 local DH networks, BIPV plants, 21 wind turbines, 13 hydropower plants, a large number of geothermal HPs. Biogas is produced from sewage sludge and food waste for usage in urban buses and organic fertilizers. 7.7% of local cars use ethanol or are electric. During the READY EU programme (2014–2019),[11] a residential district and a tertiary sector building have undergone retrofit, and this

[9]http://www.go100percent.org/cms/index.php?id=19 (accessed 23/10/2019).

[10]http://www.go100percent.org/cms/index.php?id=19 (accessed on 23/10/2019).

[11]http://www.smartcity-ready.eu/about-ready/ (accessed on 23/10/2019).

was tested on a reduction of the supply flow temperature of the DH from 80 °C to 70–75 °C (the long-term goal is a further decrease to 60 °C).

The Italian city of **Brescia**, within the EU programme PITAGORAS[12] has programmed the installation of a waste heat recovery unit (nominal capacity of 16 MWth) to reuse heat, otherwise lost, from a local steel mill. The pilot plant produces heat (nominal thermal power of 10 MWth) that is delivered via the DH network uses the waste heat to produce electricity through an Organic Rankine Cycle (ORC) (with a nominal power of 1.8 MWel and an efficiency of 0.19).

The city of **Galanta**, Slovakia, was already a pioneer in the area in terms of RES exploitation. Within the European Concerto programme GEOCOM,[13] it has defined the connection of additional utilities to the existing geothermal loop and to improve the energy efficiency of existing buildings already connected, through energy retrofit including the installation of BIPV on roofs.

In the Swiss town of **Cernier**, during the SOLUTION EU programme (2009–14),[14] renovation activities covered the refurbishment of a small mixed-use buildings cluster, achieving a reduction by nearly half of energy consumption, and the renewal of the construction of one building. At individual building level, RES integration has been promoted through the implementation of HPs, wood boilers, solar thermal collectors and PV panels. Additionally, the DH network, connected to a biomass CHP, was extended and integrated with a heat exchanger for cooling demand. A CHP biogas plant, fuelled by agricultural organic waste (7500 tons per year), was also installed.

Among the pilot cities of the EU programme SMARTER TOGETHER (2016–21),[15] the urban redevelopment area of **Lyon-Confluence (France)** is proceeding with renovation of existing buildings and doubling of solar electricity capacity with 4 PV plants (1 MW peak). A wood biomass CHP plant (of 2 MWel and 4 MWth) is included as well as the promotion of smart mobility with smart charging stands, an electric vehicle car-sharing system and an autonomous driverless electric shuttle service.

Other examples of RES integration in the Italian urban context were explored. A first selection regards cities for which a 100% RES supply of their electric and thermal energy civil uses has been estimated (Legambiente 2019) and whose features are summarised in Table 10.2. Legambiente individuated 3054 municipalities, with different sizes and distributed throughout the country, considered as able to fully cover the electricity demand with RES, plus 1200 able to cover over the 50% and 50 able to fully cover the thermal energy demand. Among these, the town of **Dobbiaco** (Bolzano, IT) includes solar thermal and PV panels, and mini-hydropower plants. Additionally, over 1000 consumers are connected to a DH network producing 53.8 GWh/year and fuelled by a biomass plant, using wood chips from pruning residues and industrial wood wastes, and a biogas plant.

[12] https://cordis.europa.eu/project/rcn/186981/factsheet/es (accessed on 23/10/2019).

[13] https://geothermalcommunities.eu/ (accessed on 23/10/2019).

[14] https://cordis.europa.eu/project/rcn/94483/factsheet/en (accessed on 23/10/2019).

[15] https://www.smarter-together.eu (accessed on 23/10/2019).

Table 10.2 Comparison of Italian urban case studies features (table elaborated based on Legambiente 2019)

	Dobbiaco	P. allo stelvio	P. S. M. di castrozza	Trento	V. di vizze
Mini-hydropower (MW)	2.46	3.73	0.31	0.94	5.49
Biogas (MW)	0.132	0.17			0.1
Bioliquids (MW)		1.62			0.05
Biomass (DH) (MW)	1.9	0.99	11.23		2.8
Geothermal energy (MW)			0.01		
DH (GWh/year)	53.82	17	37		57
Wind power (MW)				0.003	
Solar thermal (m^2)	1350	2200	176	670	36
Solar PV (MW)	1.57	7.08	0.96	1.33	4.38
Smart mobility			x	x	
ICTs towards smart grid		x			
RES share on final electricity and thermal energy consumptions of resident population (%)	100	97	100	100	100
Investments type	Public	Public	Public	Public	NA

The town of **Prato allo Stelvio** (Bolzano, IT) maintains hydropower and PV plants, together with 3 DH networks, fuelled by biogas and producing 18.3 GWh/year for 742 consumers.

Additionally, the municipality is developing fibre optics also for energy management in a smart grid perspective. The town of **Primiero—San Martino di Castrozza** (Bolzano, IT), together with the largest historical hydropower plants, benefits from 2 DH networks serving 40 GWh/year to 1250 buildings and fuelled by wood biomass from residues of local woods maintenance. The municipality is also boosting sustainable mobility by means of 17 electric vehicle and related charging stations.

The city of **Trento (Italy)** is promoting the distributed installation on the roofs of public service buildings of solar thermal collectors and PV panels as well as the 100% RES supply of the city's electric vehicle charging stations. The energy supply of the town of **Val di Vizze** (Bolzano, IT) is based on hydropower plants, PV panels installed on the roofs of public service and private buildings, a biomass DH and solar thermal collectors. Other plants based on local bio-energies complete the local energy system.

10.6 Challenges and Conclusive Remarks

Despite the several substantial contributions of RES to the enhancement of the environmental, economic, and social aspects of urban sustainability, several barriers are being experienced at global level for their diffusion. These include cost-effectiveness, technical and market barriers (such as inconsistent pricing structures), institutional, political and regulatory barriers, and social and environmental barriers.

When planning energy strategies in cities, one cannot neglect that they are complex energy systems in which a series of consumers with different habits determine different energy demand profiles during the course of the day (and according to the day of the week). In addition, most of RES are characterised by an un-programmable and intermittent behavior that results in a similar variable availability. Therefore, their availability and the related energy produced quantity are not obvious and/or do not match the simultaneous energy demand from consumers.

However, it is possible to ensure a combination of multiple sources to cover a possible lack in renewable energy, by providing storage technologies to smooth fluctuations or accumulate excess renewable energy, and by promoting energy balancing and demand response strategies, especially in mixed-use districts (and installing backup, fast-start generation units based on biofuels). In addition, there is the possibility of reusing cascade or dissipated energy.

To be effective, energy planning needs proper supporting tools for accomplishing energy and techno-economic feasibility assessments towards decision-making. Tools for simulating and mapping the energy features, in order to identify potential areas where accomplishing energy strategies and forecasting related effects, as well as tools for monitoring achieved outcomes are all fundamental.

To that end, the provision of data and maps from the local administrations and involved stakeholders for governing such processes is a priority, although quite complex to obtain. Within this framework, the GIS represent a valuable and proven tool for storing, visualising, as well as spatially and analytically correlating geo-referenced data. Even more, in the energy planning field, the use of GIS is particularly fruitful since it provides the possibility of accomplishing a spatial analysis of the urban energy density and the distribution of local sources (e.g. energy demand of several buildings, transport fluxes, RES potential production, DHC networks and location of energy plants in general, etc.).

Moreover, the management of these smart energy systems is quite complex and requires an accurate assessment of the energy profiles involved. For that purpose, going against the common practice, which is based on evaluating the energy amounts on a seasonal or annual base, the evaluation of the energy profiles with a proper time resolution (i.e. at least hourly based) has a prominent role.

Dealing with the flexibility of energy systems requires the development of robust methods and tools for balancing the energy supply and demand, establishing how to choose the different options of renewable energy sources and assessing the relative technical and economic effects of these choices, in order to minimise the impacts. According to a previous study, there is a large variety of existing tools, featured by heterogeneous characteristics in terms of scale of application, performed analysis objective, required level of detail of data, necessary user skills, etc. The most proper tools are the ones capable of assessing the energy balance with at least an hourly time step, to model several energy services and to evaluate not only the technical aspects but also the economic and environmental implications of defined energy strategies. In Table 10.3, a non-exhaustive summary is presented: this also explains the related targets and scales of application.

The full integration of renewables in urban energy systems is constrained by both economic and technical potential: in fact, the overall technical potential is rarely fully utilised since it would not be economical. First, the climate conditions affect the energy consumptions, determining the share of the different RES available and suitable for the specific situation. In cold climates of developed countries, space heating and electricity for appliances typically represent 70% of total consumption. Conversely, in warm climates, mainly in developing countries, the energy consumed for cooking and domestic hot water accounts for the 68%[16]. In temperate climates, an increase in summertime electricity demand for space cooling is observed and a global electrification of final uses is in progress. The proximity and/or availability of local RES sources (e.g. hydro reservoirs, wood forests and agricultural areas, sea, etc.) obviously drives the choice of which one to exploit and affects the capacity obtainable.

The technical capability of the system to accommodate large shares of renewables is a relevant parameter, too: this is particularly affected by the urban size and texture, in terms of building density, shape, and mutual shadowing, as well as proximity to infrastructures.

For instance, it has been demonstrated that only a half of façade surfaces are suitable for solar system installation, giving rise to a feasible covering of the total energy needs of the buildings from PV of between 8 and 44%, depending upon the urban morphology and adopted technology (Lobaccaro et al. 2019).

Summarising, despite cities being in the forefront of RES integration, the attainment of 100% RES cities is still very challenging. By the end of 2018, of 1146 cities around the world,[17] about 200 had committed to achieve 100% RES in the electricity sector, although 64% of a sample of over three hundred cities reached a RES share in electricity not greater than the 50% in 2017. The situation is even more difficult where the heating and cooling and transport sectors are concerned (REN21 2019). Looking at the fixed target by selected cities worldwide, shares of RES on total energy consumptions in the range 10–70% are reported for cities in cold climates, of 30–80% in hot climates, while RES shares on electricity are targeted in the range 30–50% in cold climates and 10–55% in hot ones. Averagely, the REN21 reports a current RES share of electricity consumptions in cities of 41%, which is reasonably greater than the global level (26%). Hence, based on reported targets for European Countries for 2020 and assuming the same increment, it can be estimated that in the EU for cities located in cold climate areas, the potential of RES share on total energy

[16]https://publications.wri.org/buildingefficiency/ (accessed on 23/10/2019).

[17]The amount includes cities with a population over 500 thousand inhabitants (UN 2018).

Table 10.3 Summarising table of RES, scale of integration, and available tools

RES	Technology	Scale		Gross final energy consumptions in EU28 (Mtoe) (EEA 2018)		Programmable	Tool[b]					
	Energy service	Building integration	District/Urban integration	Actual 2017	2020 target	[Y/N]	Energy plan	Energy PRO	Homer	Ihoga	Siren	Webopt
Wind	El.	X[a]	X	5.2	11.7	N	X	X	X	X	X	
Hydro	El.		X	30	31.8	Y/N	X		X	X	X	
Solar	El.	X	X	9.5	7	N	X	X	X	X	X	X
	Th.	X	X	2.2	6.5	N	X	X				X
Biomass	El.		X	10.1	13.5	Y	X				X	
	Th.	X	X	84.2	80.8	Y	X	X				X
Biogas	El.		X	5.5	5.5	Y	X				X	
	Th.		X	3.8	5.1	Y	X					
Biofuels	El.		X	0.4	1.1	Y	X				X	
	Th.		X	0.3	4.4	Y	X	X				X
Geothermal	Th.	X	X	0.7	2.7	Y	X					X

[a]Only for mini and micro wind turbines
[b]Based on Ferrari et al. (2019)

consumptions is 35%, of electricity consumptions 52%, and of heat consumptions 40%; whereas in warm areas, the values are, respectively, 31%, 53%, and 31%.

Urban energy planning must consequently face up to the above-mentioned challenges by proposing a combination of measures: assurance of a plentiful supply of different energy sources and technologies; flexibility of the energy system; integration with programmable RES, storage in addition to demand side management measures in order to fully meet the local energy needs.

References

British Petroleum (BP) (2019) Statistical review of world energy 2019, London

Buffa S, Cozzini M, D'Antoni M, Baratieri M, Fedrizzi R (2019) 5th generation district heating and cooling systems: a review of existing cases in Europe. Renew Sustain Energy Rev 104(C):504–522

Butera Federico M (2013) Zero-energy buildings: the challenges. Adv Build Energy Res 7(1):51–65

Caputo P, Ferla G, Ferrari S (2019) Evaluation of environ-mental and energy effects of biomass district heating by a wide survey based on operational conditions in Italy. Energy 174:1210–1218

Day T, Lim D, Yao R (2013) Renewable energy for buildings. Design and management of sustainable built environments. Springer, London, pp 203–224

EEA (2018) Renewable energy in Europe—2018. Recent growth and knock-on effects. EEA Report No. 20/2018

EU (2008) Communication from the commission to the European parliament, the council, the European economic and social committee and the committee of the regions, 20 by 2020—Europe's climate change opportunity

EU (2018a) Directive 2018/2001 of the European parliament and of the council of 11 December 2018 on the promotion of the use of energy from renewable sources (recast)

EU (2018b) Directive 2018/844 of the European parliament and of the council of 30 May 2018 amending directive 2010/31/EU on the energy performance of buildings and directive 2012/27/EU on energy efficiency

European Climate Foundation (2010) Roadmap 2050: a practical guide to a prosperous, low-carbon Europe

Ferla G, Caputo P, Colaninno N, Morello E (2019) Development of a GIS-based dataset of urban green areas for the energy exploitation of tree pruning biomass. In: Proceedings of SDEWES conference, the application to the case study of Milan, Dubrovnik

Ferrari S, Zagarella F, Caputo P, Bonomolo M (2019) Assessment of tools for urban energy planning. Energy 176:544–551

GSE (2019) Fonti rinnovabili in Italia e in Europa—2017. Verso gli obiettivi al 2020 e al 2030. Available at http://www.gse.it. accessed on 23/10/2019

Intergovernmental Panel on Climate Change (IPCC) (2014) Climate change 2014: mitigation of climate change. In: Edenhofer O, Pichs-Madruga R, Sokona Y, Farahani E, Kadner S, Seyboth K, Adler A, Baum I, Brunner S, Eickemeier P, Kriemann B, Savolainen J, Schlömer S, von Stechow C, Zwickel T, Minx JC (eds) Contribution of working group III to the fifth assessment. Report of the Intergovernmental Panel on Climate Change. Cambridge University Press, Cambridge, United Kingdom and New York, NY, USA

IRENA (2016) Renewable energy in cities, Abu Dhabi. Available at: https://www.irena.org. (accessed 23/10/2019

Legambiente (2019) Comuni rinnovabili 2019

Lobaccaro G, Lisowska Malgorzata M, Saretta E, Bonomo P, Frontini F (2019) A methodological analysis approach to assess solar energy potential at the neighborhood scale. Energies 12(18):3554

Lund H, Werner S, Wiltshire R, Svendsen S, Thorsen Jan E, Hvelplund F, Mathiesen Brian V (2014) 4th generation district heating (4GDH): integrating smart thermal grids into future sustainable energy systems. Energy 68:1–11
MiSE (2012) Decreto 15 marzo 2012. Definizione e qualificazione degli obiettivi regionali in materia di fonti rinnovabili e definizione della modalità di gestione dei casi di mancato raggiungimento degli obiettivi da parte delle regioni e delle provincie autonome (c.d. Burden Sharing)
MiSE (2017) Strategia Energetica Nazionale (SEN)
MiSE (2019) La situazione energetica nazionale nel 2018
REN21 (2019) Renewables 2019 Global Status Report, REN21 Secretariat, Paris
Saretta E, Caputo P, Frontini F (2019) A review study about energy renovation of building facades with BIPV in urban environment. Sustain Cities Soc 44:343–355
Springer Tim L (2012) Biomass yield from an urban landscape. Biomass Bioenerg 37:82–87
UN (2015a) Adoption of the Paris agreement
UN (2015b) Transforming our world: the 2030 agenda for sustainable development A/RES/70/1
UNEP (2015) District energy in cities—Unlocking the potential of energy efficiency and renewable energy
UN-Habitat (2018) Energy and resource efficient urban neighbourhood design principles for Tropical Countries. a practitioner's guidebook
Vigna I, Pernetti R, Pasut W, Lollini R (2018) New domain for promoting energy efficiency: energy flexible building cluster. Sustain Cities Soc 38:526–533
Werner S (2017) International review of district heating and cooling. Energy 137:617–631
Wolman A (1965) The metabolism of cities. Sci Am 213:179–190

Part III
Sustainable Mobility for Cities and Communities

Chapter 11
The Infrastructure for Sustainable Mobility

Renato Mazzoncini, Claudio Somaschini and Michela Longo

Abstract The objective of this chapter is to indicate in which direction cities must invest in order to become smart and sustainable in terms of mobility. On the basis of the current trends and considering technological innovation in the mobility sector, the chapter will address those aspects which the planners will have to take into account not only in the design of the new neighbourhoods, but also in the redevelopment of the recovery of the existing ones. The authors discuss both collective mobility and individual mobility, thus ranging from underground railways to electric scooters. In the final part, general criteria for sustainable mobility planning are provided.

11.1 Sustainable Mobility in Smart and Sustainable Cities

Collective mobility will be increasingly integrated and multimodal, whilst individual mobility will become increasingly shared, connected and electric. This is the new urban mobility that is slowly taking hold in all the advanced metropolises of the planet, from Europe to the USA, but also strongly in China, in India and in the most advanced countries of the Far East.

The Paris agreements on climate and the need to define road maps at 2030 and 2050 are contributing to the definition of policies to reduce smog, to face the challenge imposed by climate change and to create and have more sustainable and liveable cities. The new generations are smarter, digital and, not least, cost-conscious (Zawieska and Pieriegud 2018; Ajanovic and Haas 2019).

R. Mazzoncini (✉) · C. Somaschini
Department of Mechanical Engineering, Politecnico di Milano, Via La Masa 1, 20156 Milan, Italy
e-mail: renato.mazzoncini@polimi.it

C. Somaschini
e-mail: claudio.somaschini@polimi.it

M. Longo
Department of Energy, Politecnico di Milano, Via La Masa, 34, 20156 Milan, Italy
e-mail: michela.longo@polimi.it

G. Dall'O' (ed.), *Green Planning for Cities and Communities*,
Research for Development, https://doi.org/10.1007/978-3-030-41072-8_11

All the demographic mark trends describe a strong shift of the population towards the cities. The forecasts indicate that in 2050, there will be as many as seven city dwellers for every 10 inhabitants of the planet.

Growing cities are becoming increasingly critical in terms of harmful emissions, pollution and CO_2, also owing to the growing demand for mobility, which on the other hand is one of the major factors of competitiveness of the cities themselves. For this reason, one refers increasingly to urban plans of sustainable mobility and, in general, to make mobility increasingly green.

So far as individual mobility is concerned, much is driven by the growing availability of electric vehicles, by the growing presence of cycle paths, by the modal share which shows a drop in the rate of motorisation in almost all the advanced metropolises that grow at a steady pace in terms of number of inhabitants but often reduce the number of cars at the same time (Civitas 2015; McKinsey & Company 2018).

A representative case among all is Milan, which over the last twenty years has lost as many as 100 thousand cars and gained as many residents returning to 1.4 million inhabitants, all thanks to ambitious local policies and the tools which follow, amongst all of which the activation of the Area B (low-emission zone) after the success of Area C (congestion).

In this chapter, an attempt is made to indicate in which direction cities must invest in order to become smart and sustainable in terms of mobility. Both collective mobility and individual mobility are discussed, thus ranging from underground railways to electric scooters.

Chapter 12 instead studies the behaviour enabled by the digital revolution, arriving at the concept of Mobility as a Service (MaaS) (Crozet et al. 2019).

Digital platforms, metadata platforms for integrated mobility, sharing in all its forms, from sharing to peer to peer will thus be the topics which will be handled in Chap. 12.

For economy of narration, railway urban mobility has been excluded: this is to be understood as the demand for mobility that originates in extra-urban areas, which is served by the train and which, more and more often, allows urban penetration via city railway stations and underground rail passers. Urban railway transport has the great merit of lightening the load on the other modes of urban transport by taking up a large slice of demand that would otherwise weigh again on the city means. However, it lies typically outside the city administrative competences, being a part of large national railway infrastructure investments. In the context of this book, therefore, which aims to provide a clear indication to local government administrators on how to direct investments, urban railways have been excluded.

A reflection must be made to understand to what extent certain solutions can work regardless of the size of the cities.

Reference is not being made only to large investments, since it is clear that in order to be able to think of an underground railway service, the demand must be that typically generated in middle/large cities, but also, for example, in sharing mobility which will be discussed in Chap. 12. Today, shared mobility is taking off well in large cities, with over half a million inhabitants.

It is probably unnecessary to remind the reader that in order to accelerate this revolution of urban mobility and zero emissions, it is essential that there are concrete policies and actions capable of making cycling, walking, public transport and electric mobility grow. One can rapidly gain the conviction that it is possible to escape from the pollution that characterises urban centres and at the same time reclaim squares and streets, making cities more liveable and safer. To start this revolution, however, courageous and systematic choices are required, national policies that also include unpopular measures such as the carbon tax. Local administrators cannot be left alone in the face of these challenges. The general framework must be defined by national or supranational entities, such as the Conference of the Parties (COP), to which local realities refer.

The market and the economy once again anticipate legislation, in many cities the number and variety of mobility providers are already capable of winning against the private mobility, both from the point of view of speed and flexibility of travel, and in terms of the travel cost.

Certainly, the clear lack of legislation and the confused regulatory framework has led to the known problems between taxis and Uber, or with free-floating fleets of bikes, or even on the rules of movement of electric scooters. It is one of the main themes to work on.

In this chapter, examining future infrastructures of cities, electric mobility in all its multifaceted meanings will be focused upon. It is now well established in the international scientific community that land transport can be completely decarbonised by passing on a large scale to electric vehicles.

The different forms of mobility can be divided into:

- Collective mobility;
- Individual mobility.

The first goal that the public regulator must set itself is to move as much as possible individual to collective mobility, to increase the overall efficiency of the mobility system, in terms of generalised costs, thus also including the costs related to pollution and road congestion.

11.2 Collective Mobility

Urban public passenger transport plays a particular role in the world of mobility as it is the sector with greater volume of production (in terms of total km covered, number of employees, number of vehicles, etc.).

The complexities of urban public transport are many, including the needs to:

- Carry large masses of people;
- Adapt transport systems to urban topography, normally already strongly configured;
- Ensure intermodality and interchange between different modes of transport.

Table 11.1 Urban transport systems

Rail	Road	Cable/conveyors
Tramway Underground rail Light rail transport Monorail	Trolleybus Electric bus	Cable cars Funiculars Elevators Escalators Travelators

In recent years, in all the most developed countries, urban sustainable mobility plans have been implemented, mainly owing to two different needs: to meet the demand for mobility in urban and metropolitan areas and to make mobility increasingly sustainable and green.

In order to address the development of urban mobility, it is necessary to know the various and different modes of transport available on the market, having clear their respective capacities and potentials, investment costs, infrastructural impact, operation and maintenance costs.

In the following paragraphs, the Key Performance Indicators—KPIs characteristic of urban public transport systems will be *illustrated, together with* the various modes of transport according to the classification normally used in the literature. One feature connoting urban public transport services is the great variety of modes, vehicles, technologies and infrastructures.

This variety is explained by the need to adapt transport services to both the territory and to the demand to be served. Table 11.1 proposes a first classification of electric land-based modes used for local public transport services, while Fig. 11.1 shows the relative weight of road and rail-based public transport modes in the European Union and in some key member states in terms of passenger per km.

Comparing the costs of different transport systems is no easy task (see Fig. 11.2). The main reason is that the cost of a transport system is made up of capital expenditure (CAPEX) and operating expenditure (OPEX). High-capacity modes usually have lower OPEX but higher CAPEX per vehicle * km and it is the contrary, for low-capacity modes.

Furthermore, CAPEX is not constituted only by vehicles, but also by infrastructure that cannot be standardised on the basis of the number of vehicle per km or the number of passengers; i.e. maintaining a metro network roughly depends only upon the number of trains or daily services. Table 11.2 tries to provide just order of magnitude values, since the actual costs depend upon the context.

On the other hand, OPEX per passenger does not change considerably between different transport modes as it does for CAPEX, and it should be noted that, as public transport (PT) is normally a labour-intensive activity, the costs for providing PT services mainly depend upon the cost and productivity of driving personnel and on the commercial speed of services.

In simple terms, the performances of a PT system crucially rely on the mobility flows they should sustain.

These parameters may vary substantially from country to country and even show a great variability within a single region. This is partially due to the specific conditions

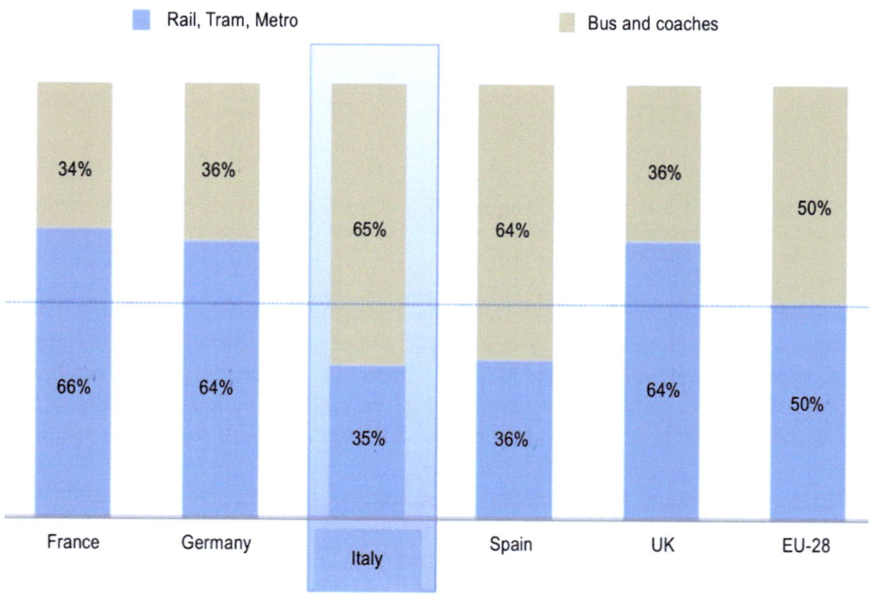

Fig. 11.1 Weight of road and rail-based public transport modes in the EU (EC 2015)

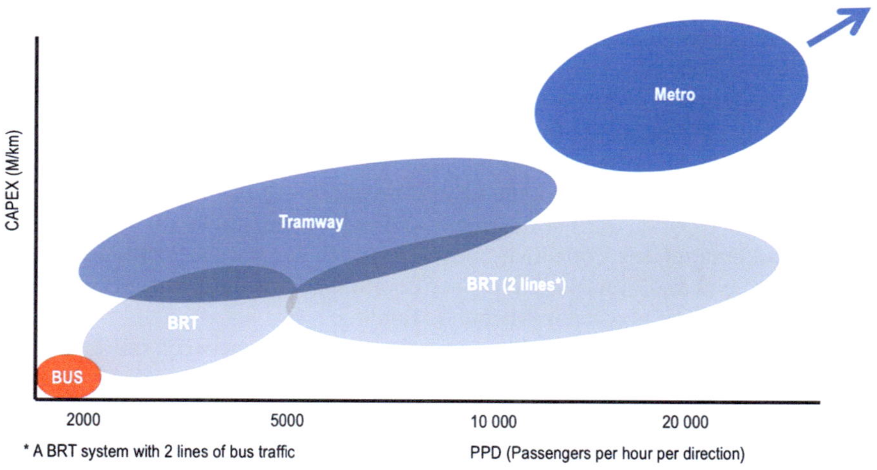

Fig. 11.2 Comparing the costs of different transport systems

Table 11.2 Average comparison between the costs of different modes of transport

Mode	Millions €/km of network
BRT[a]	5–15
Tram	15–40
Metro	80–200

[a]A Bus Rapid Transit (BRT) is a transport system based on high-frequency bus routes, segregated lanes and other forms of prioritisation over private traffic (i.e. at junctions)

in which services are operated but also reflects different efficiency levels. This is the reason for which, particularly in markets, where competition is not so extensive, universities, consultancies and public sector bodies have been struggling to calculate, through statistical, econometric, empirical models, the so-called standardised cost, that is the cost for providing a non-specified PT service with a given productivity level.

However, it should be noted that these models must be handled carefully, as their results strictly depend on the input parameters in the model itself. Table 11.3 shows the main urban local public transport vehicles classified according to their performance, first of all, the capacity in terms of passengers per hour per direction.

As described above, the main challenge for the regulators is to increase the share of collective mobility. To do this, it is necessary to bear in mind that mobility is, by nature, intermodal, since the door-to-door connection can hardly be done with a single collective means. By intermodality it is also meant: on foot + collective means, bike + collective vehicle, car or scooter + collective vehicle.

The main reason why in most cities today the individual transport largely exceeds the collective transport is precisely because individual transport is congenital door to door. To compete, the collective transport must be widespread and extensive, frequent and continuous over time so that the main differences will be only costs and speed, on which a good public transport system is normally competitive.

The backbone of any collective transport system in metropolitan cities is rail, underground and tram transport. The poor development of these networks depends also upon the lack of non-repayable public funding.

In almost all EU countries, there is a notable asymmetry between railway infrastructures, normally 100% funded by the widespread state taxation, and metropolitan infrastructures that normally receive only partial funding from the state and whose costs therefore remain a charge on the local authorities.

However, since local authorities rarely have sufficient tax revenues, the result is a significant slowdown in the construction of networks compared to the need for mobility. In essence, it would be necessary that at the level of the EU and of individual states, the extra-urban railway and the urban railway should be considered in the same way to avoid the paradox of underfunding precisely the networks with the highest utilisation rate.

Table 11.3 Collective transport—features and performances

	Unconstrained 12 m Promiscuous	Unconstrained 18 m Protected	Constrained Tramways Promiscuous	Constrained Tramways Protected	Constrained Classic Underground	Constrained Automatic Underground	Services Not programmed
Potentiality	<2500 p/h/d	2000–4000 p/h/d	2000–6000 p/h/d	3000–15,000 p/h/d	8000–45,000 p/h/d	6000–25,000 p/h/d	– p/h/d
Capillarity accessibility (m)	200	200	200	200–300	500–800	500–800	High
Regularity	Very low	Good	Very low	Good	Great	Great	Low
Frequency	Average low	High low	Average low	High low	High low	Medium high	Good
Speed (max.) (km/h)	50	50	50–80	80	50–80	80	50
Commercial speed (km/h)	10–15	15–25	20–30	30–40	20–30	30–40	10–20
Travel speed							
Short dist.	Low	Average	Low	Medium high	Medium-high	Medium-low	–
Long dist.	Very low	Medium-low	Very low	Low	Low	Very high	–

It should be noted that in the last few decades, the non-repayable public funding failure has been replaced by private investments in project finance, often financed by the construction companies themselves which have taken over the financing in order to build.

The result for the balance sheet of the local authorities is however often disastrous since the interest rates that the builders obtain from the banking system are very high, often having such companies a low credit rating, and these interests enter the overall debt that must be repaid by the local authority transport companies. Local authorities almost always pay remuneration to project finance with a fixed fee, completely absorbing the traffic risk. The overall result is an infrastructure cost that grows considerably owing to interest on the debt and makes it even more untenable for the local authority to start new investments.

11.2.1 Trams

It is considered worthwhile to spend few words on the tram, probably the oldest means of collective urban transport, but which is experiencing a new rejuvenation throughout the world.

There are several innovative technologies with which it is possible to create new-generation tram lines that are, unlike traditional ones, without an overhead contact line. These technologies are well suited to overcome some constraints typical of the historic centres of large cities, contributing to the development of eco-sustainable mobility. The innovative "catenary free" systems, which saw the first important applications in Europe especially in Spain and France, have the following strong points:

- reduction of aesthetic, environmental and urban impact;
- reduction of the zone of respect with greater possibility of overcoming obstacles (e.g. bridges);
- reduction of impact on traffic in case of adverse weather conditions (snow, ice).

A classification can be made based on the methods of energy withdrawal and transfer:

1. Pick up from the traditional contact line (existing only on part of the track) through a traditional pantograph, placed on the roof of the tram;
2. Withdrawal from a third track (third rail); in this case, the pantograph becomes a skate placed on the bottom of the tram and applied to its bogies, so-called conduction system;
3. Withdrawal by electromagnetic induction between metal loops, sunk into prefabricated panels placed on the ground, with a skate present on the vehicle, so-called induction system.

The vehicle can be equipped with on-board supercapacitors and batteries.

11.2.2 Electric Buses

For public transport even on the street, electrification seems the obvious alternative to keep pace with urban growth and to care for the city environment at the same time, using full electric or electric hybrid buses (Lin et al. 2019). With the right charging technology, the advantages of electrified buses can be utilised: less energy consumption in comparison with buses with internal combustion engines, use of renewable energy, less noise, lower particle emissions, less CO_2, lower life cycle costs and reliable service.

In recent years, all the major bus manufacturers, in addition to some newcomers, have developed and tested full electric buses, with some interesting innovations such as the Alstom Aptis, in some respects a cross between a bus and a tram.

Despite the increasing supply and capacity of batteries, up to 350 kWh, the autonomy in operation barely reaches that average of 250 km necessary to guarantee the entire daily shift of the single bus. Thus, the need has arisen to integrate the network with recharging infrastructures along the route (see Fig. 11.3). The two main solutions currently used are:

- *Opportunity charge*: high-power charging systems at the terminus with a pantograph system.
- *Electric roads*: moving charging systems inserted in the road infrastructure with catenary, conduction or induction technologies.

Both of these solutions significantly reduce the use of on-board batteries and are therefore convenient where one must manage large fleets of several hundred or thousands of vehicles.

Different companies are at the forefront of this technology to help operators to find a solution tailored to their individual challenges. Table 11.4 reports different possible solutions applied to electric buses.

Fig. 11.3 Different modeS to recharge the electric buses. **a** Plug-in charging and **b** pantograph charging (ABB 2019)

Table 11.4 Different solutions applied to E-buses

Off-board top-down pantograph	On-board bottom-up pantograph	Charging via connector
The buses need to be equipped with contact rails, a WiFi antenna and the necessary control and switchgear units	This solution is a fast-charging system mainly used in cities with the existing DC networks, such as for tramways. The most apparent feature of this charging solution is the bottom-up pantograph mounted on the bus roof	This solution is designed for fast charging of electric vehicles (EV). Plug-in DC chargers are available as single or twin chargers the last-mentioned with the advantage of reduced infrastructure expenditure. Based on proven technology, DC chargers are highly reliable and user-friendly
Opportunity charging at terminal station via 4-pole off-board pantograph	Opportunity charging at terminal station through 2-pole catenary	Fast and efficient charging solution from 30 to 150 kW
Fully automated charging process with charging power levels of 150/300/450 kW/600 kW	750 DC power input	Safe, robust, durable, stable and user-friendly
Charging power: 110/300/450 kW/600 kW	60 kW or 120 kW of charging power for minimising the required charging time	Charge any CCS or GB/T-compatible vehicle
Grid connection: 400 V to 20 kV AC, 50/60 Hz	Roof-mounted pantograph and DC/DC inverter	Battery charging status, power consumption and autonomy are displayed
Wireless communication via Wi-Fi IEEE 802.11a based on ISO 11118		Charging cycle finishes automatically or can be interrupted by user command
Remote access, service and control		Remote access, service and control

11.3 Individual Mobility

Until a few years ago, individual mobility coincided with private transport, bicycles, scooters and cars, with the only exception being the public service provided by taxis.

The digital revolution and sharing are changing habits. Certainly, the main individual means remain the private one, but one is witnessing dynamics of growing supply and demand in the various sharing sectors, with giants such as UBER and BlaBlaCar[1] or Share Now[2] driving the sector now populated by many providers.

---

[1]BlaBlaCar is a Web car-pooling platform that operates in 22 countries. With 80 million users, it is the most used in the world (https://www.blablacar.com).

[2]Share Now is a car-sharing provider of the automobile companies BMW and Daimler and thus one of five mobility service providers that emerged from the joint venture between the two companies

Since these are individual 2- or 4-wheel vehicles, the various electrical technologies available and the consequent necessary infrastructural developments will be analysed in the following paragraphs.

11.3.1 Cars

All the large car manufacturers have gone so far as to create electric models, at the moment giving the impression more of the following regulatory obligations than the real will to turn towards green mobility. Figure 11.4 shows some models of electric cars currently on the market.

The EU, Japan, several US states and several regions of China are forcing manufacturers, via legislation, to rapidly bring emissions below 100 g/km of CO_2. To avoid heavy penalties, all manufacturers now have several EV models in their catalogue. However, the marketing and price strategy still rewards those cars with internal combustion or hybrid engines. According to the writers, this depends on the absolute lack of batteries, but there is also the difficulty of battery capacity. At the global level, battery production can currently cope with the maximum of 3–4% of electric vehicle production, equivalent to 2–3 million vehicles per year.

Certainly, cities are the main beneficiaries of the transition to electric cars, both in terms of CO_2 and local pollution. To stimulate demand and supply, cities can introduce congestion charges based on the level of vehicle emissions, but at the

Fig. 11.4 Radar for the development of the electric vehicle (Ancitel Energia e Ambiente 2016)

---

founded in 2019. With Share Now, the previously competing providers car2go and DriveNow form a joint car-sharing service.

same time, they must invest in public and private charging infrastructure. A more systematic and broader use of fair and efficient tolls based on the "user/polluter pays" principle will direct users towards more sustainable transport choices.

Battery

The coming of electric mobility today is intimately linked to two strategic factors.

The first is that of the infrastructure of the territory through sufficiently widespread recharging points. The second is that of the charge capacity of the batteries and the rapidity of charging, to which an ever-increasing autonomy is required to allow the electric vehicle to equate, in a not too far distant future, to the conditions of use of a traditional endothermic engine vehicle also on long-distance routes. The batteries that today offer the greatest prospects in this regard are those operating on the basis of lithium chemistry, which have been used for many years in the power supply of consumer electronics (PCs, smartphones, small appliances) (Report Linker 2019).

The very strong diffusion that this technology has encountered in energy storage, replacing the previous chemical types, is essentially due to two fundamental characteristics. The first is the charge density, that is, the amount of electric charge which, for the same useful surface area for the accumulation, is able to absorb, which, in lithium accumulators, is significantly higher than that of all other types of batteries. The second, instead, is represented by the significantly lower weight per unit volume of these batteries, when compared to the other types existing on the market.

These two features make lithium accumulators particularly suitable for the consumer electronics sector, where many devices (in particular PCs, tablets and smartphones) require ever more energy (due to the many functions they perform) in less space, in the need for the use of batteries that do not make the equipment too heavy. For similar reasons, lithium batteries are now the only chemical type used in pure electric vehicles, since the energy storage capacity and lightweight are the two essential elements to make this form of sustainable mobility feasible.

When an electric vehicle battery has exhausted its life cycle and no longer guarantees the same original performance, it must be replaced. These batteries, however, when disposed of from electric vehicles, may still have a second useful life in their reuse as energy storage batteries for the most diverse uses, as they still retain a charge capacity equal to 70% of the original one. The theme is currently very much in the limelight. The interest, and new business activity in the reuse of these batteries for energy storage are growing. Additionally, car manufacturers are also strongly interested in the "second life" of the batteries used in their cars, since the lengthening of their useful life and the creation of a secondary business can have a positive impact on battery management costs, thereby facilitating the success of the electric market. The issue of recycling lithium batteries is thus "a current event", as is the large spread of this type of batteries on the market. For example, there is a British start-up company, operative since 2015, which gathers in these used lithium-ion batteries to create battery packs as backup power for off-grid lighting in fragile economic areas.

The *handling as waste* of these batteries does not present, in principle, particular critical aspects, although their entry on a relatively recent market, that has not yet generated large waste streams until a few years ago, has not even justified the study and consolidation of technologies. For this reason, research on the best treatment and recycling processes is still open, presenting hypotheses and sometimes very different approaches.

One of the most important aspects to be faced is represented by the first phase of grinding up these accumulators and the need to guarantee this activity in absolute safety, to avoid explosions or fires. The subsequent phases must then address the issue of recovering materials and metals, from the simplest, i.e. casings and electrical circuits, to those that are less simple such as the chemically active part of the accumulator, also called "black mass". The real challenge is to be able to define a treatment and recycling process whose costs can be sustained by the sale of secondary raw materials recoverable from the process itself.

Returning to cars, even in the case of city cars, manufacturers are focusing on medium/large batteries, currently 40/50 kWh. These capacities allow a city car with the average typical daily journey of 30 km, a need for weekly, and not daily, charging. This therefore positively impacts the city's charging columns, the use of which is subsequently optimised.

From the point of view of parking infrastructure there are no restrictions, such as those already in force for gas-driven vehicles, for the use of electric vehicles even if the extinguishing of a fire of a battery requires specific technical knowledge on the part of the fire brigade.

Hydrogen

With a hydrogen car, one means a vehicle that converts the chemical energy of this element into mechanical energy. Hydrogen can burn in an internal combustion engine, and in this case, one refers to a hydrogen internal combustion engine vehicle (HICEV), or alternatively a reaction can be caused with oxygen in a fuel cell, thus producing electricity. In this case, fuel cell electric vehicle (FCEV) is the term employed.

The main advantage of the hydrogen car is precisely that of not producing any greenhouse gases since only absolutely non-polluting water vapour is emitted from the car's exhaust pipe. Analysing the peculiarities of hydrogen cars, both the advantages and, inevitably, the disadvantages associated with this type of car must be evaluated.

The first advantage is that these cars allow harmful emissions to be reduced by 100%, since only water vapour is released from the exhaust pipe. Moreover, thanks to the electric energy coming from the fuel cells, the batteries of the electric motor can be recharged more quickly than a totally electric car. Lastly, the time needed to refuel is much lower, from 3 to 5 min.

The negative aspects of hydrogen cars include, first of all, the purchase price, currently not exactly in the popular car bracket, secondly, the almost total absence worldwide of dedicated filling stations and, last but not least, the considerable weight of these cars due to the cylinders in which the hydrogen is stored. Once again, the question is strategic. If battery solutions are chosen, a network of charging stations has to be developed, whereas if hydrogen is favoured, it is a network of hydrogen distributors which must be implemented.

At the moment, only a few countries, amongst which the most important is Japan, are investing in production and distribution of hydrogen.

Autonomous Drive

The development of autonomous driving mainly for cars and light commercial vehicles should be considered in urban planning. Investments in this technology are truly enormous, both on the part of traditional manufacturers and on the part of the digital giants and the new economy, such as Google, Apple or Amazon. All the most recent estimates predict that as early as 2040, autonomous vehicles will be between 40 and 70% of the total (Daviex 2018; Union of Concerned Scientists 2018).

From the infrastructural point of view, there are no specific requirements, except that the cities must be dotted with well-maintained *signposts and markers*.

From the point of view of traffic flows, several studies show that autonomous driving will lead to a significant reduction in the number of cars in movement on the roads.

A recent MIT study in the Manhattan area shows that once all vehicles have passed over to autonomous driving, the number of cars on the road can be reduced by 50% (AGELAB–MIT 2019). Obviously, this data must be considered in long-term urban planning.

11.3.2 Micro-mobility

Micro-mobility refers to the set of vehicles and modes used for short journeys and for the transport of a maximum of one or two people (Heineke et al. 2019; Zarif et al. 2019). City planners know that they cannot always simply add more routes and stops into the conventional public transport system, because that would cost too much or result in other undesirable outcomes. At the same time, the first/last mile problem discourages many commuters from using mass transit at all. An integrated transportation system which includes public transportation and such flexible alternatives as micro-mobility vehicles can help to fill the gap (Dutta and Sutarwala 2019; Chang et al. 2019).

This type of mobility includes all those vehicles such as electric scooters, segways, monowheels and electric bikes (see Fig. 11.5). Additionally, this also means all those services of rental of electric bikes or scooters which are becoming more widespread in the major cities.

These services offer people the possibility of renting a vehicle to move around, reserving and paying for it via a related mobile app. The *typical provisional municipal regulations* foresee an "experimental operation" in the cities of this kind of means of transport. It clearly refers to a type of intermodal mobility and therefore is intended to interface with public transport but also requires of mayors to create all the opportunities to implement in their cities those infrastructures which are necessary, being first and foremost, dedicated lanes. In fact, electric scooters, segways, monowheels

Fig. 11.5 Different symbols of micro-mobility

and hoverboards will require new updates of the Highway Code: here, one is talking about the introduction of a new category dedicated to "eco-sustainable vehicles".

Although it is true that these vehicles can be driven without a driving licence, nonetheless, they will certainly have to abide by the rules. Undoubtedly, the compulsory use of the helmet will be introduced, and a specific insurance will probably also need to be provided. A whole series of very specific limits will arise: new-generation electric vehicles, even the smallest ones, must have reflectors and position lights and cannot in any way exceed 30 km/h.

These small and practical battery-powered vehicles experienced a real boom between 2018 and 2019, and mostly, the youngest sections of the population have contributed to the spread of this phenomenon. The main reasons that led to the success of these "alternative" means of transport are that they:

- allow short trips to be made quickly and practically (home–school or home–work);
- provide the possibility to avoid both car traffic and crowded public transport, thereby saving time;
- are ecological and do not pollute;
- are a very economical alternative when compared to using a car or a motor-driven vehicle;
- can even be fun to drive (see Table 11.5).

It is very interesting to note that the average speed of travel with micro-mobility over a distance of up to 2–3 km is often even greater than that for an underground railway trip, all access times being avoided. Micro-mobility can therefore be integrated in a significant manner with collective transport.

All the micro-mobility means are also connected, and it is therefore possible to reconstruct from their displacements the origin/destination matrices for the first and last mile, which in turn can suggest changes to the collective transport system.

It is therefore suggested that local authorities authorise micro-mobility providers to request constant data transmission. If it transpires that there are major source/destination relationships in the volumes and constancy thereof during the hours of the day, it could be the case to implement a collective service.

As mentioned in the introduction, micro-mobility, in particular electric vehicles such as scooters, segways or assisted and non-assisted bicycles, is opening a relevant regulatory issue that can be divided into two macro-categories:

Table 11.5 Benefits and problems of micro-mobility

Benefits and problem	Description
First/last mile access to transit	Small. Lightweight vehicles increase access to transit stations
Satisfy short trip making	Micro-mobility can satisfy trips under 5 miles with scooters, bicycles and e-bikes. In addition, e-mobility works in hilly areas where biking can be a challenge
Low-impact transportation	These no or low power vehicles have a smaller carbon footprint and make more effect use of parking and infrastructure as compared to an automobile trip
Expansion of mobility to traditionally underserved areas	Early data shows an expansion of options to mobility deserts (not well served by transit and/or with low auto ownership)
New avenues for transportation data	Shared use mobility and opt-in data collection from riders through mobile apps provide information on the performance of the mobility system, in particular when trips are carried out with multiple modes

- road safety;
- damage management and related insurance coverage.

As far as road safety is concerned, very clearly these vehicles are particularly suited to the cycle paths, thus separated from car and larger vehicle traffic. One would imagine that the overwhelming development of micro-mobility will finally lead to a greater development of urban cycle paths even in countries where the culture of the bicycle use has never really taken off. On this point, it is hence advisable that local governments/councils continue to invest in their cycle path networks, bearing in mind that they will serve as an infrastructure for the entire micro-mobility system.

With regards to damage management, understood as damage resulting from accidents between people moving by any means of transport or on foot, the authors believe that the only solution, for which indeed, among other things, some insurance companies already envisage coverage, is to move from insurance of the means of transport to insurance of the person by the mere fact that he or she moves, therefore with any means of transport.

Obviously, dependent upon the profile of each one, the insurance will assess the level of risk that will depend on whether they have a license or not, the type of driving license and therefore the vehicles which they are permitted to drive, the fact that they use *the bicycle, the scooter or the car*, whether they fly or go by train. In short, how their mobility is assured. This seems to be the only way to avoid the grotesque insurance of an electric scooter or bicycle.

11.3.3 Charging Infrastructure

From an infrastructural viewpoint, in order to correctly outline how electric mobility can develop, it is necessary to put oneself in the shoes of the user and analyse the different usage scenarios, from which different requirements in terms of necessary infrastructures are established (Longo et al. 2018a; Engel et al. 2018; Lunz and Sauer 2015). Three different scenarios for the development of a national charging network should be considered:

- **Scenario 1—Long stay**. This scenario identifies the places where cars stop for long periods such as, for example, parking spaces at the workplace, individual parking lots in garages, condominium parking lots and parking lots for company fleets occupied for overnight stops or throughout the day. The charging systems suitable for this service are those with low power in AC (slow type charging up to 6 kW) for a prolonged period of time (even 4–10 h). This type of recharge meets the "primary" needs of the user of electric vehicles and, with some exceptions, guarantees the daily journey of the vehicle (if it is less than 150 km/day and meets 95% of the needs of European daily journeys); individual recharge wall boxes in the garages of a block of flats are usually rated at 3 kW which, in some countries, involves increasing the householder's base supply from 3 to 4 kW.
- **Scenario 2—Short stopover**. This scenario describes the typical stop in shopping centres, cinemas and restaurants where parking typically varies between 30 min and 2 h. This type of recharge meets the need to supply the vehicle with additional mileage during the day to give the user more flexibility. This category includes accelerated charging systems such as roadside columns. These charging infrastructures involve higher costs than the battery charger on the vehicle but extend the range of action of the EV.
- **Scenario 3—Stop for recharging** or stopover less than 30 min for the sole purpose of recharge. The "stop for recharging" scenario does not respond to the needs of the home–workplace, but to those who must quickly recharge (a few tens of minutes) the EV (taxis, delivery vans, buses, etc.). It is also the scenario that allows the autonomy of exceeding 450 km to be significantly increased. Fast charging is also useful in cities because it gives reassurance to the driver and guarantees flexibility in case of unforeseen needs.

These scenarios are not intended to be limited to vehicles for passenger transport but can also be applied to freight vehicles (vans, etc.) to allow the development of the so-called "last mile". Electric vehicle charging technologies are divided into: conductive and inductive. The conductive recharge allows the charging of the battery of the electric vehicle through the connection to the power supply grid in alternating current (AC) of the charger on boarding the vehicle.

An alternative method for charging the vehicle is to use an external battery charger that supplies direct current (DC) to the vehicle. Both charging methods are characterised by a physical connection (through the power cable) between the vehicle and the charging infrastructure.

Within the conductive recharge technology, a fairly complex articulation is then to be found between slow recharging (single-phase or three-phase) and fast recharging (in AC or DC), strictly related to the technological evolutions that are characterising this sector. With inductive charging, the energy transfer to the battery takes place through the electromagnetic coupling between two coils: one mounted under the vehicle and the other supported or even buried in the vehicle parking area (Arancibia and Collins 2018; Foiadelli et al. 2017).

This technology is divided into stationary, which can also be activated automatically at the time of the stop, or dynamic, an application perhaps a little more futuristic which consists of charging the vehicle while driving (i.e. smart road). The first two types of recharging are accompanied by the battery swap technique, i.e. battery replacement, which can be considered comparable to a recharge technology. In Fig. 11.6, the charging technologies currently used are represented in a schematic way, in relation to the time taken for recharging.

Cities and towns tend to favour the first two recharge scenarios, that of low power in residential areas and workplaces and that of medium power for fast recharges. Ultra-fast recharges, for very long-distance needs, can be met outside the city, particularly in motorway service areas. Exceptions are recharge islands dedicated to electric taxis or electric commercial vehicles which need to minimise downtime from service on the road.

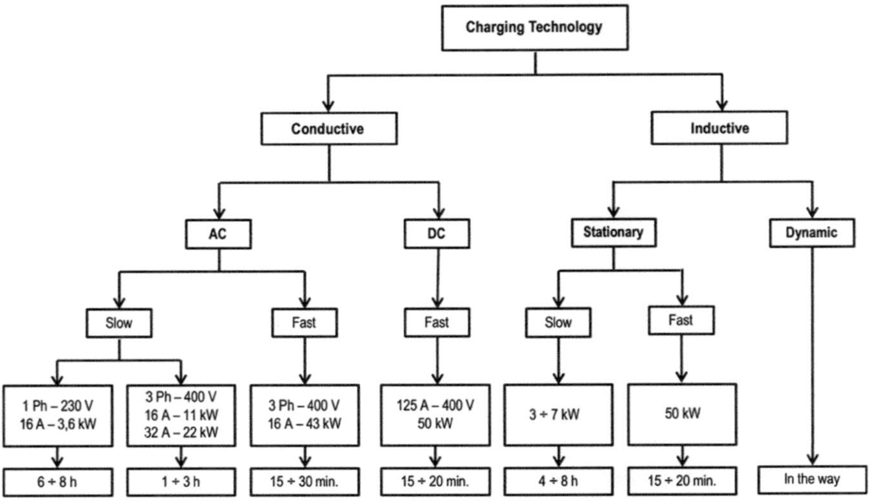

Fig. 11.6 Schematisation of the different charging technologies (Ancitel Energia e Ambiente 2016)

11.3.4 Charging System Requirements

Given that the infrastructure locations must privilege the intermodality criterion, the municipalities must ensure that the public access charging systems present on their territory adopt payment systems which are:

- functional;
- simple to use;
- readily available;
- updated and based on the latest technological solutions available on the market, with particular reference to Web/smartphone applications and digital payment systems;
- "open", or that allow the provision of the service to occasional customers through contextual payment as defined by Directive 2014/94/EU, in order to allow the use by occasional users (e.g. using common credit cards) without the need to enter into contracts.

It is also important to ensure that the charging stations are not illegally occupied by internal combustion engine vehicles or electric vehicles that are not under charge (Longo et al. 2017a, b, 2018a), having them removed when infractions occur.

11.3.5 Criteria for Planning the Charging Infrastructure Available to the Public

The charging infrastructure accessible to the public must be characterised by "high" or "high power" charging systems (greater than 22 kW), in order to guarantee the best coverage of the multiple needs of users and, in particular, a reduction in waiting times for vehicle being recharged, maximum 10–15 min. However, the charging infrastructure for private usage must be characterised by an adequate combination of "standard" or "normal power" (7.4–22 kW), or indeed household (3 kW), since, in this case, the recharge will be done in the night hours.

Standard recharge
It is preferable to install the "standard" "accelerated" charging systems in the following locations:

- *ALONG THE ROADSIDE*: with the main objective of providing the possibility of charging from the network during the overnight stop to those users of rechargeable vehicles who are not equipped with private parking/garage with a wall-box recharge.
- *PARK AND RIDE*: the main objective is to provide the possibility of charging users during daytime parking who reach the park-and-ride hubs of large urban centres by private car. They enable home-to-work commuting by electric vehicles even to those who do not have company parking lots equipped for recharging.

- *POINTS OF INTEREST*: the main objective is to provide the possibility to recharge during the stopover (average duration 1–2 h) at particular points of interest characterised by a strong influx of people.

High recharge
It is necessary to favour the installation of charging systems accessible to the general public in the following places:

- *SERVICE AREAS*: with the main objective of providing the ability to quickly recharge the vehicle on a route to be completed which is greater than the vehicle's maximum range. Wherever, due to problems of competence, it is not possible to intervene directly on the motorway, recharging service areas located near the entrance/exit toll gates should be the preference. It is also believed that the installation of recharging columns in such locations would ensure excellent capillarity but, above all, would lead the consumer to refuel in the same places previously used, presumably when he/she made trips with an internal combustion engine car. This system would improve the availability of charging stations without the need to have a specific App or interactive maps, also facilitating tourist activities, guaranteeing those who come from other towns and cities the certainty of being able to recharge their electric vehicle;
- *RAILWAY STATIONS, AIRPORTS AND KNOTS OF LOCAL AND EXTRA-URBAN PUBLIC TRANSPORT*: the main objective is to provide the possibility of quickly recharging the vehicle with access modalities such as to be able to guarantee the mixed usage of the recharging installations (by taxis, fleets in sharing, private vehicles);
- *GOODS LOADING/UNLOADING AREAS*: here, the main objective is to provide the possibility to quickly recharge from the network those rechargeable vehicles destined for the delivery of goods in urban areas, using parking bays compatible with commercial vehicles with loading up to 3.5 tons.

11.3.6 Analysis of the Actual State of Mobility

In order to plan the development of electric mobility as close as possible to the needs of anyone in possession of vehicles powered by electricity (be they residents, tourists, traders, taxi drivers or local public transport operators or other public services), a careful analysis of the existing scenario is advisable. An evaluation methodology is proposed hereunder.

First of all, it is necessary to carry out a georeferenced mapping of all the electric columns present on the municipal territory, distinguishing them on the basis of:

- locations;
- type of vehicles for which the infrastructures are destined (i.e. cars, quadricycles …);

- type of socket supplied in each column;
- distinction between "standard" or "normal power" charging systems and "high" or "high power" charging systems (CSs);
- if the CSs are intended for the public or if they are totally or partially intended for specific subjects (i.e. car-sharing operators, buses, taxis) and to what extent.

In order to allow the adequate development of electric mobility, in line with the objectives set by the European Union at 2030, which in urban transport plans to halve the use of cars "powered by traditional fuels" by that year and eliminate it by 2050 altogether, each municipality should estimate the growth volumes of the quantity of electric vehicles in circulation (private, public fleets, companies and sharing services) and calculate a recharging infrastructure requirement, including a green zone in a strategic area, well-connected with other types of mobility.

Of these top-up points, it will be noted that the majority, roughly 85% will be installed on private land (business, shopping centres, parking lots, garages and private garages/boxes) and the remaining 15% on public sector land. It is suggested to consider the following main indicators: population density (population and active population); density of economic activities and business demography; density of demand for private mobility; density of services of public interest.

If possible, and compatible with the size of the municipal territory and its morphological characteristics, it is suggested to aggregate the information, represented on a cartographic basis, by geographical area.

The process of analysis must first consider the resident population, both total and active: this first step, using the data related to the surface of each territorial area, allows the definition of the areas characterised by the highest population density. A similar criterion must be followed for production activities and their location. Subsequently, an analysis must be made of the displacements (origin -> destination) in order to determine the areas of greatest affluence. Thereafter, the location of public utility services has to be considered, in particular schools and universities, hospitals and other health facilities, sports facilities, theatres and auditoriums, museums, community centres, pharmacies, places of worship, post offices, public transport stations and stops. Finally, by assigning the same weight to each indicator, a summary table is drawn up indicating, for each selected areas, the average of the values corresponding to the classes defined above, identified as overall indicators. This synthesis, elaborated by comparing each theme, makes it possible to identify the areas for which it is estimated that there may be greater demand for infrastructure.

Considering the recharging network within the municipality of reference, it is considered appropriate to define the percentages, divided according to reference criteria normally employed (i.e. districts), to be used for the location of the new recharging infrastructures, in order to guarantee the service on the territory where, today, it is less present.

Finally, relying on the above analyses and considering the total needs of recharging points previously calculated, it is possible to proceed by making hypotheses of how to distribute them in the medium-long period among the various municipal areas.

In order to guarantee an effective use of the public top-up system by users, the infrastructure hypothesised must include a mandatory mix of AC recharging points and DC recharging points.

11.4 Conclusions

It is now fully recognised and a well-established viewpoint that terrestrial mobility can be 100% decarbonised by 2050. In metropolitan cities, the good starting point is the historical presence of many electric means of transport: the underground, trams and trolleybuses. The extension of the use of these vehicles, together with the widespread electrification of buses, commercial vehicles and cars is now a matter of time but no longer of technology.

References

ABB (2019) Pantograph down for electric buses. Available at: https://new.abb.com/ev-charging/products/pantograph-down. Accessed 1 Dec 2019

AGELAB–MIT (2019) The Advanced Vehicle Technology consortium (AVT). Available to the site: https://agelab.mit.edu/avt. Accessed 1 Dec 2019

Ajanovic A, Haas R (2019) The impact of energy policies in scenarios on GHG emission reduction in passenger car mobility in the EU-15. Renew Sustain Energy Rev 68:2017

Alstom (2015) Whitepaper from mainline signaling to digital automation. Available to the site: https://www.alstom.com/whitepaper-mainline-signalling-digital-automation. Accessed 1 Dec 2019

Ancitel Energia e Ambiente S.p.A. (2016) Guida per la mobilità elettrica: la rete infrastrutturale comunale. Available at: https://www.ancitelea.it/public/assets/uploads/flies/files/LINEEGUIDA%20MOBILITA%20I.pdf. Accessed 1 Dec 2019

Arancibia A, Collins R (2018) Electric vehicle charging infrastructure 2019–2029: forecast, technologies and players. Available at: https://www.idtechex.com/en/research-report/electric-vehicle-charging-infrastructure-2019-2029-forecast-technologies-and-players/633. Accessed 1 Dec 2019

Chang AYJ, Miranda-Moreno L, Clewlow R, Sun L (2019) Trend or fad? Deciphering the enablers of micromobility in the U.S. Available at: https://www.sae.org/binaries/content/assets/cm/content/topics/micromobility/sae-micromobility-trend-or-fad-report.pdf. Accessed 1 Dec 2019

Civitas (2015) Smart choices for cities. Making urban freight logistics more sustainable. Available at: https://civitas.eu/sites/default/files/civ_pol-an5_urban_web.pdf. accessed 1 Dec 2019

Crozet Y, Santos G, Coldefy J (2019) Shared mobility & MaaS—the regulatory challenges of urban mobility. Report, Centre on Regulation in Europe, Sept 2019

Daviex A (2018) The WIRED guide to self-driving cars. Available at: https://www.wired.com/story/guide-self-driving-cars/. Accessed 1 Dec 2019

Dutta D, Sutarwala A (2019) Micromobility: what does it mean for the future of transportation. Available at: https://dataconomy.com/2019/08/micromobility-what-does-it-mean-for-the-future-of-transportation/. Accessed 1 Dec 2019

Engel H, Hensley R, Knupfer S, Sahdev S (2018) Charging ahead: electric-vehicle infrastructure demand. Available at: https://www.mckinsey.com/industries/automotive-and-assembly/our-insights/charging-ahead-electric-vehicle-infrastructure-demand. Accessed 1 Dec 2019

European Commission—EC (2015) Statistical pocketbook 2015. Available to the site: https://ec.europa.eu/transport/facts-fundings/statistics/pocketbook-2015_en. Accessed 1 Dec 2019

Foiadelli F, Longo M, Delfino F, Bracco S, Spina D, Dhaene T (2017) Electric vehicle use in public fleets: the case of the Genoa University. In: International conference on energy and environment (CIEM 2017), Bucharest, Romania, 19–20 Oct 2017

Heineke K, Kloss B, Scurtu D, Weing F (2019) Will the micromobility market boom or bust? With billions already invested, here's an assessment of its potential. Available at: https://www.mckinsey.com/industries/automotive-and-assembly/our-insights/micromobilitys-15000-mile-checkup. Accessed 1 Dec 2019

Lin Y, Zhang K, Shen ZJM, Miao L (2019) Charging network planning for electric bus cities: a case study of Shenzhen, China. Sustainability 11(17):2019. https://doi.org/10.3390/su11174713

Longo M, Zaninelli D, Lutz NM, Daniel L, Pruckner M (2017a) Towards an impact study of electric vehicles on the italian electric power system using simulation techniques. In: RSTI 2017, Modena, 11–13 Sept 2017

Longo M, Zaninelli D, Lutz NM, Daniel L (2017b) Towards an impact study of electric vehicles on the italian electric power system using simulation techniques. In: 5th international conference on models and technologies for intelligent transportation systems—IEEE MT-ITS 2017, Naples, 26–28 June 2017

Longo M, Yaici W, Foiadelli W (2018a) Electric vehicles integrated with renewable energy sources for sustainable mobility. New trends in electrical vehicle powertrains—reliability and resilience. InTech

Longo M, Maffezzoni P, Lu X, Daniel L, Lutz N (2018b) An optimization algorithm for charging stations actually installed in North Italy. In: 18th IEEE international conference on environment and electrical engineering (EEEIC 2018), Palermo, Italia, 12–15 June 2018

Lunz B, Sauer DU (2015) Electric road vehicle battery charging systems and infrastructures. Advances in battery technologies. Available at: https://www.sciencedirect.com/topics/engineering/charging-infrastructure. Accessed 1 Dec 2019

McKinsey & Company (2018) Smart cities: digital solutions for a more livable future. Available at: https://www.mckinsey.com/~/media/mckinsey/industries/capital%20projects%20and%20infrastructure/our%20insights/smart%20cities%20digital%20soltions%20for%20a%20more%20livable%20future/mgi-smart-cities-full-report.ashx. Accessed 1 Dec 2019

Report Linker (2019) view trends, analysis and statistics on battery. Available at: https://www.reportlinker.com/market-report/Battery/7614/Battery?utm_source=adwords4&utm_medium=cpc&utm_campaign=Energy_And_Environment&utm_adgroup=Battery_Reports&gclid = Cj0KCQiAw4jvBRCJARIsAHYewPOdMIfSKNsE16lPI_-_v5Q-_fATbV6E4-_dC6_MHssmRBJChO7Y_ksaAqSuEALw_wcB. Accessed 1 Dec 2919

Union of Concerned Scientists (2018) Self-driving cars explained: how do self-driving cars work—and what do they mean for the future? Available at: https://www.ucsusa.org/resources/self-driving-cars-101. Accessed 1 Dec 2019

Zarif R, Pankratz DM, Belman B (2019) Small is beautiful: making micromobility work for citizens, cities, and service providers. Available at: https://www2.deloitte.com/us/en/insights/focus/future-of-mobility/micro-mobility-is-the-future-of-urban-transportation.html. Accessed 1 Dec 2019

Zawieska J, Pieriegud J (2018) Smart city as a tool for sustainable mobility and transport decarbonization. Transp Policy 63:39–50. https://doi.org/10.1016/j.tranpol.2017.11.004

Chapter 12
New Behaviours and Digitalisation for Sustainable Mobility, Mobility as a Service (MaaS)

Renato Mazzoncini, Claudio Somaschini and Michela Longo

Abstract This chapter will deal with changes in the behaviour of city dwellers and digital technologies which can facilitate such behaviour. Mass transportation must certainly remain the backbone of local public transport, both in terms of travel within cities and for that towards cities. However, this chapter will focus on the transformations which are taking place and which will have to be favoured in order to reduce and optimise private transport, a real scourge in large towns and cities. The final part of the chapter covers the concept of *mobility as a service*, being the evolution of the concept of ownership towards that of service.

12.1 Introduction

The current transport landscape is undergoing radical changes: the emergence of a multitude of new applications, business models, and specialisations, in addition to the entry of "new players", bears witness to this change.

Innovation in the transport industry is expected from the integration of new technologies and the development of new concepts of mobility. Smart mobility is the vision of a revolutionised individual and collective mobility, framed as a personalised "service" available "on demand", with individuals having instant access to a seamless system of clean, green, efficient, and flexible transport to meet all of their needs, as a result of emerging technologies such as automated vehicles, sharing mobility, and the Internet of things. Smart mobility is something more than the intelligent transport system (ITS) concept, the latter consisting of the application

R. Mazzoncini (✉) · C. Somaschini
Department of Mechanical Engineering, Politecnico di Milano, via La Masa 1, 20156 Milan, Italy
e-mail: renato.mazzoncini@polimi.it

C. Somaschini
e-mail: claudio.somaschini@polimi.it

M. Longo
Department of Energy, Politecnico di Milano, Via La Masa, 34, 20156 Milan, Italy
e-mail: michela.longo@polimi.it

G. Dall'O' (ed.), *Green Planning for Cities and Communities*,
Research for Development, https://doi.org/10.1007/978-3-030-41072-8_12

of high technology and computer power to increase the safety and efficiency of the transportation system.

In the previous chapter, the state of the art of technologies which will contribute to the development of sustainable mobility in cities was illustrated. This chapter will deal with changes in the behaviour of city dwellers and digital technologies which can facilitate such behaviour.

Mass transportation must certainly remain the backbone of local public transport, both in terms of travel within cities and for that towards cities. However, this chapter will focus on the transformations which are taking place and which will have to be favoured in order to reduce and optimise private transport, a real scourge in large towns and cities. To do this, a common thread, connecting some obvious macrotrends will be followed.

This chapter will start from the *sharing mobility* that is part of the broader field of the *sharing economy* of which Airbnb[1] is perhaps today the maximum representative. It will pass through the *Internet of Vehicles*, as the technological hardware enabling the whole technological revolution, to arrive at the *autonomous vehicles*, which in the name itself embodies the true revolution: the separation of the vehicle, which becomes autonomous, from its owner/driver. The final part of the chapter covers the concept of *mobility as service* (*MaaS*), being the evolution of the concept of ownership towards that of service.

12.2 From Individual to Shared Mobility

The great availability of cars with ever better performances, of ever-expanding road infrastructures, and of a very widespread supply network as well as the increase in the average revenue—just to mention some of the most important factors—has led, from the 1960s to today, ever more people to own and use individual and personal vehicles, more particularly cars.

The increase in the population which lives and works in cities, combined with the progressive dispersion of activities on the surrounding territories, the so-called *urban sprawl*, has further increased the need for this type of individual mobility. Today, in almost all the developed countries of the world, the social and territorial organisation is based on the widespread availability and regular use of cars (or other personal motor vehicles such as motorcycles and scooters but also trucks for the movement of goods). Today, road transport covers around 90% of land transport (Eurostat 2019).

The main features that have made the use of the private car becoming increasingly widespread are (ONSB 2019):

- **Accessibility**—The road network is by far the most widespread transport infrastructure in the world.

[1]Airbnb is an online portal that connects people looking for accommodation or a room for short periods, with people who have extra space to rent, generally private apartments (https://www.airbnb.com).

- **Continuity**—A private vehicle allows one to reach points that are isolated and even thousands of kilometres away without changing means of transport. It is unbeatable in the so-called door-to-door connections.
- **Availability**—Moving independently, cars allow one to move and stop in total freedom without having to buy tickets or respect timetables and connections.
- **Versatility**—The car is the most versatile means of transport, it can be used for short or long journeys, and it can be used for necessity, work, or leisure, on advanced high-capacity road infrastructures or on simple connecting rural lanes.

On the other hand, in recent years, one has been witnessing a change of trend: in heavily urbanised areas—owning a private car no longer seems strictly necessary, and indeed it is becoming an issue, in favour of collective transport services or sharing mobility, which is increasingly popular (McCarthy and O'Keeffe 2017).

This transformation of habits, which is sweeping the world of mobility, derives both from local authority decisions like congestion charging (Fulton et al. 2017) in Singapore, London, or Stockholm and from the continuous spreading of *sharing economy* models such as co-working, crowdfunding, house sharing, social shopping that are revolutionising some markets which were, historically, based on individual private property. The interest in collaborative consumption is becoming increasingly widespread, especially among the new generations: today's young people are progressively less attracted to possessing a property because they are more interested in being able to use a service.

The shared use of the same mobility service is naturally not a new concept, but it is a common feature of all the forms of transport that do not need the use of a vehicle owned by the company. "Shared mobility services" include both the more traditional collective transport services such as trains, undergrounds, trams, buses, or taxis, and the new services for sharing an "individual" means of transport such as bike sharing, car sharing, carpooling, and other innovative services enabled by the use of digital platforms. In any case, all the shared mobility services require that there be an organisation which provides the mobility service and a plurality of subjects that make use of this service, sharing vehicles and journeys.

Shared mobility services can be classified into two macro-families based on their accessibility and availability:

- Services with fixed itineraries and schedules.
- On-demand services with itinerary to be established each time.

All collective transport systems, whether or not, are part of the first family (trains and subways, trams, buses, etc.). Instead, all the modern sharing mobility services fall within the second family, together with taxis and classic rental services, with or without driver. Among these, the most well known and widespread is undoubtedly car sharing but there are different forms of sharing that can be grouped into three categories (Table 12.1).

- **Vehicle sharing** (car sharing, bike sharing, scooter sharing, and other vehicle sharing) In its most basic form, car sharing is a form of car rental with some key differences: usually, there is no requirement to return a car to a rental station

Table 12.1 Different types of shared mobility

Type	Examples	Descriptions
Commercial car sharing	Car2go	Offers a "floating" fleet that can be located on an app and rented for one-way trips within defined city areas
	Enjoy	Offers a "floating" fleet that can be located on an app and rented for one-way trips within defined city areas
	Sharengo	Offers a "floating" electric fleet that can be located on an app and rented for one-way trips within defined city areas
	DriveNow	Offers a "floating" fleet that can be located on an app and rented for one-way trips within defined city areas
Peer-to-peer car sharing	auting	Matches drivers with passengers for short-time drives, when the vehicle is not in use by the owner
Ride hailing	Uber	Matches private drivers with passengers for intra-city trips on demand
	mytaxi	Matches taxi drivers with passengers for trips on demand
Ride sharing	BlaBlaCar	Matches drivers with passengers for intercity drives
	Jojob	Corporate carpooling

(vehicles that can be rented are spread out over a region), car sharing services are charged by minute of use (vs. days of use in car rental), and all the actions of reservation, pickup, and return are self-service.
Providers include commercial entities as well as private individuals who rent out their own vehicles through peer-to-peer car-sharing programmes. These services make available to consumers all the benefits of car ownership without its attendant costs, including purchase cost, insurance, maintenance, and parking. Nowadays, they are all supported by mobile apps.

- **Ride hailing** (also called ride sourcing or on-demand ride service) These are online platforms (with mobile apps) developed by transportation network companies that allow passengers to "source" or "hail" rides from a pool of drivers that use their personal vehicles.
- **Ride sharing** Classic ride sharing is simply what was previous called carpooling: two or more travellers sharing common, pre-planned trips made by private automobile. In recent years, thanks to GPS and mobile technologies, ride sharing has evolved into a real-time or dynamic ride sharing that can match drivers with riders in real time without advanced planning. This ride-matching process is conducted through mobile apps that connect drivers with passengers travelling along similar routes, in real time, at predesignated pickup locations (commonly called casual carpooling or "slugging").

In recent years, in response to this change, an increasing number of car manufacturers, in addition to leasing and rental companies, are investing in car-sharing services. The internal reorganisation is transforming the whole automotive sector.

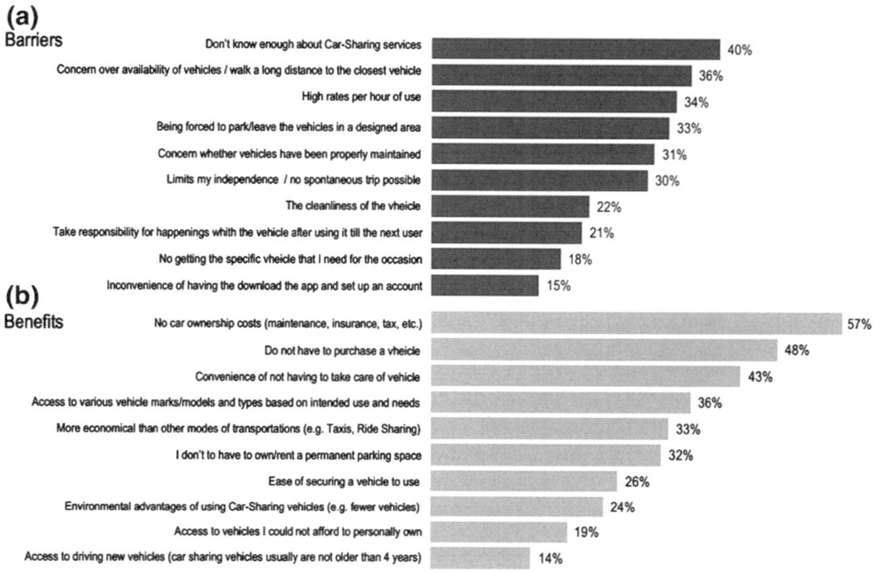

Fig. 12.1 Barriers (**a**) and benefits (**b**) of using car sharing (Yakovlev and Otto 2018)

Leading car-sharing companies such as car2go or DriveNow (recently merged into ShareNow[2]), Maven, are fully owned by car manufacturers which are changing their traditional business models where they earned money from only selling and maintaining cars: by providing car-sharing services, they also earn from operating the car itself.

Thanks also to these investments, vehicle-sharing services are increasingly diffuse despite some initial difficulties (Yakovlev and Otto 2018) owing, above all, to scepticism towards the new form of mobility (a). With the further development and spread of vehicle sharing, these obstacles will soon be overcome and progressively more people will use these services, also because the convenience of vehicle sharing in large cities is widely recognised by consumers (b). No car ownership costs (57%), no car purchase costs (48%), and the convenience of not having to take care of the vehicle (43%) are considered (by the percentages of people asked, shown in brackets) to be the three major benefits of car sharing (Fig. 12.1).

Undoubtedly, all forms of vehicle sharing are developing only in large metropolitan areas. In fact, the success of this mode of transport is achieved exclusively by making large fleets available in order to provide a truly capillary service which is competitive with the private car. However, large fleets require many customers and those numerous customers are more easily found in large urban areas. Nevertheless when examining situations such as that of Europe, where there are also many small- or

[2]Share Now is a car-sharing provider of the automobile companies BMW and Daimler. Emerged from the joint venture between the two companies, car2go and DriveNow were founded in 2019 (https://www.share-now.com).

medium-sized urban centres, it is necessary to understand how to accompany shared mobility services. A possible solution is that the sharing company which starts a service in a medium–small city immediately declares its customer break-even and that, in the growth phase towards break-even, there are public contributions to the deficit of the sharing companies, in a similar manner to what happens with local public transport.

12.3 Internet of Vehicles and Autonomous Vehicles

All the solutions that can be offered by shared mobility are and will be made possible thanks to the latest technological innovations which are rapidly and drastically changing the transport sector (Fig. 12.2). To ensure a widespread and effective service, all fleets of different vehicles (machines, scooters, bicycles, scooters, etc.) must be connected to the Internet of things to ensure both optimal management of the fleet by the service provider and access by users. The shared mobility services are in fact possible thanks to the support of mobile applications that allow access to the service, use the service, and pay for the service without any intervention by the staff of the company that manages the service.

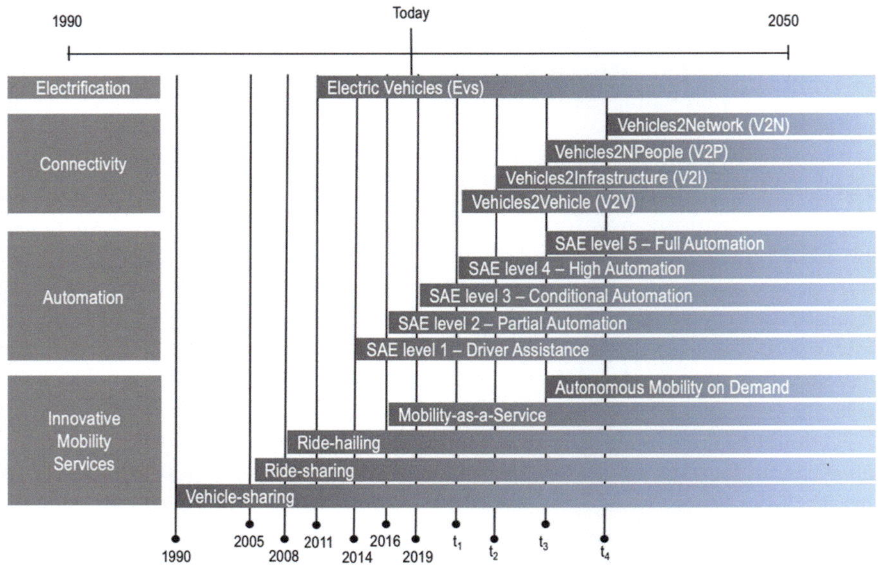

Fig. 12.2 Timeline of the maturity of vehicle technologies and mobility services (Coppola and Esztergàr-Kiss 2019)

The digitalisation of all these activities was possible thanks to the development of information and communication technologies (ICTs) which are continuously applications for better and more targeted use of vehicles and infrastructures in order to optimise the performance of the entire mobility network.

The continuous connection of the vehicles to the service management network is the first step in what is called the *Internet of Vehicles*, a subgroup of the much more famous *Internet of things*. *IoT* is a concept that represents one of the possible evolutions of the use of the Internet that is taking hold in these years: objects can communicate data about themselves and access information shared by others.

If we apply this concept to vehicles, connected vehicles will have to be not only able to exchange an increasing amount of information with a central server but also capable of communicating with all the different elements of the mobility system so as to create a much more complete connection system, usually designated *vehicle-to-everything* (V2X).

An important goal of the V2X is to allow real-time communication between a vehicle and all the entities that may affect the vehicle or vice versa. There are many levels of communication which can be grouped into:

- **Vehicle-to-driver** (V2D) systems for exchanging data between a vehicle and the driver.
- **Vehicle-to-network** (V2N) systems for exchanging data between a vehicle and a control centre.
- **Vehicle-to-infrastructure** (V2I) systems for exchanging data between a vehicle and infrastructures and supporting roadside units.
- **Vehicle-to-vehicle** (V2V) systems for exchanging data between a vehicle and surrounding vehicles. This can be done by creating the so-called vehicular ad hoc networks (VANETs) that is the spontaneous creation of a wireless network between nearby vehicles.
- **Vehicle-to-people** (V2P) systems for exchanging data between vehicle, smartphones, or other dedicated devices that support awareness for vulnerable road users such as pedestrians and cyclists.

If large cities are being considered, V2X is only possible thanks to advanced communication technologies and data management logics.

So far as communication is concerned, the speed required for V2X can only be achieved by exploiting the potential of 5G technology. The spread of this technology will introduce a new generation of high-speed, low-latency communications which will enable the implementation of V2X (Chen et al. 2017).

The management of the enormous volumes of data that will be quickly generated by these systems requires the technologies and analysis algorithms of the *big data* field. For example, traffic data from individual vehicles must be analysed in real time in order to be able to check the infrastructure for better management of the mobility network (Talebpour and Mahmassani 2016).

For example, a control room can monitor traffic lights or impose dynamic limits of optimal speeds to facilitate the flow of vehicles within the city: it can even monitor

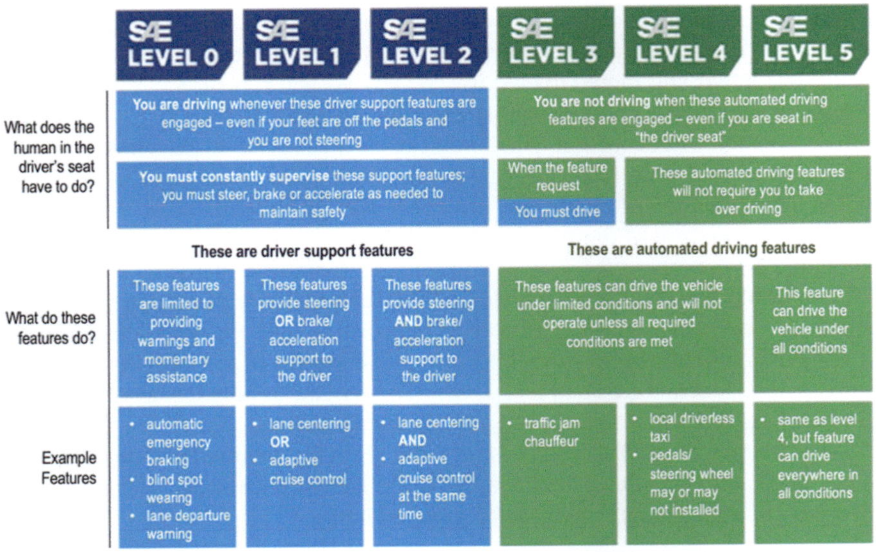

Fig. 12.3 Levels of driving automation by SAE international's J3016 standard. (https://www.sae.org/)

all the parking areas in real time and control the situation by directing the vehicles into assured parking lots.

Finally, all will have to fulfil the highest levels of computer security. *Cybersecurity* is a key aspect in the development and effectiveness of all the different V2X communications which must be protected from any kind of intrusion by outsiders.

Thanks to the implementation of all these technologies, vehicles connected with V2X systems will thereby be permitted to interface with the surrounding environment, not only facilitating a better management of the mobility network but also allowing the vehicles themselves to have access to useful information in order to foresee many dangerous situations that often lead to road accidents (Talebpour and Mahmassani 2016).

The amount and speed of information that a single vehicle could receive are such that it cannot be managed by a human driver and it is naturally necessary to have *artificial intelligence* (AI) which can exploit the information received and, based thereon, condition the driving of the vehicle itself.

The interventions to which AI has access can be very different, from the simple creation of warnings for the driver to complete control of the vehicle. The classification most commonly used to define the level of vehicle automation is that proposed by the International Society of Automotive Engineers (SAE).

SAE defines a taxonomy and definitions for on-the-road motor vehicle automated driving systems, distinguishing between six levels of automation.[3]

[3]https://www.sae.org/.

In the first three levels (see Fig. 12.3), the degree of automation is low, and one refers to assisted driving, while, starting from level 3, the characteristics of an autonomous vehicle are defined. In simple terms, "an *autonomous vehicle* is one that can drive itself using various in-vehicle technologies and sensors, including adaptive cruise control, active steering (steer by wire), anti-lock braking systems (brake by wire), GPS navigation technology, lasers, and radar".[4]

The high level of automation and communication present in the AV will surely bring advantages both for those who manage the mobility systems and for users, but there is still great uncertainty about the benefits, costs, impact on travel, and the growth of demand (Litman 2019).

For example, in an autonomous vehicle the driver is not expected to monitor constantly the roadway, allowing passengers to rest or work during the trip, reducing stress, and improving productivity.

As already indicated in the previous chapter, several studies are simulating the impact of autonomous driving in city traffic and in particular on the estimate of the minimum car fleet necessary to meet the current transport demand. For example, the result of a study by MIT is an estimate, rather surprisingly, of a possible fleet reduction from 30%, using a real-time optimisation logic, up to 40%, having the full knowledge of daily trip demand (Vazifeh et al. 2018).

Regarding safety, since human error contributes to 90% of crashes (Road Safety Authority 2013), it could be argued that autonomous vehicles will reduce crash rates and insurance costs by 90% (McKinsey 2016).

This overlooks additional risks that these technologies can introduce (Koopman and Wagner 2017), such as hardware and software failures, malicious hacking, increased risk-taking due to "over-trusting" technology (Millard-Ball 2016), additional risks to non-auto travellers, PBIC (2017), and reduced investment in conventional safety strategies (Lawson 2018).

At present, there is much interest in and curiosity about AVs since one is in a phase in which some car manufacturers are offering the first models with SAE Level 3 systems (e.g. Tesla Autopilot, Volvo Drive Me). On the other hand, much still needs to be done from the point of view of the technologies, from that of the infrastructure, and moreover from that of the legislation.

Finally, if the costs are considered, owing to the technological hardware necessary for the AV, the cost of maintenance and of necessary mappings, the prices of the AV will be on average higher than the human-driven vehicles (Litman 2019).

The AV can be competitive on a large scale only if widely implemented in sharing services thanks to which the AVs can guarantee a door-to-door service.

The autonomous driving technology will therefore have an important effect on the urban planning of cities, which has always been strongly influenced by the presence of cars, both when moving and when parked. In fact, it will allow a more rapid

[4]https://www.gartner.com/it-glossary.

development of the *sharing mobility* which, as already mentioned, will lead to a great reduction in the number of vehicles with consequent recovery of city spaces for new or different activities.

12.4 Mobility as a Service—MaaS

MaaS is a relatively new concept, presented at the European ITS Congress in Helsinki in 2014, which is mainly based on the integration, in a single application, of services generally offered by different mobility providers. Already today, technological advances have made available a wide range of digital tools related to transport and mobility, permitting the appearance of different categories in the field of intelligent mobility (El Kaim 2016). Trying to tidy up the tangle of apps related to mobility now currently available, it is useful to divide them into four broad categories, delineated by the apps' primary function (Shaheen et al. 2016):

- **Mobility apps** are apps with a primary function to assist users in planning or understanding their transportation choices and may enhance access to alternative modes.
- **Smart parking apps** provide information on parking cost, availability, and payment channels. These apps are often paired with smart parking systems.
- **Vehicle connectivity apps** are apps that allow remote access to a vehicle through an integrated electronic system that can be used in cases of emergency (e.g. locked out of a car, requesting assistance after an accident, etc.); they are generally developed by automobile manufacturers.
- **Courier network services** (also referred to as flexible goods delivery) are apps that provide for-hire delivery services for monetary compensation using an online application or platform to connect couriers using their personal vehicles, bicycles, or scooters with transport demand for light freight (e.g. packages, food).
- **Complementary apps** can be interesting for mobility, such as health apps, environment and energy consumption apps, and insurance apps.

All these apps, with basic functions, can then be incorporated and managed by applications of a higher level which can access them. These applications, which can be managed by third parties, are therefore "smarter" since they can integrate more than one service. These smart digital mobility entities can be in turn grouped in three main families:

- **Smart Mobility Providers**: Entities which enable or own a mode of transport and offer an access to its services by means of computer applications, mobile apps, API, etc. Examples: shared mobility providers, public transport companies, private bus companies, car manufacturers (Table 12.2).
- **Smart Mobility Integrators**: These provide digital platforms which link to the right combination of various modes of transportation in order to offer an integrated, multimodal door-to-door mobility solution using a mobility platform by leveraging

Table 12.2 Smart mobility provider's mobility apps

Type	Examples	Descriptions
Business-to-consumer (B2C) sharing apps	CAR2GO, SHARE'NGO, enjoy	Apps that sell the use of shared transportation vehicles from a business to an individual consumer, including one-way and round-trip car sharing
Peer-to-peer (P2P) sharing apps	auting share your car	Apps that enable private owners of transportation vehicles to share them peer to peer with others, generally for a fee
Taxi apps	mytaxi, it Taxi	Apps that supplement regulated city taxicabs
E-hail apps	lyft	Apps that supplement street hails by allowing location-aware, on-demand hailing
Ride sharing	Bla Bla	Apps that provide a platform for sourcing rides. This category is expansive in its definition so as to include "ride-splitting" services in which fares and rides are split among multiple strangers who are travelling in the same direction
Public transit apps		Apps that enable the user to search public transit routes, schedules, near-term arrival predictions, and connections. These apps may also include a ticketing feature, thereby providing the traveller with easier booking and payment for public transit services

technological expertise, operational excellence, infrastructural advancements and innovative business propositions. Examples: platform providers, B2C providers, API providers (Table 12.3).

- **Smart Mobility Aggregators**: These propose a selection of mobility services as their core business, either as stand-alone providers or through partnerships. They provide both digital platforms (including on-the-go payment and ticketing systems) and physical mobility facilities (e.g. parking zone for bike or car sharing), with the aim of becoming the mobility hub for any transportation need and of offering single payment solutions for multi-mode transportation.

Table 12.3 Smart mobility integrator's mobility apps

Type	Examples	Descriptions
B2C apps/journey planners	Citymapper urbi FREE2MOVE here moovit	Apps that route users by considering multiple modes of transport and providing the user with travel times, connection information, and distance and trip cost
Travel planners (visitors)	Google Trips	Apps that support the organisation of a travel, including transport, hotels, and so on
Complete platforms	OpenTripPlanner Rome2Rio	Complete platforms consisting in mobile apps, website, routing algorithm, API
API	*Navitia.io (Keolis, FR), OpenTransportNet (EU project), Transitland (from Mapzen), TransitFeeds (open source, USA), Transitlime (open source, USA), TransportAPI*	Components for apps building

The MaaS concept refers to the last category and is the newly emerging concept involved with mobility integration, providing a new way of thinking in terms of how the delivery and consumption of transport (or mobility) are managed (Fig. 12.4). It uses ICT integration, ticket integration, and payment integration in order to offer to users a tailor-made multimodal seamless mobility service, by means of a public or private mobility aggregator (MaaS Alliance 2017).

MaaS is based on a disruptive digital business model, offering services on subscription in the form of mobility packages, similar to a monthly telephone contract, although it could accept pay-per-ride micro-transactions. This bundling of mobility modes presents a shift away from the historic ownership-based transport system towards an access-based system. The contract could also include other complementary services, such as trip planning and reservation, hotel, events, and payments, through a single interface (Kamargianni and Matyas 2017).

The first to experiment with this type of service were the Finns and, specifically, the inhabitants of Helsinki who, since 13 June 2016 ("a groundbreaking day for urban travel"),[5] have the opportunity to use the Whim app. Thanks to this app, it is possible to search for the best route, consult timetables, choose travel options,

[5] https://whimapp.com/.

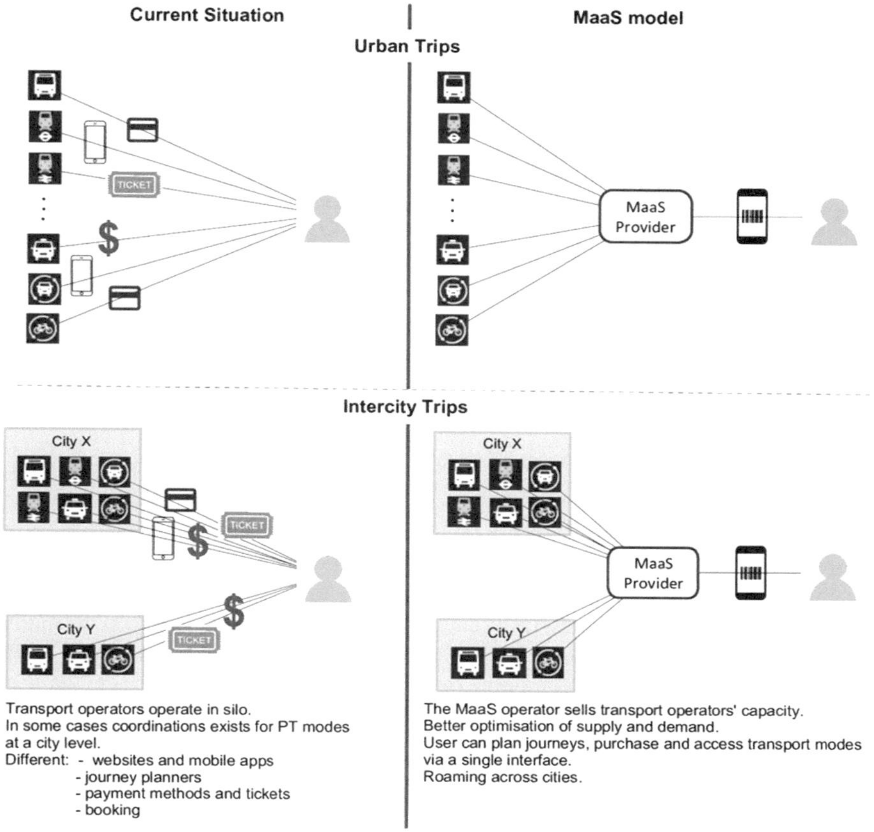

Fig. 12.4 With and without MaaS from user's perspective (Kamargianni and Matyas 2017)

and book and use the various services by purchasing a single ticket (single-ticket principle), choosing between local public transport, taxis, sharing mobility services, and rental cars (see Fig. 12.5). After the launch of Whim, it was found that among the MaaS users the modal share of public transport was increased from 48 to 74%.[6]

Following the success of Whim today, there are many examples of MaaS applications and pilot projects[7] that have shown that MaaS can be a valid tool for promoting shared modes (Matyas and Kamargianni 2018).

Among these, we mention the Communauto/Bixi project started in Quebec, Canada, where some municipal transport companies have decided to offer packages that include in their offer of public transport the possibility of using bike-sharing and car-sharing services. In Gothenburg, Sweden, UbiGo ("We want to simplify your

[6]https://whimapp.com/.

[7]https://maas-alliance.eu.

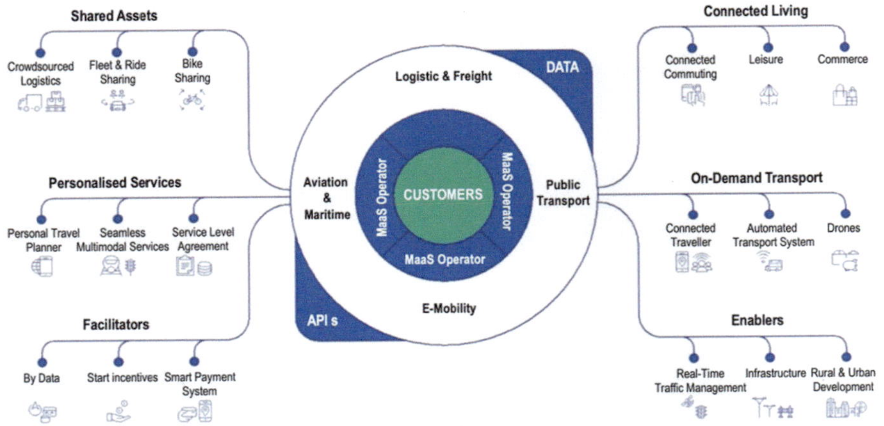

Fig. 12.5 Mobility as a service framework (Huhtala-Jenks 2016)

journey from A to B")[8] adds, to the previous types of transport, the possibility of renting a car and using taxis through a single application which, among other features, has the interesting possibility of assigning bonuses to users if they make choices of sustainable movement by exploiting shared transport services with low environmental impact. In Italy, the NUGO application, developed by the Ferrovie dello Stato group, today already integrates more than 200 mobility providers, including intercity and regional rail transport, urban transport in many Italian cities, various car-sharing and bike-sharing providers, parking lots, car rentals, and more.

In general, the business model of MaaS apps allows to access different transportation options with just one account: within the application, one can plan, book, and pay trips using the services of the partners (Table 12.4).

Typically, a digital mobility solution makes use of a *Journey Planner*, i.e. a search engine aimed to find optimal routes and ways of moving between a set of locations (Shaheen et al. 2016).

Trip Planner, Travel Planner, and Route Planner are often used as synonyms of Journey Planner:

- *Route Planners* are typically thought of as using only one mode of private transportation (such as driving a vehicle, walking, or cycling).
- *Journey Planners* usually make use of at least one public transport mode which operates according to published schedules.
- *Trip Planner and Travel Planner* are often used as a trip organiser, i.e. for tourist or visitor travel.

The process of planning a journey from one location to another mainly involves decisions on the mode of transport (e.g. car, bicycle, public transport, on foot) and potential routes to reach the destination. The choice of mode of transport and of routes

[8] https://www.ubigo.me/.

Table 12.4 MaaS core characteristics (Jittrapirom et al. 2017)

Journey planning and integration of transport modes	A goal of MaaS schemes is to encourage the use of collective transport services, by bringing together multimodal transportation, allowing the users to choose, and facilitating them in their intermodal trips. Urban modes may be included: public transport, taxi, car sharing, ride sharing, bike sharing, car rental, on-demand bus services; but also long-distance buses and trains, flights, and ferries
Tariff option	MaaS platform offers users two types of tariffs in accessing its mobility services: "mobility package" and "pay-as-you-go". The package offers bundles of various transport modes and includes a certain amount of km/minutes/points that can be utilised in exchange for a monthly payment. The pay-as-you-go charges users according to the effective use of the service
One platform	MaaS relies on a digital platform (mobile app or web page) through which the end-users can access to all the necessary services for their trips: trip planning, booking, ticketing, payment, and real-time information. Users might also access to other useful services, such as weather forecasting, synchronisation with personal activity calendar, travel history report, invoicing, and feedback
Multiple actors	MaaS ecosystem is built on interactions between different groups of actors through a digital platform: demanders of mobility (private customer or business customer), a supplier of transport services (public or private), and platform owners (third party, PT provider, authority). Other actors can also cooperate local authorities, payment clearing, telecommunication, and data management companies
Use of technologies	Different technologies are combined to enable MaaS: devices, such as mobile computers and smartphones; a reliable mobile Internet network (Wi-Fi, 3G, 4G, LTE); GPS; e-ticketing and e-payment system; DBMS and integrated infrastructure of technologies (i.e. IoT)

(continued)

Table 12.4 (continued)

Demand orientation	MaaS is a user-centric paradigm. It seeks to offer a transport solution that is best from customer's perspective to be made via multimodal trip planning feature and inclusion of demand-responsive services, such as taxi
Registration requirement	The end-user is required to join the platform to access available services. An account can be valid for a single individual or, in certain cases, an entire household. The subscription not only facilitates the use of the services but also enables the service personalisation
Personalisation	Personalisation ensures end-users' requirements, and expectations are met more effectively and efficiently by considering the uniqueness of each customer. The system provides the end-user with specific recommendations and tailor-made solutions on the basis of his/her profile, expressed preferences, and past behaviours. Additionally, they may connect their social network profiles with their MaaS account
Customisation	Customisation enables end-users to modify the offered service option according to their preferences. This can increase MaaS' attractiveness among travellers and its customers' satisfaction and loyalty. They may freely compose a specified chained trip or build their mobility package with a different volume of usage of certain transport modes to better achieve their preferred travel experiences

could take into account several factors. The most important factors are journey time, journey cost, number of interchanges, type of transport means; but other factors can be taken into account such as sustainability issues (Fig. 12.6).

The core component of a journey planner is the (shortest) path search algorithm. For road route planning, one can compute driving directions in milliseconds or less, even at continental scale. The techniques for solving the vehicle routing problem include basic approaches, but also heuristic search algorithms aimed at solving classic

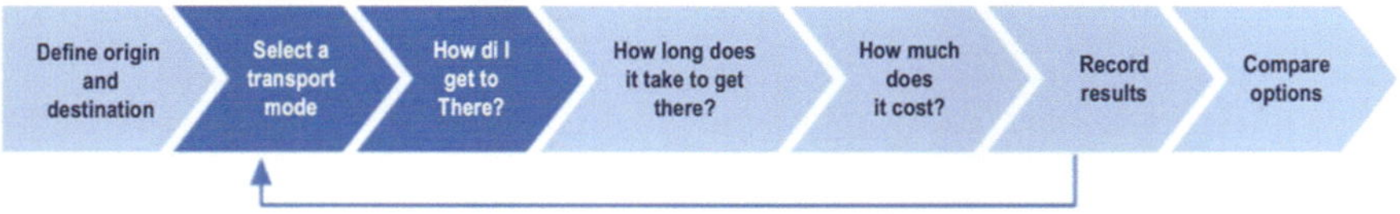

Fig. 12.6 Mobility as a service, conceptual scheme

operational research problems such as the travelling salesman problem (TSP), its generalisation given by the dial-a-ride problem (Cordeau and Laporte 2007), and the integrated dial-a-ride problem that includes the integration of demand-responsive services with fixed route services.

Journey planning on public transport systems, although conceptually similar, is a significantly harder problem owing to its inherent time-dependent and multicriteria nature. The multimodal route planning problem, which seeks journeys combining schedule-based transportation (buses and trains) with unrestricted modes (walking and driving), is even more difficult, relying on approximate solutions even for metropolitan inputs (Bast et al. 2015). The ultimate goal, a worldwide multimodal journey planner, must take into account real-time traffic and transit information, historic patterns, schedule constraints, and monetary costs. Moreover, all these elements should be combined in a personalised manner (World Economic Forum 2014).

MaaS will certainly become the most widespread way to move, in the same way that Apple Music and Spotify revolutionised the world of music or Netflix revolutionised the world of films. Cities and local government regulators have an important task: to prevent mobility providers from slowing down this unstoppable process by refusing commercial agreements for the development of a single mobility supplier.

Over the interests of individual companies, the general interest of the citizen/customer must prevail. Indeed, the public service regulators could impose on both historic and new suppliers the integration into a public platform which, in order to keep the fees low, must be regulated at national, European, or international level.

12.5 Conclusions

Given that the goal of the public administration bodies (local and regional councils) which promote mobility solutions is to ensure that the use of private cars must be gradually abandoned, there are many advantages provided by MaaS for private individuals, for companies that offer transport services and for the community as a whole.

First of all, there is a reduction in traffic and pollution (both atmospheric and acoustic). Operators in the transport world, having a vast amount of data on which to build their strategies, will be able to optimise their services by responding in a timely manner to the needs of their interlocutors. In this regard, even traffic managers will be able to opt for solutions and modular prices based on usage, timetables, and the current traffic situation.

Moreover, thanks to the offer of much more efficient planned travel solutions, service users will be able to have cheaper, faster, and safer journeys thanks to always having access to information updated in real time on the situation of the route they are taking.

Figure 12.7 shows the concepts expressed in this chapter concerning the mobility as a service and the new paradigms of sustainable mobility:

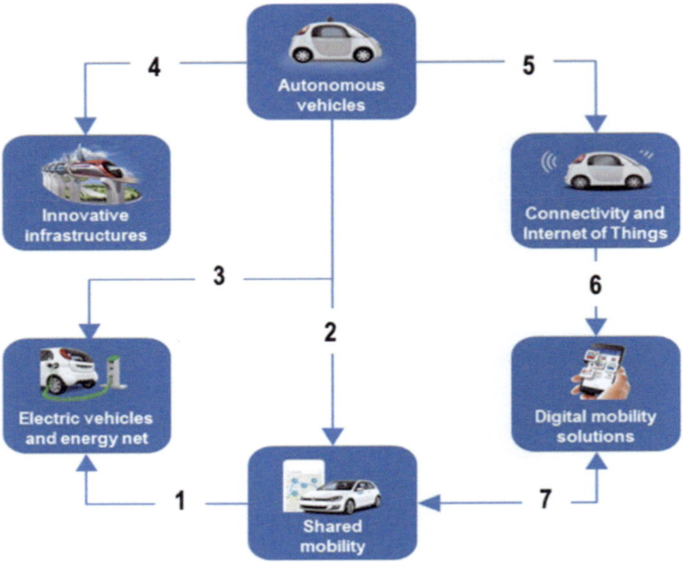

Fig. 12.7 Mobility as a service, new paradigms of sustainable mobility

- An uptake in shared mobility will accelerate electrification, since greater utilisation is favourable for the economics of electric vehicles.
- Self-driving could cause the merger of shared mobility business models into a single proposition, competitive with private car ownership and public transport.
- Self-driving vehicles, both private and shared, will probably increase demand for, and consumption of mobility, in which case electric vehicles offer lower total cost of ownership.
- Self-driving electric vehicles will have a different usage and hence have different requirements for charging infrastructure.
- Self-driving vehicles might accelerate the uptake of IoT applications.
- Connectivity and Internet of things (IoT) will enable progressively more digital mobility applications.
- The availability of digital mobility solution can support shared mobility and vice versa.

Finally, if one considers the possibility of having electric, autonomous, and shared vehicles then it is possible to outline a scheme in which all the major mobility trends have an effect of mutual reinforcement in favour of a more integrated, multimodal, and thereby more efficient and effective mobility.

References

Bast H, Delling D, Goldberg A, Müller-Hannemann M, Pajor T, Sanders P, Wagner D, Werneck RF (2015) Route planning in transportation networks. Technical report published by Microsoft Research. Available on https://arxiv.org/pdf/1504.05140.pdf. Accessed on 1 Dec 2019

Chen S, Hu J, Shi Y, Peng Y, Fang J, Zhao R, Zhao L (2017) Vehicle-to-everything (V2X) services supported by LTE-based systems and 5G. IEEE Commun Stand Magazine 1:70–76. https://doi.org/10.1109/MCOMSTD.2017.1700015

Coppola P, Esztergàr-Kiss D (2019) Autonomous vehicles and future mobility. AET, Elsevier, Amsterdam. ISBN: 978-0-12-817696-2

Cordeau JF, Laporte G (2007) The dial-a-ride problem: models and algorithms. Ann Oper Res 153:29–46

El Kaim W (2016) Mobility as a service. Enterprise architecture and governance. Available on https://www.slideshare.net/welkaim/mobility-as-a-service-maas. Accessed on 1 Dec 2019

EUROSTAT (2019) Passenger transport statistics. Available on https://ec.europa.eu/eurostat/statistics-explained/index.php/Passenger_transport_statistics. Accessed on 1 Dec 2019

Fulton L, Mason J, Meroux D (2017) Three revolutions in urban transportation. How to achieve the full potential of vehicle electrification, automation and shared mobility in urban transportation systems around the world by 2050. UC Davis ITDP Joule 2(4):575–578

Huhtala-Jenks K (2016) Digital services and mobility as a service at ministry of transport and communication, Finland. Available on http://sunergos.ee/wp-content/uploads/2018/10/From-Transport-to-Mobility-as-a-Service-Krista-Huhtala-Jenks.pdf. Accessed 1 Dec 2019

Jittrapirom P, Caiati V, Feneri A, Ebrahimigharehbaghi S, González M, Narayan J (2017) Mobility as a service: a critical review of definitions, assessments of schemes, and key challenges. Urban Plan 2(2):13–25

Kamargianni M, Matyas M (2017) The business ecosystem of mobility-as-a-service. In: Conference: transportation research board. Available on https://www.researchgate.net/publication/314760234_The_Business_Ecosystem_of_Mobility-as-a-Service. Accessed on 1 Dec 2019

Koopman P, Wagner M (2017) Autonomous vehicle safety: an interdisciplinary challenge. IEEE Intell Transp Syst 9(1). https://doi.org/10.1109/mits.2016.2583491

Lawson S (2018) Tackling the transition to automated vehicles, roads that cars can read report III. European Road Assessment Association (www.eurorap.org)

Litman T (2019) Autonomous vehicle implementation predictions. VTPI. Available on https://www.vtpi.org/avip.pdf. Accessed on 1 Dec 2019

MaaS Alliance (2017) White paper. Guidelines and recommendations to create the foundations for a thriving MaaS ecosystem. Available on https://www.drivesweden.eu/en/white-paper-maas-ecosystems

Matyas M, Kamargianni M (2018) The potential of mobility as a service bundles as a mobility management tool. Transportation 1–18. https://doi.org/10.1007/s11116-018-9913-4

McCarthy J, O'Keeffe D (2017) Autonomous, connected, electric and shared vehicles. Reimagining transport to drive economic growth

McKinsey (2016) Automotive revolution—perspective towards 2030: how the convergence of disruptive technology-driven trends could transform the auto industry. Available on https://www.mckinsey.com/~/media/mckinsey/industries/high%20tech/our%20insights/disruptive%20trends%20that%20will%20transform%20the%20auto%20industry/auto%202030%20report%20jan%202016.ashx. Accessed on 1 Dec 2019

Millard-Ball A (2016) Pedestrians, autonomous vehicles, and cities. J Plan Educ Res 1–7, https://doi.org/10.1177/0739456x16675674

ONSB, Osservatorio Nazionale Sharing Mobility (2019) 3° Rapporto nazionale sulla sharing mobility. Available on http://osservatoriosharingmobility.it. Accessed on 1 Dec 2019

PBIC (2017) Automated and connected vehicles, pedestrians, and bicyclists, pedestrian and bicycle information center. Available on http://www.pedbikeinfo.org/topics/automatedvehicles.cfm. Accessed on 1 Dec 2019

Road Safety Authority (2013) Road safety strategy 2013–2020

Shaheen S et al (2016) Smartphone applications to influence travel choices. US department of transportation, federal highway administration. Available on https://ops.fhwa.dot.gov/publications/fhwahop16023/fhwahop16023.pdf. Accessed on 1 Dec 2019

Talebpour A, Mahmassani HS (2016) Influence of connected and autonomous vehicles on traffic flow stability and throughput. Transp Res Part C: Emerg Technol 71:143–163. https://doi.org/10.1016/j.trc.2016.07.007

Vazifeh MM, Santi P, Resta G, Strogatz SH, Ratti C (2018) Addressing the minimum fleet problem in on-demand urban mobility. Nature 557(7706):534–538. https://doi.org/10.1038/s41586-018-0095-1

World Economic Forum (2014) Connected world. Hyperconnected travel and transportation in action. Available on http://www3.weforum.org/docs/WEF_Connected_World_HyperconnectedTravelAndTransportationInAction_2014.pdf. Accessed on 1 Dec 2019

Yakovlev A, Otto P (2018) The future of mobility: shared mobility. Game changers, IPSOS. Available on https://www.ipsos.com/it-it/future-mobility-shared-mobility. Accessed on 1 Dec 2019

Part IV
Rating Systems for Green Planning of Cities and Communities

Chapter 13
Green Protocols for Neighbourhoods and Cities

Giuliano Dall'O' and Alessandro Zichi

Abstract The rating systems which certify the environmental sustainability of neighbourhoods or cities contribute greatly to enhancing the quality of built environments, also from a social and economic point of view, thus improving the quality of life of the inhabitants. For this reason, in a proper vision of green planning of cities and communities they are essential tools. This chapter focuses on the most widespread protocols used for the certification of urban space sustainability: BREEAM® Community, LEED® V4 for Neighbourhood Development and LEED® for Cities and Communities. The analysis confirms the validity of these instruments which, in addition to certifying the environmental, social and economic sustainability of neighbourhoods and cities, provide indicators that facilitate an objective comparison between different urban realities. The completeness of the areas covered by these protocols certainly places them among smart city protocols.

13.1 Sustainable Rating Systems for Neighbourhoods and Cities, General Concepts

If the rapid growth of cities and communities requires strong and incisive actions to direct development towards sustainable models not only from an environmental but also from an economic and social point of view, then tools that enable the management of sustainable planning in a long-term perspective, contributing to the verification of the effectiveness of the choices, become indispensable supports.

Some organisations, generally non-profit associations, have developed systems for assessing the sustainability of buildings such as BREEAM® (United Kingdom 1990), LEED® (USA 1998), CASBEE® (Japan 2001), GREEN STAR® (Australia

G. Dall'O' (✉) · A. Zichi
Architecture, Built Environment and Construction Engineering, ABC Department, Politecnico di Milano, Via G. Ponzio, 31, 20133 Milan, Italy
e-mail: giuliano.dallo@polimi.it

A. Zichi
e-mail: alessandro.zichi@polimi.it

G. Dall'O' (ed.), *Green Planning for Cities and Communities*,
Research for Development, https://doi.org/10.1007/978-3-030-41072-8_13

2002) or DGNB® (Germany 2008), used and recognised internationally. These systems provide a set of indicators to evaluate the sustainability of a building: the said indicators can help the designer involved in the management of the impacts of the project. Although the number of indicators and metrics may differ, they all define a set of criteria to assess a construction project based on their environmental, social and economic dimensions (Pedro et al. 2019).

A common feature of these sustainability assessment systems, also called sustainability rating systems, is that the measurement of the sustainability of the building also takes into consideration the territorial context in which it is located.

Taking account of the link between a building and the surrounding territory is fundamental and should be viewed as a response to the need to reduce the environmental impacts of the built environment at a territorial level.

13.1.1 The Certification of Sustainability at the Territory Scale

The environmental certification of buildings in accordance with the most widespread international protocols is meeting with considerable success, helping to increase the awareness of environmental issues in the real estate market. To certified buildings, for example according to the LEED® or BREEAM® protocols, a greater value is attributed which generally compensates for the greater effort, also in economic terms, that is made for a more sustainable design and construction. In terms of greater energy efficiency of certified buildings: in particular, this can lead to a greater reduction in operating costs over time.

In recent years, some sustainability certification systems have evolved by shifting interest from the building to the territorial area (neighbourhood or city); it is the case, for example, of the neighbourhood sustainability assessment (NSA), such as BREEAM® Communities, LEED® Neighbourhoods (ND), CASBEE® Urban Development (UD), DGNB® Urban Districts (UD), GBC® Quartieri (I) and finally LEED® for Cities and Communities (see Fig. 13.1).

These new certification protocols contribute greatly to improving the sustainability of cities: at the neighbourhood scale, buildings become a component of an urban system, interrelated with the other components, such as public transport, green areas and green infrastructures and services (e.g. schools, shopping centres, hospitals, etc.).

Figure 13.1 shows the rating system timeline including the systems of building sustainability assessment (BSA) and the systems of neighbourhood sustainability assessment (NSA) which are highlighted. One can observe that the certification protocols for neighbourhoods have been published in more recent times, starting from 2006.

The management of the sustainability certification protocols of neighbourhoods is more complex when compared to that of buildings: the number of indicators is greater and the indicators can relate to public services (e.g. the mobility system

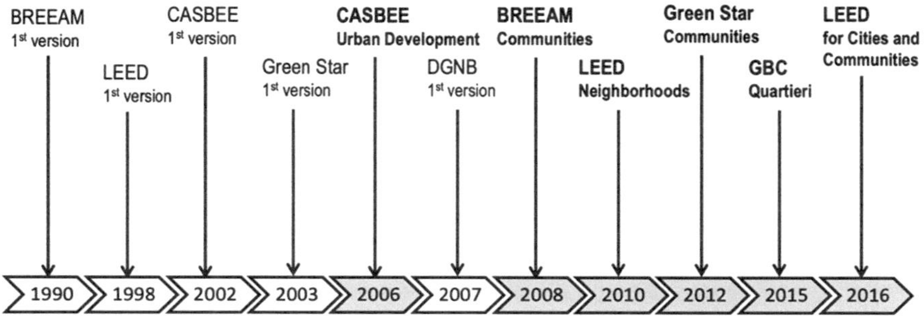

Fig. 13.1 Neighbourhood sustainability assessment (NSA) timeline

or urban district heating networks) with the involvement of a greater number of stakeholders.

Then there is the question concerning the person who promotes the protocol: whilst, in the case of a building, the promoter is normally a private individual (e.g. the owner of the building or the real estate), in the case of a neighbourhood, the property can be publically owned (e.g. the municipality/local authority) and this leads to greater complexity surrounding the decisions that must be taken. An in-depth analysis of this aspect is provided by Dall'O' et al. (2013). For these reasons, there are still very few certified neighbourhood projects compared to those at building scale. For instance, sustainable assessment systems such as BREEAM® have already passed the 570,400 building certificates mark, but only summed up to 54 neighbourhood projects by December 2019; this is followed by LEED® system with over 104.000 certified buildings and only 247 neighbourhood projects.

Before going into detail on the issues related to the criteria common to rating systems for sustainability, it is appropriate to reflect upon the concepts underlying the assessment of sustainability on an urban scale.

The classic approach considers the sustainability of a territory, from the district to the city, as the integration of three areas (see Fig. 13.2):

- *Environmental Sustainability* defined as the responsible interaction between man and the environment to avoid the depletion or degradation of natural resources and allow long-term environmental quality.
- *Social Sustainability* understood as the capacity of a community to develop processes and structures for shared and inclusive development.
- *Economic Sustainability* understood as the adoption of strategies which support long-term economic growth without generating a negative impact on the social, environmental and cultural aspects of the community.

The indicators and strategies contained in a sustainable certification protocol at the territorial scale should consider all the aspects linked to sustainability, understood in its most inclusive definition of environmental, economic and social aspects. Some authors highlight the need to broaden the areas of interest relating to the issue of sustainability on the territorial scale.

Fig. 13.2 Sustainability seen as an integrated approach between environmental, social and economic needs

Turcu (2013), for example, argues that little is known about whether the "integrated" sets of sustainability indicators normally used work in practice and effectively reflect local perspectives, values and understandings of the sustainability which they are supposed to represent. In his paper, the author concludes that sustainability indicators are not isolated pieces of information, but manifestations of local underlying processes and interconnections that can be mapped and which have the potential to expand the understanding of local sustainability.

The author, therefore, proposes an enlargement of the areas of sustainability including *institutional sustainability*, according to the diagram shown in Fig. 13.3.

The topic of indicators for the environmental sustainability of cities and communities is dealt with in detail in Chap. 16 of this book (Indicators and rating systems for sustainable smart cities).

13.1.2 How a Sustainability Assessment Systems Works

The sustainability assessment systems for the building scale and those for the territory scale are structured in an almost similar manner.

Figure 13.4 provided as an example shows procedure provided by the LEED® for Neighbourhood Development (ND) V4 protocol.

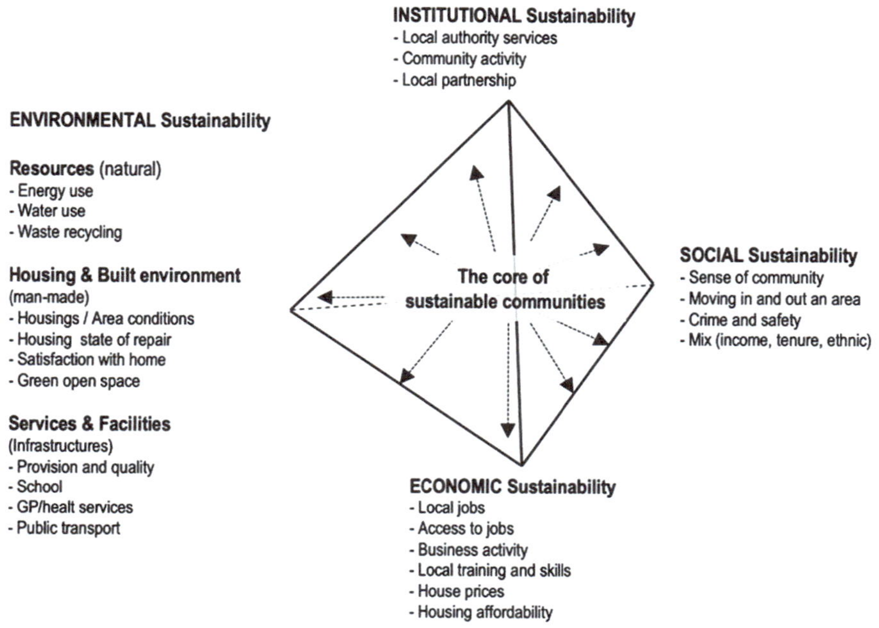

Fig. 13.3 List of indicators for assessment of sustainable communities (Turcu 2013)

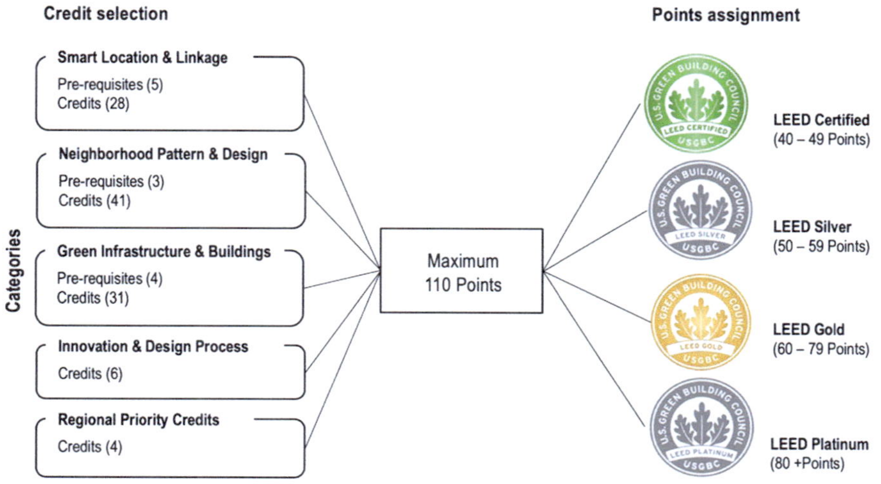

Fig. 13.4 Prerequisites and credits management scheme in the LEED® ND V4 protocol

First of all, the protocol defines some categories or areas of evaluation (in this specific case 5): within each category, there are prerequisites and credits.

The prerequisites define the characteristics of the project that must be respected to allow access to the certification path. The prerequisites do not provide credits and do not affect the final score but constrain the project to meet minimum environmental quality requirements.

The credits of each category provide the project with design options, obviously aimed at improving the sustainability of the project: if the project then complies with those options, points can be earned.

A sum is made of the points obtained subsequently and, based on the overall score obtained, different rating levels can be obtained: in the case of LEED® ND V4, for example, four levels have been defined: LEED® certified (from 40 to 49 points), LEED® Silver (from 50 to 59 points), LEED® Gold (from 60 to 79 points) and set LEED® Platinum to 80 and more points.

From the diagram in Fig. 13.4, one can observe how the budget of the points obtainable with credits (110) is higher than the points required to obtain the highest rating (more than 80 for LEED® Platinum).

This demonstrates the flexibility offered by the protocol that provides the designer with various options to achieve important sustainability levels. This aspect, which characterises all protocols, is positive for two reasons:

In some cases, the characteristics of the project, or of the territorial context, do not allow it to obtain credits; furthermore, the protocol gives the designers the freedom to follow different paths (different strategies) to reach the final objective.

The conceptual simplicity of the diagram in Fig. 13.4 must not lead one to believe that the certification process is simple. In fact, even obtaining just one credit can sometimes be complex. A design choice involves the implementation of strategies, sometimes complex, which must be demonstrated in an objective way: almost all credits require calculations that must be made in compliance with national and international technical standards based on the *prerequisite and credit description sheets* contained in the manual provided by the body that manages the protocol.

The credibility of a certification protocol also depends very much upon the role exercised by the external control body, acting independently: in the case of the protocol of the LEED® family, the supervisory body, which then issues the certification, is the GBCI® (Green Business Certification Inc.).

The relationship between the client (neighbourhood property), supported by an experienced professional (e.g. LEED® AP—Accredited Professional for LEED® protocol), and the certification body is normally managed through an online platform. Any credit that is considered to be obtained must be demonstrated by filling in online forms with ample supporting documentation that must be provided to prove that the credit itself is in fact deserved.

From a procedural point of view, the protocols for the rating systems may be different. Figure 13.5 shows, as an example, a certification path.

First of all, it is necessary to choose the rating system which is most suitable for the certification of the project (*Choice Phase*). The subsequent administrative phase

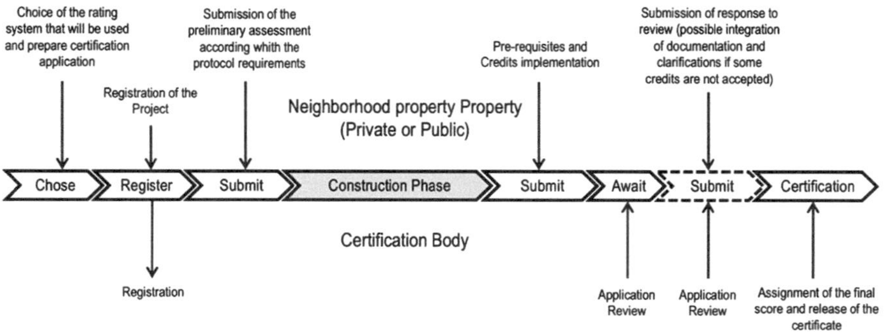

Fig. 13.5 Example of a typical certification process

is the registration of the project which normally requires the payment of a fee to the certification body.

Registration allows access to the online platform which will become the way to communicate until the end of the process (*Register Phase*).

The next phase involves the deposit of the project documentation accompanied by a pre-assessment (*Submittal Phase*).

The owner, with specialist professional support, compiles a checklist in which they hypothesise the planning strategies to which correspond the hypothetical credits that could be attained.

The next phase involves the execution of the project: in the case of a building, this phase can last a few years, whilst in the case of a neighbourhood, it can clearly last several years (*Construction Phase*). The executive phase for the construction of a neighbourhood, even if small, involves the construction of various buildings, services and infrastructures.

In the development of the project and the consequent certification process, it may happen that unexpected critical decisions cannot be developed and some credits cannot be respected and, in this case, other credits can be added to compensate, to maintain the level of certification previously hypothesised.

All this to explain that there may be differences between the situation assumed in the pre-assessment and the final situation.

Once the construction of the district has been completed, the updated documentation in support of the requested credits must be transmitted to the certification body (*Submittal Phase*).

This is also an opportunity to make changes. In the submission phase, it is necessary to demonstrate, using supporting documentation, that the prerequisites required by the protocol are assured and that the credits provided in the preliminary assessment submitted before implementation are merited.

At this point, the certification body analyses in depth the documentation provided (*Await Phase*). After a waiting period whose duration is established by the procedures, the certification body could:

- approve the submitted assessment confirming both the prerequisites and the credits;
- request clarifications regarding prerequisites and/or credits (for example, integration of documentation);
- consider some prerequisites or some credits to be non-compliant, thus blocking the certification process or lowering the final rating of the project.

If supplementary documentation is required, the next phase is performed (*Submittal Phase*). In the event that some claims are disputed, it is possible, in a spirit of democracy, to act by submitting a complaint report.

The certification process ends with the last phase (*Certification Phase*) which provides for the emission of the certification.

13.1.3 An Overview of the Main Sustainability Assessment Systems for Neighbourhood and Cities

The following is a brief description of the main sustainable assessment systems for neighbourhoods and cities mentioned in Fig. 13.1. Three of these rating systems, namely BREEAM® Communities, LEED® ND and LEED® for Cities and Communities, have also been applied beyond the borders of the country that proposed them and therefore can be considered international rating systems: they will be studied in depth in the following paragraphs of this chapter.

BREEAM® Communities was the first rating system to be developed for sustainable certification in urban areas. Its first version, which dates back to 2012, was launched by the the UK Building Research Establishment (BRE). BREEAM® Communities can be used to assess and certify the performance of medium to large scale developments, including new communities and regeneration projects.

The system is structured in five categories (governance, social and economic wellbeing, resources, land use and ecology, transport and movement) which include 40 credits subject to a score of which 12 are mandatory (prerequisites). It also provides bonus credits for innovative features. By meeting the criteria, the projects earn points which determine the level of the certification: pass (30 points), good (45 points), very good (55 points), excellent (70 points) and exceptional (80 points). In December 2019, the BREEAM Communities certified projects are 54 (BRE Global 2019).

The first version of **LEED® ND** (Neighbourhoods Development) was launched by the US Green Building Council in 2014. The rating system, engineered to inspire and help create better, more sustainable, well-connected neighbourhoods, evaluates sustainability performance at the urban scale by addressing three main categories: smart location and linkage, neighbourhood pattern and design, and green infrastructure and buildings. It also accounts for two extra categories: innovation and regional priority. Distributed under these categories, there are 40 credits subject to scoring and 12 prerequisites that must be respected but do not count in the score. By fulfilling credits, projects earn points that determine their certification level: Certified (40 points), Silver (50 points), Gold (60 points) and Platinum (80 points). In December 2019, the LEED ND certified projects are 247 (USGBC 2019a).

CASBEE® UD (Urban Districts) was launched by the Japan Sustainable Building Consortium in 2008 [JaGBC 2014]. The rating system considers the improvement of three main urban dimensions: environment, society and economy. The evaluation attributes a score to each of these dimensions based on two factors: built environment quality (Q) and built environmental load (LR); the assessment scale for Q and LR ranges from 1 to 5. Then, these two factors are used to calculate the final score for Built Environment Efficiency (BEE). For CASBEE® UD, there are no mandatory requirements. The calculated score for projects determines their certification level: poor (BEE 0.5), fairly poor (BEE 0.5–1), good (BEE 1–1.5), very good (BEE 1.5–3) and excellent (BEE 3). In 2018, there were 5 CASBEE® UD projects, all of them in Asia (Pedro et al. 2019).

DGNB or German Sustainable Building Council is the main German organisation that promotes eco-construction projects. The members of the DGNB are architects, entrepreneurs, designers, construction engineers and experts in the real estate and construction sectors. **DGNB® UD** (Urban Districts), the rating system for measuring the sustainability of urban districts, was published in 2012.

The scheme includes all areas relevant to sustainable building: from the location and energy supply of the district, to public space amenities and mixed use, sustainable mobility and reducing costs throughout the entire life cycle (DGNB 2019).

The system includes the five main categories of assessment: environmental quality, economic quality, sociocultural and functional quality, technical quality and process quality. Distributed under these categories, there are 30 criteria subject to scoring. By fulfilling credits, projects earn points which determine their certification level: Silver (55 points), Gold (65 points) and Platinum (80 points). In 2019, there were 18 certified DGNB® UD projects, all in Germany (DGNB 2019).

GBC® Quartieri, launched in 2015, is the certification protocol developed by the Green Building Council Italia for the projects of areas subject to redevelopment or new expansions, which promote among its primary objectives the environmental sustainability performances of the territory, infrastructures, equipment and sustainable buildings. GBC® Quartieri recognises the value of interventions that promote an integrated approach to quality of life, public health and respect for the environment. The certification encourages the best practices orientated to the analysis of the territory, the choice of areas in relation to environmental preservation, the promotion of the connection to public transport, the relationships of areas with preexisting structures, the creation and development of services and social functions.

GBC® Quartieri was not created as an urban planning tool but can be a useful instrument for municipalities/local councils which could adopt it for the promotion and provision of incentives for the sustainable redevelopment of their urban territory or for the drafting of development guidelines. The rating system assesses performance at the urban scale by addressing three main categories: smart location and linkage, neighbourhood pattern and design, and green infrastructure and buildings. It also takes into account two additional categories: innovation and regional priority. Distributed under these categories, there are 42 credits which count towards scoring and 12 prerequisites which must be respected but do not count in the scoring. By fulfilling credits, projects earn points that determine their certification level: Certified (40 points), Silver (50 points), Gold (60 points) and Platinum (80 points). Currently, several districts are under study and a neighbourhood is undergoing certification (GBC Italia 2019).

LEED® for Cities and **LEED for Communities**, launched in 2016 by the US Green Building Council, provide cities and communities with a globally consistent way to measure and communicate performance. LEED® v4.1 for Cities and LEED® v4.1 for Communities help cities and communities develop responsible, sustainable and specific plans for natural systems, energy, water, waste, transportation and many other factors which contribute to the quality of life.

The Plan and Design option is available to new cities and communities to plan, design and develop sustainable strategies across six LEED® credit categories: natural systems and ecology, transportation and land use, water efficiency, energy and greenhouse gas emissions, materials and resources and quality of life. To start, one must document the sustainable strategies implemented across the six LEED® credit categories during the planning, design and development phase.

The Existing option is for existing cities and communities to measure the performance of social, economic and environmental conditions at a city-wide scale or at a community level. To start, they must share data across six LEED® credit categories: natural systems and ecology, transportation and land use, water efficiency, energy and greenhouse gas emissions, materials and resources and quality of life. On December 2019, more than 90 cities and communities around the globe have certified through LEED® (USGBC 2019b).

13.2 BREEAM® Communities

BREEAM® Communities rating system integrates the sustainable design into the master planning of new communities and urban regeneration projects. The BREEAM Communities rating system is useful to assess and certify the energy and environmental performance of medium to the large-scale developments, including new communities and regeneration projects.

BREEAM® Communities International is internationally applicable; the case studies are developed across Europe, the Middle East, Africa and Asia. The simple and flexible approach to measure and certify the sustainability of the development can be tailored on the criteria to better reflect a country's or a region's specific environmental, political, economic, social and cultural development.

The BREEAM® rating system if is applied during the early master planning process, with an integrated and holistic process, helps to save time and money throughout the project and facilitate the planning process.

The BREEAM® Communities integrates all the aspects of sustainability such as energy provision, carbon reduction, local economic impact, ecological enhancement, housing provision, transport networks, community facilities, social and economic factors.

To use the BREEAM® Communities, one needs to apply the approved standards list to locate the appropriate best practice guidance in the country of assessment. If this does not exist, then one can use one of the locally approved standards, for which the applicability must be demonstrated by verifying the minimum requirements as set out in the list of approved standards: the proposed documents must then be submitted to BRE Global for approval. Thereafter, the design team should demonstrate compliance with the UK or European standards as listed in each relevant country reference sheet.

The certification process expected three phases:

- Step 1: Establishing the principle of development
- Step 2: Determining the layout of the development
- Step 3: Designing the details.

In the first step, the BREEAM® Communities tool assesses the constraints and opportunities relating to the sustainability at the site and requires consideration of how the development will impact on the wider community. At this stage, is important to conduct surveys regarding flood risk, ecology, energy, transport and social and economic aspects.

Step 2 focuses on designing the layout of the landscape for the safeguard of biodiversity and habitat and to develop the transport and circulation planning for pedestrian, cyclist and vehicular movement, public transport.

The third step in BREEAM® Communities focuses on the detailed design of the development. This step covers the development of the project of the landscaping, the management and long-term stewardship of facilities and services, the building design, and an inclusive design for the entire lifecycle of the building and indeed also includes the period of its construction. The strategy for the construction materials and the typology of the buildings are defined in this phase.

Whilst in Steps 1 and 2 there are requirements and mandatory strategies, in Step 3, none of the issues are obligatory.

The minimum requirements for the development of BREEAM® certification are distributed across the different categories focused upon, to be applied in an integrated approach, from social and economic aspects to landscape design, through to energy and transport aspects.

The most interesting mandatory credits are briefly described below.

13.2.1 Consultation Plan

The members of the local community and appropriate stakeholders are identified for consultation. If the consultation takes place early enough in the process, then the community and stakeholders can readily influence the key decisions.

13.2.2 Economic Impact

When an economic analysis is completed and clearly identifies the needs and opportunities within the local area and surrounding economy, it can be used to characterise the development of the neighbourhood. This study should be focused on understanding how the proposed development can enhance the economic well-being of future occupants. It should also ensure that the development complements and enhances existing economic activity in the local area.

13.2.3 Demographic Needs and Priorities

The development plans for the provision of housing, services, facilities and amenities are based upon the local demographic trends and priorities. The scope of the proposed development, including the housing mix, community facilities and employment opportunities, is to have an overview of the current demographic profiles and future trends of the local area.

13.2.4 Flood Risk Assessment

A site-specific flood risk assessment is carried out in accordance with current best practice and planning policy and includes as a minimum:

- risk and consequences of flooding from all sources on the site and from the site to the surrounding area. How the risks will be managed;
- changes in flood risk due to climate change;
- consultation with appropriate statutory bodies;
- knowledge of possible flood risk held within the local community.

The flood zone or zones for the development are determined in accordance with current best practice and planning policy.

13.2.5 Noise Pollution

This credit is mandatory to ensure, in the site, the mitigation of the noise impacts. This includes mitigation from existing sources of noise, reducing potential noise conflicts between future site occupants and protecting nearby noise-sensitive areas from noise sources associated with the new development.

The mandatory requirements require a noise impact assessment carried out by a suitably qualified acoustician to determine the sources and nature of existing noise on and around the site. The suitably qualified acoustician will have prepared a report setting out recommendations for addressing all identified site issues and, if necessary, attenuation of on-site noise to prevent disturbance to neighbouring noise-sensitive areas.

13.2.6 Energy Strategy

To recognise and encourage developments designed to minimise the operational energy demand, consumption and carbon dioxide emissions, the mandatory requirements provide an energy strategy development plan.

13.2.7 Existing Buildings and Infrastructure

To reduce the embodied carbon impact, BREEAM encourages, where possible, the reuse of the existing buildings and infrastructure. The mandatory requirements require demonstration of compliance with the following impacts:

- heritage and local identity;
- location and condition of buildings and infrastructure;
- embodied carbon in existing materials;
- potential uses of buildings and infrastructure;
- possible use of materials (on- or off-site);
- community and local authority knowledge and opinion;
- local or regional authority; best practice guidance, policies or strategies.

A decision is made and justified with evidence regarding the use and/or demolition of all existing buildings and infrastructure at the site.

13.2.8 Water Strategy

To ensure that the minimised water demand and maximum water efficiency strategy is applied in the project, it is necessary to demonstrate an integrated analysis of the water management starting from the current water availability and demand in the area and covering the predicted future usage.

To demonstrate the strategies applied, a water management plan of demand must be prepared which includes:

- targets for water use in residential and non-domestic buildings in the development site;

- strategies to minimise the predicted use of the development, and maintain this in the future in accordance with the ownership and future facilities management company;
- design options to reduce the water demand in landscaping, any other predicted water use and on-site collection/storage opportunities.

13.2.9 Ecology Strategy

The ecological impact assessment (ECIA) is required to minimise and mitigate the impact of the development on the biodiversity of the site. The suitably qualified ecologist confirms that the master plan conforms with the ecology strategy based on the preexisting biodiversity and the best strategy to avoid damage to any valued ecological features on or near site.

The ecology strategy, covering both the construction and operation phases, will have been drawn up by a suitably qualified ecologist in such a way as to avoid damage: where it is unavoidable that damage to a valued ecological feature will occur, a mitigation plan or compensation plan will have been agreed by the appropriate statutory body and the suitably qualified ecologist to ensure there is no net loss of any of the valued ecological features.

13.2.10 Land Use

To avoid the use of land which has not been previously disturbed, the use of terrain previously developed or contaminated is encouraged. A preliminary investigation is required to define the existence of contaminated areas and to identify whether decontamination work is necessary. Where the preliminary investigation has identified potential land contamination, the contaminated land specialist must elaborate a site investigation and risk assessment to determine the presence and levels of any contamination affecting the site and make recommendations on any remediation required.

13.2.11 Transport Assessment

It is mandatory to produce a transport and movement assessment to define the best strategies to reduce the impact of the transport infrastructure. The scope of the proposed development includes housing mix, community facilities and employment opportunities, information for which derives from a review of the current demographic profiles and future trends of the local area.

The transport assessment or transport statement is developed with the local authority and local, national and regional authorities responsible for roads and other transport networks and aims at the improvement of environmental and social sustainability through appropriate transport. A travel plan is developed by a travel plan coordinator for the site, outlining the design methods used to encourage and implement sustainable transport and movement. To ensure the needs, ideas and knowledge of the community are used to improve the quality of stakeholder engagement, throughout the design, planning and construction process.

13.3 LEED® V4 for Neighbourhood Development

LEED® for Neighbourhood Development (LEED® ND) System of rating is focused on development of neighbourhoods. It is engineered to develop more sustainable, well-connected neighbourhoods in consideration of the city and communities exigencies. There are two options:

- *ND Plan Certification* for the planning and design phases of a neighbourhood. Is the best solution to help the design team and developers' market and, by affirming the intended sustainability strategies, give an immediate image of the project among prospective tenants, financiers, public administration officials and the population.
- *Built Project* designed for neighbourhood-scale for projects in their completion phase or where the construction has been completed within the last three years.

LEED® v4 for Neighbourhood Development Plan requires one to demonstrate that the projects are in the phase of planning and design with no more than 75% of the project's total floor area constructed, and if the project is in the early conceptual phase, the project has the option of pursuing a letter of support review, prior to the plan review. This letter of support is the guideline for the evolution of the project, and consequentially, the design team will in the final phase need to demonstrate accomplishment of the strategies designed in the preliminary analysis.

LEED® v4 for Neighbourhood Development Built Project focuses on the completion of a neighbourhood project or on one that was completed within the last three years.

A project may be registered under LEED® v4 ND Built Project at any time but must be substantially complete before submittal for review. If the project previously earned the LEED ND—Plan Certification, the adaptation of the certification can be requested.

The quality of the project is represented by the following certification levels:

- LEED Certified: 40–49 points earned
- LEED Silver: 50–59 points earned
- LEED Gold: 60–79 points earned
- LEED Platinum: 80+ points earned.

To guarantee a standard of quality of the neighbourhood projects, the system of rating provides three strategies including prerequisites and credits as described in the following paragraphs. Other credits can be obtained considering other areas (innovation, pilot credits, pilot credit registration, innovation catalogue and regional priority).

13.3.1 Smart Location and Linkage

The protocol requires a preliminary analysis to define the characteristics and the impact of the construction activities in them imperilled species and ecological communities, agricultural land conservation, wetlands and water body conservation and floodplain avoidance. To ensure the correct long-term conservation of the ecologic assessment needs to demonstrate the strategies and conservation management strategies of habitat or wetlands and water bodies.

From the perspective of transforming existing cities in terms of sustainability, it is recommended to insert the neighbourhood development system in locations already previously developed or where Brownfield remediation is necessary.

To preserve the ecological system, the rating suggests a steep slope protection and an integrate approach for the sustainable construction plan.

To assure the correct lifestyle of the population, the analysis is requested of the public transportation and the integration into the project of bicycle facilities, for example, illustrating the distribution of the bicycle tracks, the parking areas close to or within the buildings and in the offices' shower services for the bicycle users.

The most successful projects provide mixed-use districts and buildings, taking into consideration the existing services in the district near the site and putting into the project development the necessary design to assure the correct proximity of the housing and jobs of the population.

13.3.2 Neighbourhood Pattern and Design

In the *neighbourhood pattern and design,* strategies are mandatory the analysis for the development of the pedestrian streets and precincts and a distribution of buildings to have a compact development. To ensure affordable development are suggested mixed-use neighbourhoods and a correct mix of housing types and affordability.

To reduce the traffic movement of private cars, a diminished parking footprint strategy with the preference for the location of parking areas for green vehicles is required. To demonstrate the correct design of the public transportation system, the strategies are resumed in the transit facilities and transportation demand management credits.

Other rewarding strategies are the access to civic and public space, neighbourhood schools, and the access to recreation facilities and the presence in the district of local food production and a marketplace.

13.3.3 Green Infrastructure and Buildings

Green infrastructure and buildings strategies are related to the quality of buildings and infrastructures. It is mandatory to have the certification of minimum three buildings with the LEED® rating system for buildings, the definition of the guidelines for all buildings and infrastructure to reduce the energy and water consumption. Particularly, rewarding are the strategies to reduce the energy and water consumption with reference to the baseline project.

In order to reduce the construction cost and optimise the bioclimatic behaviour, in this category the heat island reduction and solar orientation credits with the related production of energy from solar panels are important. The use of the distributed renewable energy production and district heating and cooling systems is promoted to reduce the energy costs.

To optimise the **water efficiency in the buildings and in outdoor spaces,** the rating system suggests the creation of an integrated water system design within which rainwater could be collected and reused, and the reuse of the grey water is promoted.

The same analysis requested for buildings is useful to define the reduction in energy consumption of the infrastructure: for example, for the lighting system, the use of lamps with reduced energy consumption and a lighting analysis is requested.

The **impact of the material construction** is focused on some credits for example with the building reuse and historic resource preservation and adaptive reuse strategies. Another contribution is the recycled and reused infrastructure credit and that for solid waste management where there is a need to create a waste management plan.

13.4 LEED® for Cities and LEED® for Communities

LEED®-certified cities and LEED®-certified communities are the instrument to plan new cities and transform those existing into healthy, sustainable cities, based on high quality of life, safety, health, education, resilience and energy for all residents.

The system of rating has been developed for Plan and Design projects and also for Existing projects. The project team will identify and prepare documentation demonstrating the LEED® credits to achieve the prerequisites and selected credits.

All the information will be entered into the Arc platform to fulfil each of the required actions, including any supporting documents and the platform returns a performance score. The certification levels are the four hereunder:

- LEED Certified: 40–49 points earned
- LEED Silver: 50–59 points earned
- LEED Gold: 60–79 points earned
- LEED Platinum: 80+ points earned.

The system of rating is based upon prerequisites and credits in the following fields:

- Integrative process
- Natural Systems and ecology
- Transportation and land use
- Water efficiency
- Energy and greenhouse gas emissions
- Material and resources
- Quality of life
- Innovation
- Regional priority.

Integrative process, Integrative Planning and Design Process is a prerequisite and helps to support high-performance, cost-effective city outcomes through an early analysis of the interrelationships among city systems. In order to demonstrate the related strategy, a comprehensive plan which includes the key goals and objectives of the city or community must be produced.

Once identified the conditions and the local resources, the sustainable design needs to evaluate the existing ecosystem conditions and services provided by ecosystems, built landscapes and other open spaces to inform the city development along with conservation and restoration efforts. The minimum information covers the topography, the topographic features, slope stability risks, the soil–natural resources, the vegetation and habitat, the hydrology and aquatic ecosystems.

This system of rating focuses on the quality of life and wellness of the community, for example, ensuring green spaces of 11.25 m^2 per person within walking distance (defined as max. 800 m).

A very interesting aspect is the *resilience planning* to ensure the strengthen resilience of the communities to climate change risks, natural and man-made hazards and extreme events (Table 13.1).

Starting from the vulnerability and capacity assessment, the climate change risks and the natural and man-made hazards and extreme events are identified in the local context.

For each risk identification, one must identify the impacts from which an area is at risk, estimate the probability of occurrence of the extreme events, study their characteristics, frequency and potential severity and conduct a socio-economic and environmental assessment of the impact. Thereafter, an *adaptation and mitigation analysis and strategy* to reduce all the impacts and the related fundamental emergency planning are required.

For the green development of the city, it is recommended to encourage a compact and mixed-use development and high level of connectivity within the city or community: walking, biking and *public transport* transit are to be encouraged. It is

Table 13.1 Classification of impacts for vulnerability and capacity assessment (USGBC 2019c)

Natural		Man-made	
Geo-physical	Earthquake	Social	Complex emergencies/conflicts
	Landside		Displaced population
	Tsunami		Cyber attack
	Volcanic activity	Technological	Infrastructure failure
Hydrological	Avalanches		Fire
	Flood	Industrial	Explosion
Climatological	Extreme temperatures		Accidents
	Drought		Accidents on road, air, railway maritime
	Wildfire	Transport	Air and water
	Extreme heat/cold	Pollution	Air and water
	Heat Island effect		
Meteorological	Cyclones		
	Storm/wave surges		
Biological	Disease epidemics		

suggested to have the construction of green building based on a local policy and with incentives.

To promote efficient operation of transport systems, user facilitation, behavioural change and reduced environmental impact through smart technologies and transportation policies are all necessary elements.

In order to reduce pollution by promoting alternatives to fossil fuel vehicles, the installation of electric vehicle charging facilities and electric vehicle supply equipment (EVSE) in 2% of all public parking spaces, or at least two spaces is requested: it must also be demonstrated that the number of planned private and public electric vehicle charging stations exceeds 1.07 per 10,000 residents. As an alternative, one must demonstrate that the total number of government and privately owned planned alternative fuel stations meets or exceeds 1.52 per 10,000 residents.

For water management, an *integrated water management strategy* to reduce freshwater consumption, increase the resiliency of the cities and encourage to move towards a net-zero water city must be demonstrated. The fundamental strategy is to develop the water balance statement for the city by projecting the total water demand based on various uses and designing the supply side to meet the total water demand. In a typical project, the ratio of water withdrawals for human usage to the total freshwater resources is less than 0.2.

To ensure the quality of the water, the provision to all sections of the society of equitable access to clean drinking water and sanitation services is required: pollution from storm water runoff must also be prevented. An integrated process is requested to demonstrate that the water and wastewater systems guarantee 100% coverage of all buildings within the boundary of the public water supply and wastewater collection system, including centralised and decentralised systems. One must additionally provide adequate infrastructure to manage storm water within the city to demonstrate reduction in runoff, maximum capture, treatment and reuse of storm water for a minimum of the sixtieth percentile of local or regional rainfall events based upon ten-year 24 h rainfall event data.

To control the efficiency of the water management system, it is necessary to verify the 100% of water connection units with smart water metres to measure the total potable water use at the point of connection and the water supply monitoring and audit.

Off-grid developments or micro-grids and strategies to reduce consumption, supported by the city development plans or policies, are required to integrate the energy and greenhouse gas emissions management. For example, the design and installation of street lighting and public area lighting should meet the efficiency related requirements of "ANSI/IESNA RP-8-14 Roadway Lighting" or internal equivalent standard. Another strategy is to develop a grid harmonisation to improve operational efficiency of the energy system, with an infrastructure that provides access to dynamic pricing for metered users to motivate load shifting.

Yet another strategy is to incorporate a district energy system (DES) to reduce the environmental and economic harm associated with fossil fuel energy and reduce greenhouse gas emissions by increasing self-supply of renewable energy, use of grid-source renewable energy technologies and carbon mitigation projects.

For a correct management of construction materials, the strategies start from a construction and demolition waste management that includes solid waste management and organic waste treatment and are completed by recycling infrastructure. Another strategy is to provide smart waste management systems using any *appropriate technology* or adopt the supporting policy to develop smart waste management systems for the city. The points are awarded based upon the percentage of total waste generated by the city which is handled by the smart systems.

During the design and construction of new infrastructures, it must be demonstrated that the city is using one or more of the following criteria for a minimum of 20% by cost of the total value of permanently installed top three materials used in infrastructures.

Furthermore, it is suggested to use products and materials for which the lifecycle information is available and helps to produce a lifecycle analysis.

In function of the reuse of the recycled construction materials, smart systems to collect the waste from the final users are particularly interesting. An example is the use of pneumatic transport systems where two pipes, one for compostable and other for recyclable waste, should run parallel underground to the central point collection. Another strategy is the smart bin and route optimisation where the plan must include information about bin location via sensors and route optimisation of waste vehicles moving within the city, or the use of sensor bins; the ultrasonic sensors installed in refuse bins with the RFID technology[1] guide the level of filling with waste and a communication system will transfer this information to the cloud for further processing and analysis.

So far as the social impact is concerned, the *demographic assessment*, the *social infrastructure* and the *affordable housing and the public health* strategies are relevant.

The demographic assessment helps to describe the population demographics and housing characteristics of the area; starting from a report on the history of development and the critical points of change for the overall area or specific neighbourhoods, a report is requested concerning the preeminent sociocultural groups present, such as immigrants, religious and linguistically isolated groups. Racial/ethnic composition should be included if applicable.

The social infrastructure provides facilities and services to citizens which help in meeting their social needs, maximise their potential for development and enhance community well-being. For education, it is necessary to demonstrate the presence of infrastructure facilities for secondary education, higher secondary education and vocational training.

In the public health credits, the purpose is to promote health and wellness in the city, for example, with outdoor air monitoring for a minimum of five pollutants including ozone (O_3, particulate matter ($PM_{2.5}$, PM_{10}) and any of the three pollutants from—sulphur dioxide (SO_2), nitrogen dioxide (NO_2), carbon monoxide (CO), ammonia (NH_3), lead (Pb) or volatile organic compounds (VOCs).

Another social strategy concerns the access to healthy food, i.e. provide permanent and accessible urban agriculture space of at least 2.5 square feet per capita (0.25 square metre per capita) allocated within all residential zones through one or more locations. These can be for example community garden, edible landscaping with labelled plants and signage containing harvesting guidelines, a small farm or orchard, private gardens and/or rooftop gardens. Another alternative is to adopt regulations with the obligation to provide supermarkets or grocery stores selling fresh produce within all residential zones. Within each residential zone, a minimum of 50% of the buildings must have access within ¼ mile (400 m) walking distance.

The active lifestyle credits provides access to a minimum of two services within all residential areas for example an athletic field and/or sports courts, trail network, play space geared towards children (e.g. playground) or an all-weather fitness zone.

For the emergency management and response, sufficient capacity and capability to respond to emergency incidents and reduce their impact on human life/health must be created. The requirements call for design the emergency management systems such that the emergency response time meets the National Fire Protection Association (NFPA) 36 1710 and 1720 requirements of response time to be within 9 min of the dispatch notification. Alternatively, it must meet the local, state or national equivalent for each of the following emergencies: fire, medical, police and special circumstances or operations as relevant for the city.

---

[1]RFID is an acronym for "radio-frequency identification" and refers to a technology whereby digital data encoded in RFID tags or smart labels that are captured by a reader via radio waves.

13.5 Implementation of a Sustainable Protocol in a Real Context: A Case History

The use of urban-scale sustainability certification systems described in this chapter greatly facilitates sustainable planning of neighbourhoods or indeed of entire cities. In fact, rating systems define sets of indicators that consider a broad spectrum of problems linked to environmental sustainability and therefore contribute to promoting spatial planning which can actively compete with climate change.

However, it is important to understand what the purpose of rating systems really is and also what may be the interferences that can arise with other planning tools such as the Town Planning Scheme (TPS), the Municipal Energy Plan (MEP), Urban Mobility Plan (UMP) and the Building Code (BC).

All the planning tools, if used correctly, can contribute to increasing the sustainability of neighbourhoods, towns and cities. However, it is useful to remember that, whilst environmental certification protocols such as LEED® ND or BREEAM® Communities are proposed on the basis of a private initiative, often aimed at giving added value to projects economically speaking, the other instruments mentioned above (TPS, MEP, UMP, BC) originate in the public sector sphere, supported by laws and regulations, and are managed by the public administration (e.g. the municipality or local council).

A study conducted by Dall'O' et al. (2013) investigated the possible conflicts that may arise from the simultaneous application of rating systems for environmental sustainability on an urban scale and the conventional planning tools managed by the public administration/local authority.

The case study analysed is the town of Baranzate, a medium-small municipality located in the north-east of the Province of Milan (Italy) (Lat. 45° 31′37″ N, Long. 9° 06′51″ E). Baranzate has a population of 11,448 inhabitants, the average population of a typical Italian municipality, and a total area of 2.78 km^2 2023.84 inhabitants/km^2. The climate is temperate (2404 Winter Degree Days-DDH), sub-continental, and is characterised during the year as follows: four months cold and wet (from November to June February) and two hot and humid months (July and August). The average annual air temperature is equal to 13.2 °C with an average temperature range (difference of warm weather, July, and coldest, January) of 21.3 °C. The design air temperature for heating systems is -5 °C.

The urban context was characterised by the presence of many industrial and service sectors areas, some of which, partly because of the economic crisis, are no longer used. From the social point of view, the municipality of Baranzate is an interesting and unique case, considering the Italian context. The population is characterised by a strong presence of foreign nationals, average to lower-middle-class workers, who have created communities through which integration and social cohesion are sought. The municipal government/local council has proposed, in recent years, a policy strongly orientated towards social cohesion but also to the improvement of energy and environmental sustainability. All the tools of sustainable planning have been activated:

- approval of a *Town Planning Scheme* aimed to recover and enhance unused urban areas;
- approval of a *Building Code* which requires high energy quality standards for new buildings (energy class A) and for buildings subject to major renovation;
- implementation of a *Urban Mobility Plan* that addresses sustainable mobility;
- signatory to the Covenant of Mayors and the approval of an *Sustainable Energy Action Plan* (SEAP);
- implementation of a *Municipality Energy Plan* in order to reduce energy consumption and greenhouse gas (GHG) emissions.

The political and administrative context of Baranzate seemed ideal to test a possible implementation of the LEED® ND Protocol and its integration with the energy planning and management tools approved or in the approval process.

Figure 13.6 shows the general layout of the town identifying the areas in which urban redevelopment was planned. The first step of analysis was to identify areas that may be certified using the LEED® ND Protocol.

Within the TPS, that is the framework for the urban programming and development, it is possible to define particular areas subject to urban regeneration; Fig. 13.6 highlights these zones by means of greyscale outlined and shaded areas.

The areas highlighted in light grey are characterised by the presence of industries and commercial areas; these areas are not compatible with LEED® ND certification.

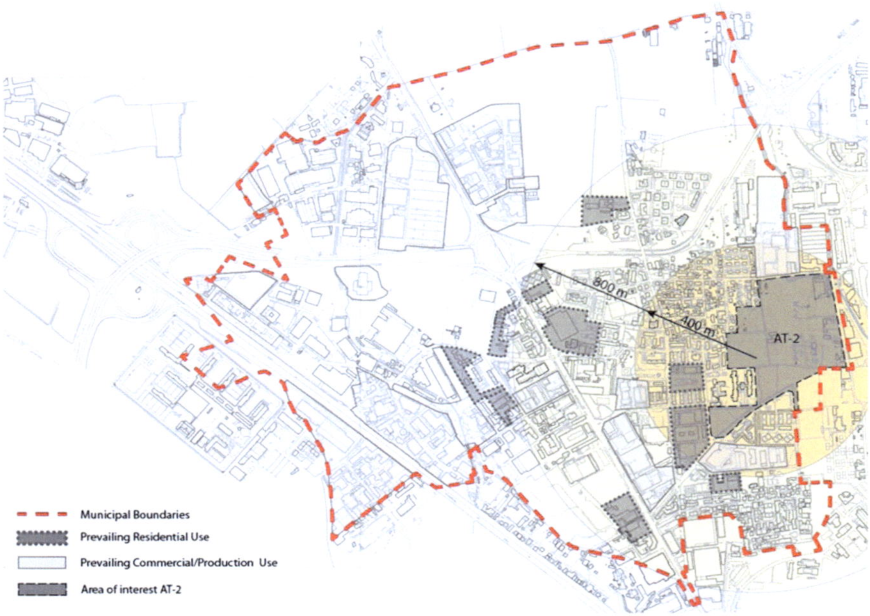

Fig. 13.6 General layout of the town highlighting the areas in which the urban redevelopment is scheduled within the Town Planning Scheme (TPS) (Dall'O' et al. 2013)

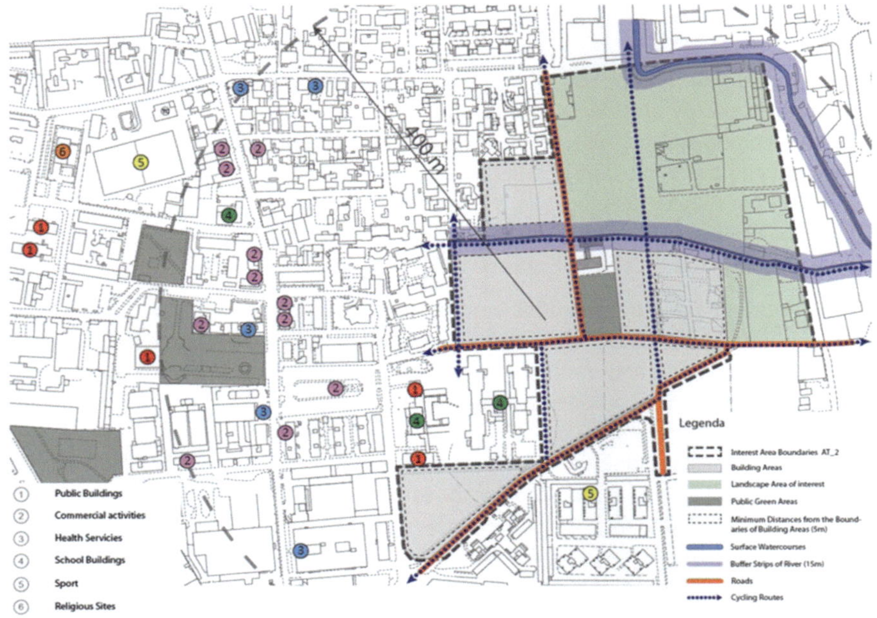

Fig. 13.7 Characteristics of the urban area covered by the application of the LEED® ND Protocol (Dall'O' et al. 2013)

The areas indicated in dark grey, however, have a strong presence of residential usage and thus could be considered. An area located to the east of the city was chosen for the test of a possible application of the LEED® ND Protocol as shown in larger scale in Fig. 13.7.

For the area under consideration, all the provisions contained in the planning and sustainable tools already approved or in the approval phase (TPS, IPA, MEP, UTP and SEAP) were carefully analysed in order to verify the compatibility and compliance with the prerequisites and credits of the LEED® ND certification scheme.

In the area covered by the study, shown in Fig. 13.7, three scenarios were considered:

- *Baseline scenario*: the prerequisites and credits in the LEED® ND were compared with the actions already provided for within the energy and environmental planning tools;
- *Silver scenario*: the energy and environmental planning tools already implemented were integrated with additional rules (prerequisites and credits) in order to achieve the Silver certification of the LEED® ND Protocol;
- *Gold scenario*: the energy and environmental planning tools already implemented were integrated with additional rules (prerequisites and credits) in order to achieve the Gold certification of LEED® ND Protocol.

Table 13.2 Results of application of LEED ND to Baranzate municipality

Rating scenario	Rating				Additional effort
	SLL	NPD	GCT	Total ([a])	
Baseline	10	13	3	26	NA
Silver	11	32	9	52	Low
Gold	13	34	15	62	Medium

([a]) The credits for Innovation and Design Process (IDP) and Regional Priority Credits (RPC) were not considered in this simulation
SLL smart location and linkage; *NPD* neighbourhood pattern and design; *GCT* green construction and technology (Dall'O' et al. 2013)

The results of the three simulations are shown in Table 13.2, which highlights the scores achieved in the three areas of evaluation (SLL, NPD and GCT).

The baseline scenario shows the impossibility of obtaining the LEED® ND certification, because the rules contained in the actual planning tools do not comply with the prerequisites that are mandatory in the protocol.

By implementing the existing planning tools with a few additional rules, and limited effort for the public administration/local authority, it is possible to aspire to the Silver certification. There is still a margin and the need for further additional rules, to obtain Gold certification.

The aim of the LEED® ND Protocol is to become an effective tool for the transition of a neighbourhood or a part of a city towards an ideal situation in which the concept of sustainability becomes a fact and not just a fascinating theory.

To provide a better understanding of the role and potential of tools such as LEED® ND, it is necessary to make a clarification. Raimi et al. (2013) define a neighbourhood as a place with its own unique character and function, where people can live, work, shop and interact with their neighbours. The most sustainable neighbourhoods tend to exhibit high levels of walkability, a sense of place, social cohesion and stability and neighbourhood resiliency amidst changing economic and socio-political conditions.

Since the LEED® ND programme is market driven, local governments have the opportunity to influence new development in a way that was only previously possible with zoning and permits. Local governments and councils can prepare themselves for the implementation of LEED® ND developments and shape the future growth of a city by considering a number of aspects related to the programme (Miller 2010).

This protocol could be the real driver for a policy strongly orientated towards the sustainability of land use, for a "greening" of our neighbourhoods and of our cities, and could be a uniting element for all the other planning tools around energy and sustainability. Additionally, foreign investors would be better ensured in the cases where land planning references an international standard.

13.6 Conclusions

The environmental sustainability certification protocols declined in the urban context can be considered to be effective tools not only to promote the sustainability of neighbourhoods and cities, but also to objectively measure, through shared indicators, the evolution of the territory in the different areas of sustainability (environmental, social and economic).

A question that emerges from the analysis of the most widespread protocols is what weight to give to these planning tools within a context in which other programming tools that are often supported by the public administration are applied.

At this point, one may ask: can the rating systems with LEED® ND, BREEAM® Communities or others be considered to be planning tools? To support this doubt, it is useful to remember that these rating systems, which are not yet very widespread on an international scale, were created with the objective of making real estate environmentally sustainable at a territorial scale and, thereby, of promoting a culture that finally can change a market that is too tied to interests far from environmental awareness.

Considering these systems simply for commercial purposes would underestimate their importance. In these protocols, in fact, systematic aspects of green and sustainable planning are considered in a methodical manner and approaches that significantly improve the quality of life of the citizens are redefined: so a new relationship is provided between the built environment and the natural environment, among the grey infrastructures and green infrastructure, just to give examples.

In the technical scientific literature, the sustainability certification protocols applied on a territorial scale have been the subject of interesting studies and research.

Some of these are reviews which compare different approaches, such as the work of Kaur and Gang (2019) that presents a comprehensive review of literature on the most widely used urban sustainability assessment tools. The aim of the paper is to understand the similarities and differences in existing urban sustainability assessment tools and identify the gaps so as to discover whether these are capable of and suitable for addressing multiple issues of urban sustainability in multiple contexts, including settlements, in a diversity of geo-climatic and ecologically sensitive regions.

Pedro et al. (2019) provide a literature review of the most widely used urban sustainability assessment systems: BREEAM® Communities, LEED® Neighbourhoods, CASBEE® Urban Development, Green Star Communities and DGNB® Urban Districts. The study revealed that there is a need for context customisation of global targets into local actionable measures, involvement of regulatory bodies to ensure the successful application of such systems and consideration of socio-economic factors since the assessment is still very focused on the environmental impact.

The transition from the protocols for the certification of the sustainability of buildings to those for the certification of community sustainability has been studied by Wu et al. (2018). The authors, in order to facilitate the transition process, in this paper review and analyse the credits obtained from LEED® ND 2009 certified plans. A total of 55 projects were identified from the LEED® project directory of the US

Green Building Council. The results indicate that there is an unbalanced allocation of credits to economic, social and environmental sustainability in the LEED® ND rating. In addition, green infrastructure and building credits, such as wastewater management, on-site renewable energy and solar orientation, have extremely low percentages of achievement, indicating that these credits should be reworked. The results also provide useful insights for developers to prepare for LEED® ND certification and for regulatory bodies to improve the performance of the current LEED® ND rating system.

Some works critically analyse rating systems to verify their effectiveness as "green" planning tools. Reith and Orova (2015) argue that neighbourhoods are the building blocks of cities, which have their own architectural, cultural and economic systems. In the scope of their study, five assessment systems, CASBEE® UD, the 2009 and 2012 versions of the BREEAM® Communities, LEED® ND, and DGNB® UD were compared and the paper presents the methodology and results of their comparative investigation.

Diaz-Sarachaga and Jato-Espino (2019) analyse the contribution of rating systems on the issue of city resilience. In their research, the authors, after selecting the most relevant community rating systems, determined their adequacy to handle urban challenges by benchmarking them against the major international efforts and some relevant tools of resilience assessment extracted from the study of eight global frameworks. The existence of prominent gaps in all the community rating systems which were screened suggests the need for the development of a new tool which involves both resilience and sustainability.

On the issue of resilience the contribution of Uda and Kennedy (2018) is interesting. The paper examines LEED® ND to determine the extent to which its supported actions also contribute to a neighbourhood's specified resilience. The authors argue that LEED® ND does in fact contribute to a neighbourhood's specified resilience, but it does not do so comprehensively nor in the optimum manner.

Several authors have analysed the aspects related to the perception that citizens have about the value of neighbourhoods designed for sustainability. For example, the study of Stanislav and Chin (2019) analyses the capability of residents and users of spaces to perceive the actual design components and theories in place which have actively created the community in which they live and which they use.

In a holistic view of the green planning of cities and communities, such as that proposed in this book, the rating systems, with their potential to affect the real estate market, by radically transforming it, are indispensable pieces of a complex mosaic that is the planning of neighbourhoods of cities and of communities.

For this reason, the environmental certification protocols of neighbourhoods and cities should be considered to be urban planning tools although it would be a mistake to consider them to be the only ones that can be used.

References

BRE Global (2019) Green book live: certified BREEAM assessments. Available on http://www.greenbooklive.com/search/scheme.jsp?id=202. Accessed 1 Dec 2019

Dall'O' G, Galante A, Sanna N, Miller K (2013) On the integration of leadership in energy and environmental design (LEED)® ND protocol with the energy planning and management tools in Italy: strengths and weaknesses. Energies 6:5990–6015. https://doi.org/10.3390/en6115990

DGNB (2019) DGNB—Urban Districts. Available https://www.dgnb-system.de/en/schemes/scheme-overview/urban_districts.php. Accessed 1 Dec 2019

Diaz-Sarachaga JM, Jato-Espino D (2019) Do sustainable community rating systems address resilience? Cities 93(2019):62–71

GBC Italia (2019) GBC quartieri. Available on http://www.gbcitalia.org/quartieri. Accessed on 1 Dec 2019

JaGBC (2014) Dissemination of CASBEE in Japan. Available on http://www.ibec.or.jp/CASBEE/english/statistics.htm. Accessed on 1 Dec 2019

Kaur H, Garg P (2019) Urban sustainability assessment tools: a review. J Clean Prod 210:146e158

Miller K (2010) Sustainable neighborhood development 2010. A Toolkit for Understanding US and European Approaches, Master's thesis. Polytechnic University of Milan, Milan, Italy

Pedro J, Reis A, Duarte Pinheiro M, Silva C (2019) A systematic review of the international assessment systems for urban sustainability. In: Sustainable built environment conference 2919 (SBE19 Graz), IOP conference series: earth and environmental science, vol 323. IOP Publishing, pp 012076. https://doi.org/10.1088/1755-1315/323/1/012076

Raimi + Associates and the Natural Resources Defense Council (NRDC) (2013) A citizen's guide to LEED for neighborhood development: how to tell if development is smart and green. Available online https://www.nrdc.org/cities/smartgrowth/files/citizens_guide_LEED-ND.pdf. Accessed 1 Dec 2019

Reith A, Orova M (2015) Do green neighbourhood ratings cover sustainability? Ecol Ind 48(2015):660–672

Stanislava A, Chin JT (2019) Evaluating livability and perceived values of sustainable neighborhood design: new urbanism and original urban suburbs. Sustain Cities Soc 47:101517

Turcu C (2013) Re-thinking sustainability indicators: Local perspectives of urban sustainability. J Environ Plan Manag 56(5). https://doi.org/10.1080/09640568.2012.698984

Uda M, Kennedy C (2018) Evaluating the resilience of sustainable neighborhoods by exposing LEED neighborhoods to future risks. ASCE Am Soc Civ Eng Infrastruct Syst 24(4):04018030. https://doi.org/10.1061/(ASCE)IS.1943-555X.0000443

USGBC (2019a) Projects US green building council. Available on https://www.usgbc.org/projects. Accessed 01 Dec 2019

USGBC (2019b) LEED for cities and communities. Available on https://new.usgbc.org/leed-for-cities. Accessed 1 Dec 2019

USGBC (2019c) LEED V4.4 cities and communities, plan and design (Beta Manual)

Wu P, Song Y, Hu X, Wang X (2018) A preliminary investigation of the transition from green building to green community: insights from LEED ND. Sustainability 10:1802. https://doi.org/10.3390/su10061802

Chapter 14
Sustainable Rating Systems for Infrastructure

Giuliano Dall'O' and Elisa Bruni

Abstract The use of any territory at various scales, from the national to the urban or neighbourhood, would not be possible if there were no infrastructures. Such infrastructure is composed of a variety of public and private physical attributions to improve that territory: these include various types of civil engineering structures, e.g., bridges, motorways, railways, airports but also energy and communication networks, just to mention a few examples. It plays a strategic role in the social, economic, environmental well-being of human beings but also of the ecosystems. To underline the importance of infrastructures, one of the Sustainable Development Goals of the U.N. (SDG 9) regards the development of quality, reliable, sustainable and resilient infrastructures. In the green planning of cities and communities, the topic of infrastructure must be examined with great attention. The purpose of this chapter is to address the issue of infrastructure by focusing on the relationship between international environmental sustainability protocols and infrastructure. The chapter will analyse in particular the international Envision® protocol, designed to certify the sustainability of infrastructures by presenting some examples of its application.

14.1 Introduction

The complexity of the topic "infrastructures" is demonstrated first of all by the different definitions that are given to this subject. Considering a territory made up of places (e.g., buildings) in which the most diverse functions can be performed, from the residential to the industrial, artisan craft or commercial, "infrastructure" can be defined as the set of connection systems between these places. This definition is clearly reductive since a railway network can be considered to be an infrastructure,

---

G. Dall'O' (✉)
Architecture, Built Environment and Construction Engineering, ABC Department, Politecnico di Milano, Via G. Ponzio, 31, 20133 Milan, Italy
e-mail: giuliano.dallo@polimi.it

E. Bruni
ICMQ S.p.A, Via G. De Castilia, 10, 20124 Milan, Italy

G. Dall'O' (ed.), *Green Planning for Cities and Communities*,
Research for Development, https://doi.org/10.1007/978-3-030-41072-8_14

but at the same time, a railway station, a physical place similar to a building, can also be considered as an infrastructure.

Attempts to define infrastructure have been made by public and private institutions, professional organizations, stakeholders associations and, of course, researchers. Although a complete overlap between the definitions is impossible, there are common elements that concern physical components that make up the infrastructure, interrelation systems and social needs understood as the need for services that must be guaranteed.

Among the different definitions, one of the most complete and the current one is the definition proposed by the American Society of Civil Engineers: "*The infrastructure supporting human activities includes complex and interrelated physical, social, ecological, economic, and technological systems such as transportation and energy production and distribution; water resources management; waste management; facilities supporting urban and rural communities; communications; sustainable resources development; and environmental protection*" (Fulmer 2009; ASCE 2009).

In the report published in 2009, ASCE (2009) supply cards describing the infrastructures considered cover water and environment (dams, drinking water, hazardous waste, levees, solid waste and wastewater), transportation (aviation, bridges, inland waterways, rail, roads and transit), public facilities (parks, recreation and schools) and Energy (Table 14.1).

Fulmer (2009) states that there are two general types of infrastructure, hard or soft:

- *Hard infrastructure* that refers to the physical networks necessary for the functioning of the modern industry (e.g., roads, bridges, railways, etc.);
- *Soft infrastructure* that refers to institutions that maintain the economic, health, social and cultural standards of the country (e.g., educational programmes, official statistics, parks and recreational facilities, law enforcement agencies and emergency services).

In the cited paper, Fulmer provides a complete list of hard infrastructures and soft infrastructures, although it must be remembered that technological evolution and possible future needs on the part of users could extend both lists (e.g., one thinks of the possible evolution of 5G and autonomous driving of vehicles).

If the purpose of green planning is also to stimulate and harmonize the relationship between urbanized areas and natural areas, we can consider a further classification between *grey infrastructures*, namely traditional infrastructures (roads, bridges, railways, etc.) and *green (blue) infrastructures*.

By this term is meant the infrastructures designed to structurally enhance the relationship between a city and nature (Table 14.2).

Green infrastructure refers to ecological systems, both natural and engineered, to provide clean water, conserve ecosystem values and functions, biodiversity and a wide array of benefits to people and wildlife.

Infrastructures significantly affect the economy of a nation as they represent an important element in public and private economic balance sheets.

Table 14.1 Infrastructures divided by sectors and functions (adapted from Fulmer 2009)

Sector	Function
Electric power	Power generation Power transmission Power distribution
Oil and gas	Exploration and production Oil and gas gathering and separation Oil storage, refining, transport and distribution Natural gas processing, transport and storage
Potable and wastewater	Water supply, storage and treatment Water delivery Wastewater collection Treatment and discharge
Transportation	Road transport Rail transport Waterway transport Sea transport Air transport Mass transport
Communication	Wireline services Cable services Wireless services Broadcast services Data processing/network management

Table 14.2 Green and blue infrastructures in urban contexts

Infrastructure	Description
Green infrastructure	Green ways and ecological corridors Urban parks and urban forests Rain gardens, streams, wetlands, floodplains Rainwater tank (or rain barrel) Cycle and pedestrian paths
Blue infrastructure	Pools, ponds and pond systems Artificial buffer basins Watercourses (rivers and canals) Urban lakes Fountains

Infrastructures create value and contribute to the economic growth of countries thanks to both direct and indirect effects, increasing GDP (worldwide annual investment in economic infrastructure is included, between 4 and 8% of GDP) and creating jobs. Infrastructures can therefore favour, in principle, the overall well-being of a territory (Secchi and Riela 2019).

If the infrastructures are so important not only from a social but also from an economic and environmental point of view and represent important (often public) investments, which features should they have?

On the subject of infrastructures, important indications come from the Sustainable Development Goals (SDGs) under the 2030 Agenda for sustainable development proposed by the United Nations in 2015.

SDG 9, industry, innovation and infrastructure, aims to build resilient infrastructure, promote inclusive and sustainable industrialization and foster innovation. To make infrastructures and industries sustainable by 2030, resources will have to be used more efficiently by clean and environmentally friendly technologies and industrial processes will have to be encouraged. Technological development, research and innovation will have to be supported in particular in developing countries. Furthermore, the access of industries and other small businesses to financial services, including credit at advantageous conditions, should be fostered, and their integration into markets and value creation chains should be increased. Finally, in less-developed countries, Internet access must be simple and widespread.

The topic of infrastructures is also considered in the SDG 11 (sustainable cities and communities) aimed at making cities and human settlements inclusive, safe, resilient and sustainable, and in SDG 15 (life and land), whose aims are to protect, restore and promote sustainable use of terrestrial ecosystems, sustainably manage forests, combat desertification, halt and reverse land degradation and halt biodiversity loss.

To achieve the objectives defined by the aforementioned SDGs, it is necessary to ensure that the design premises related to sustainability, inclusiveness and resilience of infrastructures are really implemented.

The environmental sustainability protocols, structured to manage all these aspects, can be used as an effective control and verification tools in all the design phases, from the conception of the projects to that of infrastructure management after their execution.

14.2 How International Rating Systems Deals with Infrastructures

International rating systems for measuring the sustainability of buildings, neighbourhoods and cities enhance the presence of infrastructures in the territory, on the one hand by defining pre-requisites to be respected in order to obtain certification and on the other by allowing them to obtain credits.

In this section, the rating systems closest to the concept of green planning will be specifically considered, i.e., whose purpose is to measure and certify sustainability on an urban scale.

Generally speaking, rating systems for sustainability value urban planning works do not require the construction of new infrastructures, for example an efficient public transport network or the presence of services for citizens, but rather exploit existing ones, thus making the new intervention of urban planning of lesser impact and more sustainable.

14.2.1 LEED® for Neighbourhood Development

LEED® for neighbourhood development (LEED® ND) is a rating system promoted by the United States Green Building Council (USGBC) that integrates the principles of smart growth, urbanism and green building into a national system for neighbourhood design. This rating system recognizes development projects that successfully protect and enhance the overall health of the population, the natural environment and the quality of life. The rating system encourages smart growth and New Urbanism best practices (Boeing et al. 2014).

While the potential of LEED® ND has already been described in Chap. 13, in this section, we examine in depth how the protocol interfaces with infrastructure and sustainable transport. The protocol is not only a certification system but also a set of sustainable standards for land development which can be useful to anyone interested in sustainable community planning and design, including community leaders and government officials (US GBC 2015).

The protocol is organized into three sections:

- *Smart Location and Linkage*—SSL—(where to build)
- *Neighbourhood Pattern and Design*—NPD—(what to build)
- *Green Infrastructure and Buildings*—GIB—(how to manage environmental impacts).

LEED® ND has the aims of connected neighbourhoods (i.e., frequent street connections and pathways to surrounding areas), neighbourhoods that use land efficiently (compact development requires less infrastructure to serve the same number of people) and bicycle-friendly design (bicycle network).

Categories which are most related to infrastructure and sustainable transport are smart location and linkage and neighbourhood pattern and design, and in particular, the credits are given in Table 14.3, where the credit requirements and objectives are explained.

14.2.2 LEED® v4.1 for Cities and LEED® v4.1 for Communities

In 2016, USGBC introduced the LEED® for Cities and LEED® for Communities to help city leaders to communicate performance across an array of objectives and to differentiate types of stakeholders.

Table 14.3 Credit requirements and objectives of LEED ND Protocol related to infrastructure and sustainable transports

Credits	Objectives	Requirements
Preferred locations	To conserve the natural and financial resources required for infrastructure	Option 1. Location type Option 2. Connectivity Option 3. Designated high-priority locations
Bicycle facilities	To promote bicycling and transportation efficiency and reduce vehicle distance travelled	Provide short-term bicycle storage Provide long-term bicycle storage Option 1. Bikeable location Option 2. Bicycle network
Compact development NPD credit 2: Compact development	To conserve land. To promote livability, walkability and transportation efficiency and reduce vehicle distance travelled. To leverage and support transit investments	Design and build the project to meet the densities specified
Connected and open community	To promote projects that have high levels of internal connectivity and are well connected to the community. To encourage development within existing communities that promote transportation efficiency through multimodal transportation	Design and build the project such that its internal connectivity is at least 54 intersections per square kilometre
Reduced parking footprint	To minimize the environmental harms associated with parking facilities, including automobile dependence, land consumption and rainwater run-off	Do not build new off-street parking lots, or locate all new off-street surface parking lots at the side or rear Use no more than 20% of the total development footprint area for all new off-street surface parking facilities

New and developing cities and communities can use LEED® v4.1 for Cities and LEED® v4.1 for Communities to plan, design and develop sustainable strategies across six LEED® categories[1]:

- Natural systems and ecology;
- Transportation and land use;
- Water efficiency;
- Energy and greenhouse gas emissions;

[1]Source: Selina Holmes on https://www.usgbc.org/articles/leed-v41-launches-cities-and-communities (Accessed 20 October 2019).

- Materials and resources;
- Quality of life.

More than 90 cities and communities around the globe have been certified through LEED® v4.1 for Cities and LEED® v4.1 for Communities (October 2019).

Most of the pre-requisites and most of the credits of these protocols involved the infrastructures (e.g., green spaces, walkability and bikeability, smart mobility and transportation policy, integrated water management, organic waste treatment, recycling infrastructures, smart waste management systems, social infrastructures, smart water systems, wastewater management, storm water management, light pollution reduction, renewable energy and grid harmonization).

The indicators collected during the protocol application process can be included in the ARC platform developed by the Green Business Certification Inc. (GBCI) and managed dynamically to assess the effects of implementing the actions and to create a comparison, on a global scale, with other cities and with other communities.

Scores generated by ARC can be used in a variety of different ways, including streamlining green building certification, tracking project performance, guiding portfolio management and supporting city-wide sustainability efforts. Arc performance scores are an integral part of the LEED v4.1 rating system.[2]

14.2.3 BREEAM® Communities

BREEAM® promoted by BRE is an international scheme that provides independent third-party certification of the assessment of the sustainability performance of individual buildings, communities and infrastructure projects.

BREEAM® Communities is a certification standard based on the established BREEAM® methodology, and the scheme covers the assessment of the designs and plans for new development and regeneration projects while post-construction is not included. There are three steps involved in the assessment of sustainability at the master planning level that interfaces also with infrastructure and sustainable transports (Table 14.4) (BRE Global Limited 2017):

1. Establishing the principle of development.
2. Determining the layout of the development includes requirements regarding how people will move around and through the site.
3. Designing the details involves more detailed design of the development including transport facilities.

BREEAM® Communities has five assessment categories; a sixth category promotes the adoption of innovative solutions. The categories are listed below:

- *Governance*. Encourages community involvement in decisions.

---

[2]Source: Chris Pyke on https://arcskoru.com/learn-more-about-arc (Accessed on 20 October, 2019).

Table 14.4 Credits and objectives of the credits related to transport and movement category

Credits	Objectives
Transport assessment	To ensure transport and movement strategies and reduce the impact of the development upon the existing transport infrastructure
Safe and appealing streets	To create safe and appealing spaces that encourage human interaction
Cycling network	To promote cycling as a leisure activity and as an alternative to vehicle use by providing a safe and efficient cycle network
Access to public transport	To ensure the availability of frequent and convenient public transport links to fixed public transport nodes and local centres
Cycling facilities	To promote cycling by ensuring the adequate provision of cyclist facilities
Public transport facilities	To encourage frequent use of public transport by providing safe and comfortable transport facilities

- *Social and economic well-being*. Considers societal and economic factors affecting health and well-being.
- *Resources and energy*. Addresses the sustainable use of natural resources and the reduction of carbon emissions.
- *Land use and ecology*. Promotes sustainable land use and ecological enhancement.
- *Transport and movement*. Directs the design and provision of transport and infrastructure to encourage the use of sustainable transport.
- *Innovation*. Promotes the adoption of innovative solutions.

Within the transport and movement categories, the credits related to infrastructures are:

- Transport assessment;
- Safe and appealing streets;
- Cycling network;
- Access to public transport;
- Public transport facilities.

In November 2015, CEEQUAL®, a sustainability assessment, rating and awards scheme for civil engineering, infrastructure, landscaping and public realm projects became part of the BRE Group. The scheme is now operating alongside BREEAM®.

14.3 Envision® Rating System for Infrastructures

In the previous section, it has been illustrated how some international rating systems relate to the structures present in the territory, enhancing their presence. This section, instead, examines an international sustainability rating system, Envision®, specifically designed for infrastructures.

14.3.1 General Concepts

The Envision® Protocol is the first rating system to create sustainable infrastructures through an analysis grid, adaptable to any infrastructure development project. Envision® is an independent assessment tool, able to concretely support companies, designers, public sector administrators and citizens in infrastructure design. In terms of investment effectiveness, respect for the ecosystem, climate and environmental risk, durability, leadership and improving the quality of life, Envision® looks at the development of the infrastructure and its sustainability in a holistic and long-term way.

The protocol was developed in joint cooperation between the Zofnass Programme for Sustainable Infrastructure at the Harvard University Graduate School of Design and the Institute for Sustainable Infrastructure (ISI) which is a non-profit organization grouping other organizations and individuals involved in the planning, design, construction, and maintenance of infrastructure. Envision® was launched in 2012.

With Envision®, it is possible to design and build roads, railways, ports, airports, power lines, power plants, communication networks, etc., based on the objective measurement of the advantages that the project brings to the community, management and maintenance capabilities throughout its entire useful life and on the opportunity of sharing investment between public and private capital.

Envision® can be used by proponents and infrastructure managers, designers, community groups, environmental organizations, builders, public sector authorities and private administrations to:

- meet sustainability objectives;
- obtain public and super-partes recognition for the levels of sustainability obtained;
- optimize the location of the project;
- assist local communities and proponents in the participatory planning process and in creating consensus;
- make decisions about the destination of investments;
- include the needs of the territory and local communities in infrastructure projects.

Envision® tools help the project team by:

- facilitating dialogue with stakeholders in order to speed up the implementation process and reduce the risk of territorial disputes;
- enhancing sustainable choices, both environmental and social, in design;
- stimulating creative discussion within the team;
- evaluating the costs and benefits throughout the life cycle of the work (location, design, implementation, operation and disposal);
- evaluating environmental benefits;
- identifying result-based goals;
- achieving higher levels of sustainability;
- valuing high-performance design choices.

Envision® has been applied widely throughout the USA and Canada but is in fact applicable worldwide. Indeed, the protocol was implemented and is being used for something in excess of 100 projects worldwide, whereas 59 projects are pending for certification, 47 of these have already been certified. These projects include two water treatment plants, one water supply network, one road, one industrial facility for fish farming, one wind farm and various plans for industrial area redevelopment. The link https://v3.sustainableinfrastructure.org/projects/map is available for viewing the map of all Envision projects "Registered" or "Verified" (ISI 2019).

In Italy, Envision credentialing and verification are administered by the joint initiative of ICMQ—a certification body specialized in the construction sector—and Stantec, a leader in engineering and architectural consulting and design. In 2019, the first project in Europe to earn Envision Award for Sustainable Infrastructure was the Frasso Telesino—Telese—S. Lorenzo section of the extension of about 21 km of the Naples–Bari railway route, obtained by Rete Ferroviaria Italiana (RFI).

14.3.2 Structure of Envision® Rating System

The structure of the protocol is based on three different levels: categories, sub-categories and criteria (Table 14.5). There are five categories, and they represent the macro areas of impact according to which to assess the sustainability of the project and are:

- *Quality of Life*. Some important questions should be asked upstream of the planning phase, including: does the infrastructure preserve and enhance local resources? Does it help local communities to develop while minimizing potential negative impacts? Is the existing transport network used? The Quality of Life category answers these questions. According to the protocol, the infrastructure project must first take into account the priority objectives of the community, defining what and how many long-term benefits it can really derive, while minimizing the negative impacts on the community.
- *Leadership*. A greater involvement of the community and the main stakeholders in the planning of an infrastructure allow the adoption of far more efficient design solutions. In fact, for a project to be considered sustainable, it is not sufficient that it is efficient from an energy point of view, because the evaluation needs a broader vision. The Leadership category enhances the actions that emphasize a real and continuous community involvement and the project teams' and clients' attention to all aspects of sustainability.
- *Resource Allocation*. In the design, construction and management of an infrastructure, the resources employed play a fundamental role in terms of influence on overall sustainability. The protocol, through the credits of this section and the three sub-categories materials, energy and water, deals with the issues of energy consumption, water and materials in general, providing an operational methodology on their use and exploitation in respect of conservation of non-renewable resources and the containment of water and energy consumption.

Table 14.5 Indicators of Envision rating system (Institute for Sustainable Infrastructure 2018)

Category	Sub-category	Credits	Max points
Quality of life	Well-being	Improve community quality of life, enhance public health and safety, improve construction safety, minimize noise and vibration, minimize light pollution, minimize construction impacts	200
	Mobility	Improve community mobility, encourage sustainable transportation, improve access and wayfinding	
	Community	Advance equity and social justice, preserve historic and cultural resources, enhance views and local character, enhance public space and amenities	
Leadership	Collaboration	Provide effective leadership and commitment, foster collaboration and teamwork, provide for stakeholder involvement, pursue by-product synergies	182
	Planning	Establish a sustainability management plan, plan for sustainable communities, plan for long-term monitoring and maintenance, plan for end-of-life	
	Economy	Stimulate economic prosperity and development, develop local skills and capabilities, conduct a life-cycle economic evaluation	
Resource allocation	Materials	Support sustainable procurement practices, use recycled materials, reduce operational waste, reduce construction waste, balance earthwork on site	196
	Energy	Reduce operational energy consumption, reduce construction energy consumption, use renewable energy, commission and monitor energy systems	

(continued)

Table 14.5 (continued)

Category	Sub-category	Credits	Max points
	Water	Preserve water resources, reduce operational water consumption, reduce construction water consumption, monitor water systems	
Natural world	Siting	Preserve sites of high ecological value, provide wetland and surface water buffers, preserve prime farmland, preserve undeveloped land	232
	Conservation	Reclaim brownfields, manage stormwater, reduce pesticide and fertilizer impacts, protect surface and groundwater quality	
	Ecology	Enhance wetland and surface water functions, maintain floodplain functions, control invasive species, protect soil health	
Climate and resilience	Emissions	Reduce net embodied carbon, reduce greenhouse gas emissions, reduce air pollutant emissions	190
	Resilience	Avoid unsuitable development, assess climate change vulnerability, evaluate risk and resilience, establish resilience goals and strategies, maximize resilience, improve infrastructure integration	

- *Natural World.* The infrastructural systems generate a high impact on the landscape in which they are located, producing both short- and long-term effects on a considerable quantity of elements such as ecosystems, habitats and geomorphological systems. The Envision protocol, through the Natural World category and its three sub-categories siting, conservation and ecology, highlights and rewards the best practices that an infrastructure project should implement in approaching the environment and the natural world.
- *Climate and Resilience.* The infrastructural system is a very complex concept that encompasses many aspects related not only to the design and engineering field, but also to the social, economic, environmental and climatic aspects. The protocol provides tools to evaluate all these aspects and in particular, through the Climate and Resilience category, offers indications to implement a preventive assessment of climate and pollution risks and a consequent optimal design.

This family of credits and its sub-categories, emission and resilience, examines fundamentally two aspects that, respectively, reflect a medium- and short-term vision and a longer-term one: the minimization of the emissions of an infrastructural element or system and the concept of resilience.

The 14 sub-categories identify the main elements of each area and group together a total of 64 sustainability and resilience indicators called "credits". Each criterion provides an indicator of sustainability related to a specific aspect of interest—environmental, social or economic.

These credits measure infrastructure sustainability and address areas of human well-being, mobility, community development, collaboration, planning, economy, materials, energy, water, siting, conservation, ecology, emissions and resilience.

Credits have multiple levels of achievement representing the possible performance goals beyond conventional practice. Following the evaluation of the results achieved in each of the credits, the project teams establish how the project is positioned with respect to the sustainability indicators and decide how to pursue higher performance.

Each of the 64 credits has multiple levels of achievement representing the spectrum of possible performance goals. Teams establish how well the project addresses the full range of sustainability indicators and are challenged to pursue higher performance:

- *Improved*: Performance that is above conventional;
- *Enhanced*: Sustainable performance that is on the right track;
- *Superior*: Sustainable performance at a very high level;
- *Conserving*: Performance that has achieved essentially zero negative impact;
- *Restorative*: Performance that restores natural or social systems.

Scores of each credit are added together to give the total score. The final Envision score is a percentage of the total achieved points compared to the total applicable points.

Finally, Envision rewards innovative methods that promote sustainable practices or demonstrate exceptional performance beyond what is expected in credits. Innovation credits include three options to earn bonus points:

- Innovation;
- Exceptional performance;
- Additional aspects of sustainability.

Envision promotes high performance across three dimensions (Fig. 14.1):

- *Performance achievement*: through the recognition that success in sustainability is incremental;
- *Project phases*: addressing the full project life cycle, beyond planning and design through construction to operations and maintenance;
- *Stakeholder collaboration*: when an inclusive, representative group of stakeholders is engaged throughout the project, the results satisfy the widest possible portion of the community.

Envision was designed to help infrastructure stakeholders implement more sustainable projects to offer benefits for each category of stakeholder: from owners

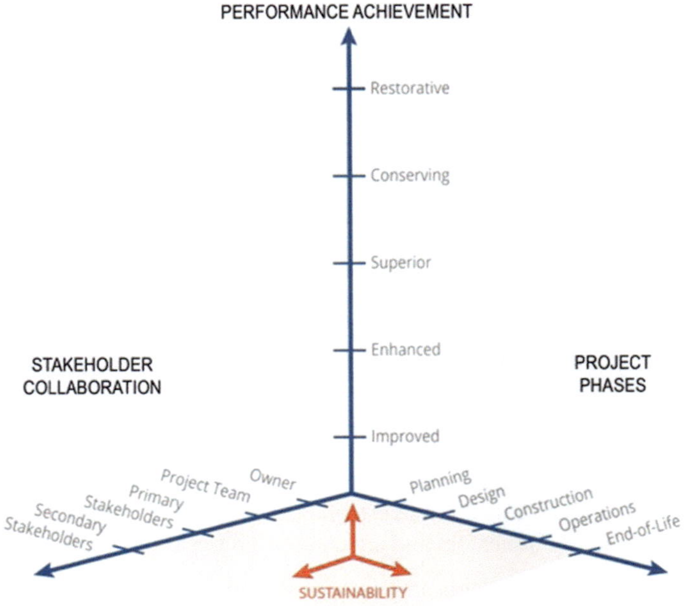

Fig. 14.1 Strategies for Envision® approach (ISI 2018)

and design teams to community and environmental groups, construction companies, regulators and policymakers.

14.3.3 Application and Verification of Envision® Rating System

The process starts with the initial self-assessment of the project using the Envision® scoresheet, after which it is possible to register the project for verification.

The Envision scoresheet is an online tool that allows the project team to jointly evaluate projects, and it also allows to calculate the scores of each credit automatically and finally to register it for third-party verification. Verification is an independent review to confirm that a project meets the evaluation criteria.

At the beginning of the verification process, ISI assigns a qualified Verifier to verify the levels of achievement and the final score submitted by the project team.

Verification enables projects to become eligible for ISI awards, as described below. To qualify for an award, a project must achieve a minimum of the total applicable points (Table 14.6).

Table 14.6 Award levels of Envision® rating system (ISI 2018)

Recognition level	Total applicable points (%)
Verified	20
Silver	30
Gold	40
Platinum	50

Fig. 14.2 Envision® verification pathways (ISI 2018)

A project team may choose to pursue verification after the design phase or after the construction phase, so projects may follow one of the two pathways of verification:

- Path A: Design + Post-Construction
- Path B: Post-Construction

The steps of the two verification pathways are shown in Fig. 14.2.

The diffusion of the Envision® approach also involves the creation of a community of professionals able to master and promote its contents, as a heritage of shared know-how. For this reason, the protocol includes training sessions and an exam for the qualification of Envision® Sustainability Professionals (Envision SP), essentially:

- Reference persons figures for the project teams, who must document the achievement of the sustainability objectives set out in the protocol.
- Guarantees of the adequacy of the project for the purposes of assigning the Envision® certification.
- The protocol also provides for an annual maintenance of the Envision SP qualification obtained.

14.4 Conclusions

Infrastructures are essential elements of the territory at different scales, from national (or international) to urban: they guarantee connections between different places but also energy and information flows. Their integration with the territory, although indispensable to guarantee the necessary services to the community and to individual users, if not managed correctly, could generate environmental impacts and risks for safety. The infrastructures must consequently be evaluated in terms of sustainability and resilience.

A green planning of cities and communities cannot ignore the careful consideration of infrastructures which, over time, thanks to technological innovation and changes in human needs, are subjected to continuous renewal. Infrastructures often require considerable investments and very long planning and implementation times when compared to those of buildings; however, they do represent a considerable added value for the economy. Their execution, in addition to stimulating technological innovation, creates investments and jobs, not only during the implementation phase but also in the subsequent phases of operation and maintenance.

The management of a complex issue such as infrastructure must be able to count on effective tools that are able to support not only the decision makers but also the stakeholders involved in making choices. International rating systems for sustainability such as LEED® or BREEAM® can contribute in this direction since they enhance the added value determined by the presence of infrastructures in the territory. A protocol dedicated exclusively to infrastructures such as Envision®, examined in this chapter, can moreover contribute in a remarkable way to a sustainable, and at the same time, resilient management of the infrastructures.

References

American Society of Civil Engineers ASCE (2009) 2009 Report cards for American's infrastructures. ISBN: 978-0-7844-1037-0

Boeing G, Church D, Hubbard H, Mickens J, Rudis L (2014) LEED-ND and livability revisited. Berkeley Plann J 27(1). http://dx.doi.org/10.5070/BP327124500. Retrieved from https://escholarship.org/uc/item/49f234rd. Accessed 20 Oct 2019

BRE Global Limited (2017) BREEAM communities technical manual. Available on https://www.breeam.com/discover/technical-standards/communities. Accessed Sept 17 2019

Diaz-Sarachaga J (2016) The role of sustainable infrastructure rating systems in urban development. In: Conference paper, international congress of civil engineering, madrid (Spain). Available on https://www.researchgate.net/publication/296647048. Accessed Sept 12 2019

Fulmer JE (2009) What in the world is infrastructure? PEI Infrastruct Invest. Available on https://30kwe1si3or29z2y020bgbet-wpengine.netdna-ssl.com/wp-content/uploads/2018/03/what-in-the-world-is-infrastructure.pdf. Accessed on 16 Oct 2019

Institute for Sustainable Infrastructure (2018) Envision: sustainable infrastructure framework guidance manual. Washington, DC. Available on www.sustainableinfrastructure.org. Accessed Sept 12 2019

Institute for Sustainable Infrastructure ISI (2018) Envision: sustainable infrastructure framework guidance manual—Version 3. Institute for Sustainable Infrastructure, Washington, DC, ISBN 978-1-7322147-0-5. www.sustainableinfrastructure.org

Institute for Sustainable Infrastructure (2019) Envision projects map. Available on https://v3.sustainableinfrastructure.org/projects/map. Accessed Sept 12 2019

Secchi C, Riela S (2019) Report—infrastructure for growth: how to finance, develop and protect it, ISPI—Italian Institute for International Political Studies. Available on https://www.ispionline.it/it/pubblicazione/infrastructure-growth-how-finance-develop-and-protect-it-22906. Accessed on 20 Oct 2019

U.S. Green Building Council (2014) LEED Reference Guide for Neighborhood Development, LEED v4 Edn. Available on https://www.usgbc.org/resources/leed-v4-neighborhood-development-current-version. Accessed Sept 13 2019

Part V
The "Smart City" Approach to Improve Urban-Scale Sustainability

Chapter 15
Greening the Smartness of Cities and Communities

Giuliano Dall'O'

Abstract The topic of smart cities has come to the fore at a time when technological innovation in the cities, owing to the ever increasing use of ICT infrastructures and services, especially the Internet, is spreading widely. The digital city seems to have taken the upper hand over the real one, and cities are joining the smart revolution aiming to become smart cities. The risk is that the most concrete meanings of the smart revolution will be lost, giving too much weight to technologies. This chapter addresses this issue by highlighting the true meanings of the smart city approach on urban planning and the necessary and useful synergy with the green aspects most linked to the physical environment of the city.

15.1 The Meaning of "Smart" Applied to a City

Ecocities, digital cities, solar cities, sustainable cities, green cities and smart cities are just some of the definitions with which cities are classified, giving them ecological aspects rather than aspects related to technological innovation.

These codified classifications, which often overlap creating a little confusion, risk directing the urban planner towards favouring certain aspects and neglecting others. The new urban settlements can be orientated towards strong choices, especially regarding the aspects linked to sustainability; however, it is frequent to make comparisons with pre-existing cities, characterized by a stratification of building interventions or infrastructures that have lasted even centuries.

How can a large town, city or a metropolis suddenly obtain the green or smart label?

The cases of excellence obviously exist: for example, the sustainable district of Vauban, Freiburg, Germany, or the eco-sustainable Krönsberg district of Hanover again in Germany, or the district Bo01, called City of Future, of Malmö in Sweden, are examples of important reference, however they are not cities as such.

G. Dall'O' (✉)
Architecture, Built Environment and Construction Engineering, ABC Department, Politecnico di Milano, Via G. Ponzio, 31, 20133 Milan, Italy
e-mail: giuliano.dallo@polimi.it

G. Dall'O' (ed.), *Green Planning for Cities and Communities*,
Research for Development, https://doi.org/10.1007/978-3-030-41072-8_15

The big challenge is to accelerate the ecological, sustainable and smart revolution of the cities and metropolises that already exist, and that will have to accommodate an ever-increasing number of new inhabitants, towards urban models that can help us fight climate change, in the spirit of SDG 11—sustainable cities and communities: make cities and human settlements inclusive, safe, resilient and sustainable.

Another example is Oslo, the icon city of sustainable mobility, where electric cars make up 16% of total cars with 9% fully electric cars and 7% plug-in hybrids.[1]

To understand the complexity of the topic, it is useful to examine the definitions of the "Smart City" concept shown in Table 15.1.

One should not state that a city is smart or that a city is green: regenerating a neighbourhood is very important but definitely not enough to assign the "smart" or "green" label to the entire town or city in which the neighbourhood is inserted.

Labels are misleading because they assign a quality in a static and absolute way: it is essential instead to put a value on the path followed and the efforts that a city is making to improve its smartness, or to improve its sustainability, or both. One should not ask oneself if a city is smart or sustainable, but how smart or green it is.

Among the definitions of the "Smart City" shown in Table 15.1, the most correct is the definition proposed by Department for Business, Innovation and Skills, UK (2013): "*The concept is not static, there is no absolute definition of a Smart city, no end point, but rather a process, or series of steps, by which cities become more "livable" and resilient and, hence, able to respond more quickly to new challenges*".

An interesting contribution to this topic of sustainable smart cities (SSC) is given by Maysoun et al. (2015). The said paper draws upon examples of sustainable smart cities (SSC) initiatives and existing SSC transformation frameworks to more fully articulate the challenges of achieving successful SSC projects, arguing that transformation towards SSC faces a set of four types of challenges that vary from one region to another and between countries within the same region:

- *Complexity challenges*: SSCs require creation of a sustainable system series of systems capable of generating opportunities for cities and their citizens.
- *Economic challenges*: a need for a sustainable financial investment to support SSC projects, which is highly dependent on the economic situation of a country, and hence of its towns and cities.
- *Social challenges*: the necessity to adopt SSC services to the specific needs of each user (i.e. citizen).
- *Governance challenges*: the need of coordination and integration between public, private and civil bodies for the purpose of making a city function as one organism in an efficient and effective manner.

In the following sections, an analysis is made of the conceptual framework for the green city approach and for the smart city approach to planning, and then, a comparative analysis is performed.

[1]Source: https://www.weforum.org/agenda/2018/08/the-oslo-model-how-to-prepare-your-city-for-electric-vehicles/m (Accessed 20 September 2019)

Table 15.1 Selected definitions of a "Smart City" concept

Definition	Source
Is a city that monitors and integrates conditions of all of its critical infrastructures, including roads, bridges, tunnels, rails, subways, airports, seaports, communications, water, power, even major buildings, can better optimize its resources, plan its preventive maintenance activities, and monitor security aspects while maximizing services to its citizens	Hall (2000)
Regional competitiveness, transport and Information and Communication Technologies economics, natural resources, human and social capital, quality of life, and participation of citizens in the governance of cities	Giffinger et al. (2007)
A city can be defined as "smart" when investments in human and social capital and traditional (transport) and modern (ICT) communication infrastructure fuel sustainable economic development and a high quality of life, with a wise management of natural resources, through participatory action and engagement	Caragliu et al. (2009)
Is a city in which it can combine technologies as diverse as water recycling, advanced energy grids and mobile communications in order to reduce environmental impact and to offer its citizens better lives"	Setis-EU (2012)
Is one that utilizes ICT to meet the demands of the market (the citizens of the city), and that community involvement in the process is necessary for a Smart city	Deakin (2013)
The concept is not static, there is no absolute definition of a Smart city, no end point, but rather a process, or series of steps, by which cities become more "livable" and resilient and, hence, able to respond more quickly to new challenges	Department for Business, Innovation and Skills, UK (2013)
Should be regarded as systems of people interacting with and using flows of energy, materials, services and finance to catalyze sustainable economic development, resilience, and high quality of life; these flows and interactions become smart through making strategic use of information and communication infrastructure and services in a process of transparent urban planning and management that is responsive to the social and economic needs of society	European Commission, (EC 2013)
A developed urban area that creates sustainable economic development and high quality of life by excelling in multiple key areas; economy, mobility, environment, people, living, and government. Excelling in these key areas can be done so through strong human capital, social capital, and/or ICT infrastructure	Business Dictionary (2014)
Is a city seeking to address public issues via ICT-based solutions on the basis of a multi-stakeholder, municipally based partnership. A Smart City is quintessentially enabled by the use of technologies (especially ICT) to improve competitiveness and ensure a more sustainable future by a symbiotic connection of networks of people, businesses, technologies, infrastructures, consumption, energy and spaces	European Parliament (2014)

(continued)

Table 15.1 (continued)

Definition	Source
Is one that dramatically increases the pace at which it improves its social economic and environmental (sustainability) outcomes, responding to challenges such as climate change, rapid population growth, and political and economic instability by fundamentally improving how it engages society, how it applies collaborative leadership methods, how it works across disciplines and city systems, and how it uses data information and modern technologies, in order to provide better services and quality of life to those in and involved with the city now and for the foreseeable future	ISO/IEC (2015)
Should be designed to achieve holistic objectives, focusing on intelligent urban ecosystem development. This research helps local government CIOs (Congress of Industrial Organizations.), urban planners and strategists assess emerging technologies and solutions for cities and governments for a sustainable societal outcome	Tratz-Ryan and Finnerty (2019)
Is a designation given to a city that incorporates information and communication technologies (ICT) to enhance the quality and performance of urban services such as energy, transportation and utilities in order to reduce resource consumption, wastage and overall costs. The overarching aim of a smart city is to enhance the quality of living for its citizens through smart technology	Techopedia (2019)

15.2 A Conceptual Framework for Green Cities

Before going into detail on the topic of *smart cities*, it is useful to examine what the available technical literature provides about *green cities* in order to verify possible synergies between the green approach and the smart approach to urban planning.

An interesting contribution on the topic of green cities is the work of Brilhante and Klaas (2018) of the Institute for Housing and Urban Development Studies (IHS), Erasmus University of Rotterdam, which illustrates more complex ways of looking at urban growth and sustainability.

Their paper introduces a Green City Conceptual Framework (GCCF) and two tools: one tool to measure city Environmental Performance (EP) over time (IHS-GCI), and the other is a Green City Action Planning methodology (GCAP) which contains innovative approaches to improve the city's EP for a chosen timeline.

The IHS-GCI tool can be used to measure EP, to explain possible factors influencing the calculated EPs over time, set targets, track achievements and help in the preparation of a city's GCAP.

The paper addresses a range of gaps in the green city field and consolidates prior research into one actionable conceptual framework and related tools, adaptable to various cities and needs.

The Authors define a green city as "*a city that promotes energy efficiency and renewable energy in all its activities, extensively promotes green solutions, applies land compactness with mixed land use and social mix practices in its planning systems, and anchors its local development in the principles of green growth and equity.*"

The Green City Conceptual Framework proposed by Brilhante and Klaas (2018) is reproduced in Fig. 15.1. The diagram shows seven flexible thematic areas:

- Green infrastructures: storm water systems;
- Green transport;
- Greening of public services: drinking water and sanitation;
- Green buildings;
- Green technologies: smart ICT;
- Environmental quality/water security;
- Greening urban architecture: vertical farms.

The conceptual framework includes four entry points or distinctive aspects:

1. Energy efficiency/renewable energy in all sectors;
2. Greenery: extensive use in all sectors;
3. Spatial planning with land compactness/mix use/social mix;
4. Green growth and equity principles.

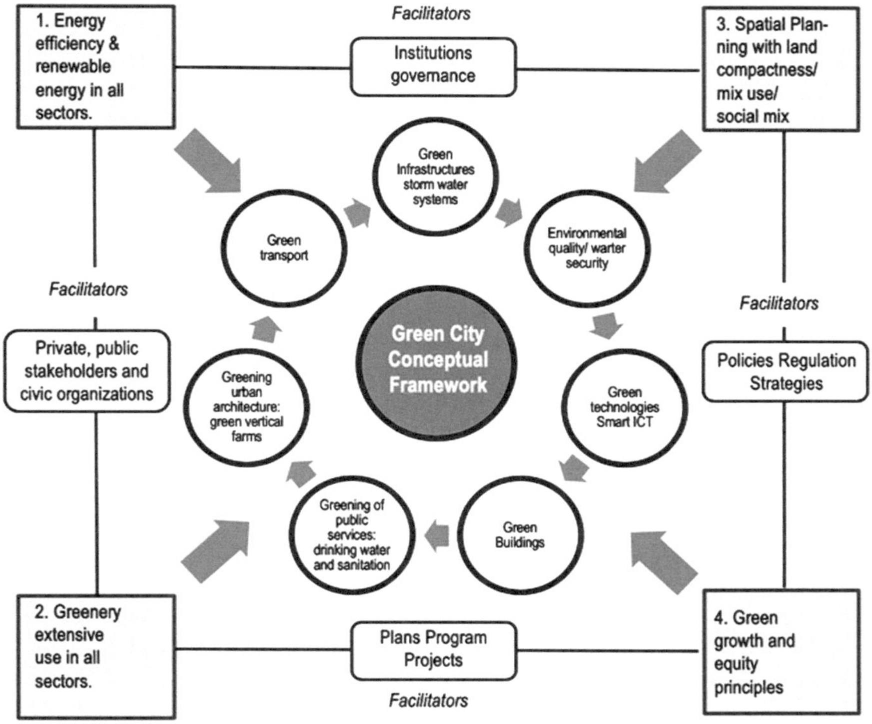

Fig. 15.1 HIS green city conceptual framework (Brilhante and Klaas 2018)

To promote urban planning according to the green city scheme, to share experiences between cities that have made this choice and to increase the critical mass of cities in Italy, the "*green city network*" was created.

An interesting document to better understand the approach adopted and the future strategies of the network is the "Statement for the climate adaptation of the Green City", signed in Milan on 16 July 2019, in the occasion of the 2nd National Green City Conference. The statement is structured into ten points which are reported below:

- **Defining and updating plans and measures for city climate adaptation**. Southern Europe and, in particular, Mediterranean countries such as Italy are already heavily affected by the impacts, which are tending to worsen, of extreme atmospheric changes generated by climate change: heat waves and prolonged periods of drought; very intense rainfalls, concentrated in short periods, with increased risks of flooding, floods and landslides; strong winds, cyclones, tornadoes and violent storms. Cities, particularly exposed to the risks of climate change, must be protagonists of policies and adaptation measures for which local action is essential. More effort is needed: the measures that cities have begun to implement need to be improved, strengthened and made more extensive and incisive.
- **Integrate policies and adaptation measures with mitigation of climate change**. The first objective of prevention for the cities can only be to help avoid the further dangerous worsening of the climate crisis. To contain the average global temperature increase below 2°, as envisaged by the Paris Agreement, it will be necessary to reach 2050 net greenhouse gas emissions close to zero and therefore to demanding reductions also in the cities by 2030. Then update the plans and measures to reduce greenhouse gas emissions in cities and integrate them with those for adaptation to climate change, enhancing positive synergies. Greater and more incisive initiatives are urgent, in particular to increase the production and use of renewable energy sources, for electricity, thermal uses and fuels, for energy savings in public and private buildings with sustainable mobility and a circular economy.
- **Update risk assessment and measures and both emergency, on medium and long terms**. It is necessary to keep up to date the knowledge of the urban impacts of climate change, to have preventive warning systems and contingency plans for risk management and to reduce the exposure of the population, in particular of the most vulnerable sectors. On the basis of the local microclimatic profiles and variables of the different urban systems, the effectiveness of the measures adopted to reduce vulnerability and exposure to short- and long-term risks, and to increase their resilience. Particular attention should be paid to the integration of adaptation measures into medium- and long-term plans and programmes for urban transformations and intervention projects.
- **Enhancing the positive returns of adjust measures and accounting the cost of the absence of such measures**. The plans and programmes for climate adaptation, according to the Green City model, in addition to reaching the specific finalities, must contribute to improving the environmental quality of the city, urban districts and built complexes as well as individual buildings. Adaptation measures, in addition to increasing the resilience of cities and territories by reducing impacts

and risks, can also generate other benefits for citizens: from better air quality to healthier living, from better resource management water resources to increase biodiversity. Adaptation measures can promote regeneration and revitalization of marginal urban districts, often peripheral and degraded. Climate adaptation, if well implemented, contributes to improving safety and stability, including economic and social, while, vice versa, failure to adapt involves significant costs, which must be accounted for and evaluated, for citizens and economic activities. In addition, the links between health protection and climate adaptation must be strengthened, and information on the risks and costs, including health, of the climate crisis must be improved along with information on the benefits generated by adaptation measures.

- **Develop adaptive skills**. The adaptation of the city is a process of change that requires flexibility, periodic updates, the ability to incorporate resilience to climate change into plans, policies and measures, investments in urban development, land management and use, resources water, buildings and infrastructure. This change, aimed at ensuring adequate safety standards and protecting the health of citizens, the functioning of services and infrastructures and the regular conduct of economic activities, must be set and managed as a real process of continuous learning. In this process, it is essential that local administrations make use of research, knowledge, good practices and the best available techniques to increase the capacity for local adaptive response, both in the interventions for existing buildings and infrastructures and for new designs and achievements as well in land management, land recovery and use.
- **Point more on nature-based solutions**. Road trees, public and private gardens, parks, roofs and green walls, vegetable gardens and peri-urban agricultural areas, natural expansion crates integrated into the urban fabric, as well as providing different ecosystem services, contribute significantly to climate adaptation, to reduce heat waves, to improve surface runoff and water absorption. Adaptation based on green infrastructure and urban and peri-urban natural capital is multifunctional and offers multiple advantages: in terms of biodiversity, mitigation, air and soil quality and social well-being. This multifunctionality should be better integrated into the assessment of adaptation options. Adaptation solutions based on nature (nature-based solutions) must be kept in mind both in urban planning and in plans to increase, manage and finance urban green spaces. Promoting the implementation of ecological corridors and green belts with the redevelopment of urban and peri-urban open spaces is important for mitigating climate risk and hydrogeological instability.
- **Reduce vulnerability and risks of very intense precipitations**. It is necessary to stop the waterproofing and the consumption of new soil and increase the interventions of de-waterproofing the floors. Green networks and infrastructures are of great importance both as a microclimatic moderator and to absorb and retain more rainwater. Squares or green areas lower than the street level, for example, can contribute to the accumulation of rainwater in the case of extreme events. The particularly intense rainwater runoff can be conveyed to specially prepared wetlands in peri-urban areas that can also be used for the expansion and rolling of river and

stream floods. Existing urban drainage networks should be better monitored, and efforts should be made to ensure separation between the sewer networks—and the relative flood spillways—and the hydrographic network of canals and rivers, even in the case of intense meteorological phenomena.

- **Face the "heat island" effect**. Advanced analysis and assessment tools must be used to adapt to the increasingly frequent waves and increasingly consolidated heat islands in cities. On the basis of these analyses and evaluations, adaptive design, technical and managerial solutions must be implemented in urban regeneration, in the redevelopment of buildings and relevant spaces. It is necessary to promote measures for the bioclimatic control of buildings, for shading and control of solar radiation, to increase natural ventilation and cooling, to improve thermal insulation even with innovative materials. It is also useful to use and upgrade green infrastructures and improve cooling by using water tables and surface water bodies.
- **Promote investments in adaptation measures**. Urban adaptation policies and measures require more public and private investments. It is necessary to promote public–private partnerships, greater use of financial and insurance instruments, direct with technical standards and favour, with cost-and-benefit analysis, private investments in adaptation projects. Adaptation measures can help create positive effects for the local economy and attract investments, including international ones, improving the safety and attractiveness of the city. Linking adaptation, mitigation and green economy can transform a risk into an opportunity by creating synergies between public interest and the private sector. Finally, it is also useful to check the availability of forms of financing such as, for example, donations and crowdfunding.
- **Strengthen governance**. To define and implement a climate adaptation strategy in the city, it is necessary to increase the political commitment of local administrations and establish a precise responsibility as well as a coordination between the different competences and the offices/departments involved. The measures of climate adaptation, prevention and risk monitoring require active citizen participation, effective information, as well as constant dialogue with the various stakeholder groups. Adaptation is a process that requires awareness and knowledge, adequate skills and technologies, defined economic resources, training of administration staff and interested parties. Fundamental to the development of policies and measures of urban adaptation is the circulation, the exchange, the knowledge of the best interventions, of the most innovative, realized and ongoing projects. Finally, the coordination of policies and adaptation measures between the different levels of government needs to be improved: local, regional and national.

To summarize, we can affirm that the planning green strategy of cities is very environmentally orientated: with this objective, a multidisciplinary strategy to combat climate change is promoted.

15.3 A Conceptual Framework for Smart Cities

Associating with *smart city,* a model of the city of the future is both too simple and reductive. Defining it in this way would mean not considering its true meaning, renouncing those positive values that the term "smart" does not always manage to highlight.

Speaking of smart cities today means referring to a model of the city in which, first of all, relations between citizens and institutions, between citizens and the world of economy are modified, and obviously among the citizens themselves (Dall'O' 2014).

The social dimension of change, therefore, must be placed at the centre of attention for those who really want to integrate the dynamics in place and perhaps in this way be active in this process.

Table 15.1 demonstrates that the concept of smart city or sustainable smart city is not completely clear, not even for the actors involved and the stakeholders who obviously force the declination of meaning and the sectoral aspects.

A company that deals with ICT emphasizes for example aspects relevant to the protection of information, both on the network and on the air, managing the related services. Similarly, a company that is involved in energy services points out aspects related to energy issues, e.g. generation of energy, use of renewable energy sources.

The spread of the *smart city* concept, on the other hand, generates curiosity and stimulates expectations first and foremost on the part of citizens, who must be put in a position to understand what is changing and what will change, on the part of those who actively work, not only for example workers, professionals and companies, but also by public service and local authority administrators, who have the delicate task of defining the political choices that underlie the current and future development of the cities they govern.

With the smart city approach, one is facing an evolution in which two elements confront each other and find a value only if they contribute synergistically in the evolutionary process. On the one hand, there is the general objectives of improving the quality of life at all levels, from that environmental to the social, from the economic to the cultural, on the other the new technologies that can lead to the creation of new needs but also of new opportunities (just think how the Internet has changed our lives).

The smart city is not a goal, but a process that must be conceived as such: we can say that it makes no sense to define that a city is "smart", while it makes more sense to be able to manage an evolution towards ever greater smartness, possibly monitoring its changes.

Given the importance of the theme, the technical–scientific literature on the topic of smart cities is consistent. Several studies provide more in-depth definitions; others investigate specific sectors such as energy, the environment or Information and Communication Technologies (ICT). Further studies have entered into the merits of indicators, tools necessary to objectively measure this evolution, examining case studies of cities, or more reasonably neighbourhoods, which have invested heavily in accelerating the process of transformation towards levels of greater intelligence.

Albino et al. (2015) identify many academic definitions of smart city highlighting how the term smart tends currently to be replaced by "intelligent" or "digital". The smart city, which is then the up-to-date city, in fact arises from the overlap between two cities: the real one that we can see and touch, made up of physical spaces, and the digital, apparently ephemeral, but which we use daily through our devices: smartphones, tablets, personal computers or urban interactive panels.

A more complete definition is given by Bakıcı et al. (2012): the authors consider a smart city as an intensive and advanced high-tech city that connects people, information and elements of the city itself: these connections are possible using new technologies that create a city which is sustainable and greener, competitive and innovative for trade, with a better quality of life.

All definitions, however, converge on recurrent concepts such as sustainability, interaction with the IoT (Internet of things), quality of life. From the analysis of these works, it is undeniable that the smart city concept is a complex and dynamic concept: Javidroozi et al. (2015) define it, not without reason, "*a system of systems*" that must be efficient, interoperable and integrated. In this vision, cities have become intelligent not only in terms of automation of services, buildings, traffic systems, but also in ways with which they allow us to monitor, understand, analyse and consequently plan the city to improve efficiency, equity and quality of life for citizens (Carli et al. 2013).

However, smart cities should not be just a concentration of innovative technologies: other issues are equally important and therefore deserve attention. In this sense, Thite (2011) underlines the importance of the social aspect, speaking of cities that aim to support a creative economy through investment in the quality of life: this is the point that attracts knowledgeable workers to live and work in smart cities. The nexus of competitive advantage has shifted to those regions capable of generating, retaining and attracting the best talent.

Komninos (2011) also put emphasis on the more creative side of smart cities, which are defined as territories with a high capacity for learning and innovation.

Carli et al. (2013) offer a not less important point of view, the economic aspect: the concept of the smart city was recently introduced as a strategic means to understand the importance of ICT, social and environmental capital in the definition of competitiveness in a common framework of cities and sustainability.

The intelligence of a smart city to assume its credibility, to become something concrete, must be able to be analysed with objective assessment tools that have a scientific basis: anything that cannot be measured can at most arouse emotions, but this may not be enough.

If the cities that already exist contain at least some of the elements that contribute to creating smartness, the goal is to have available tools, objective and necessarily shared, through which it can be measured. If we can define these "indicator" tools, their combination will finally allow us to develop a ranking or ranking among the cities and to establish the location of each of them between the two extremes in which the poorer cities are considered, on the one hand, from the evaluation system relatively the worst, and on the other those with more points considered the best.

A global ranking, which is what one tends to do, makes no sense: every city, in fact, can be innovative in certain aspects and present shortcomings in others: the important thing is to develop this analysis in detail in order to grasp the elements that can guide governance political and social in the right direction.

There is no single and shared classification system, but several projects with approaches that may be different. Nonetheless, a common element is that they permit the definition of a single classification but at the same time they allow partial evaluations to be made within the individual evaluation areas.

If the smart city, as Javidroozi et al. (2015) states, is "*the system of systems*", it can be useful to analyse its complexity through segmentation into areas of evaluation as shown in Fig. 15.2. Dall'O' et al. (2017). Coherently with the European guidelines and with the certification schemes already used, seven areas can be defined:

Smart Economy

The smart economy highlights the aspects that identify the competitiveness of a city in economic and productive terms. The city has the task of stimulating and promoting this scenario, enhancing the possibilities of synergy between private companies, public bodies and research institutes, engages in the harmonization and promotion of virtuous enterprises, thus raising the general level of all activities and creating a stimulating environment.

Smart Environment

Environmental sustainability is an aspect that strongly characterizes a smart city. This area defines all the strategies that a city can put in place to improve the quality

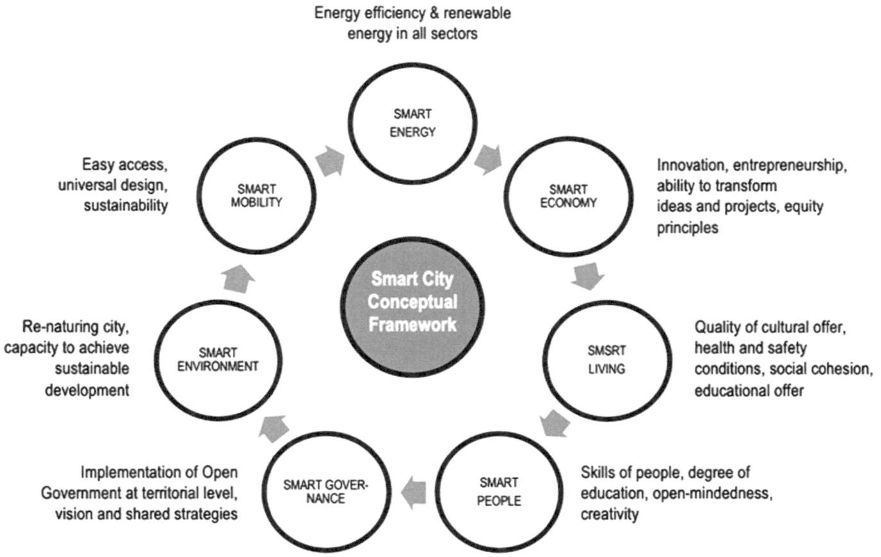

Fig. 15.2 Smart city conceptual framework (Dall'O' et al. 2017)

of the environment. In this area, the aspects linked to the re-naturing of the city can converge, aspects that characterize the green approach (green roofs, green facades, parks, greenways, etc.).

Smart Energy

A city with more energy-efficient buildings is a city that greatly reduces the environmental impact. The energetic redevelopment of the existing building stock, the incentive to build low-consumption buildings, the rationalization of public lighting and the correct management of green areas to limit the "heat island" effect in the summer months are all actions that contribute to making a smart city come true.

Smart Mobility

Within a city, mobility is often a very critical element, but this does not happen if its management is not based on the principle of smart mobility that can arise from the synergy between public transport infrastructures, private mobility and ICT.

This intelligent approach facilitates travel through an efficient public transport infrastructure offering at the same time a reduced environmental impact.

Smart Governance

A new and stimulating relationship between the citizen and institutions in a smart city is the object of smart governance. Through smart governance, transparency of government procedures is promoted and data flows are opened and shared (open data).

Smart Living

A smart city is also a city in which its citizens live well and therefore the quality of life is guaranteed, and this is precisely the meaning of smart living, the implementation of a city model in which all aspects the quality of living is best guaranteed.

Smart People

The smart people evaluation area enhances human capital by stimulating creativity, professional growth, flexibility, cosmopolitanism and open-mindedness.

Smart city is based on a rediscovered awareness and participation in public life, stimulates and raises the qualification of citizens, the coexistence of different stakeholders and communities, interaction and continuous dialogue to detect concrete needs and make efficient and effective response.

15.4 Greening Smart Cities, a New Paradigm

As a result of the growing interest in smart cities over recent years, the scientific community has been faced with the problem of assessing the differences between "smart" approaches or "green" approaches, or those that are mainly linked to promoting sustainability, and the possible synergies to be developed.

Bibri and Krogstie (2017), in their paper, provide a comprehensive overview of the field of smart (and) sustainable cities in terms of its underlying foundations and assumptions, state-of-the art research and development, research opportunities and horizons, emerging scientific and technological trends, and future planning practices.

As to the design strategy, the paper reviews existing sustainable city models and smart city approaches. Their strengths and weaknesses are discussed with particular emphasis being placed on the extent to which the former contributes to the goals of sustainable development and whether the latter incorporates these goals.

As a result of the research, an integrated approach is proposed based on an applied theoretical perspective to align the existing problems and solutions identification for future practices in the area of smart sustainable urban planning and development.

Comparing the conceptual frameworks of the *green* approach and the *smart* approach to urban planning, it is possible to highlight some overlapping common aspects (Table 15.2 considers also a comparison with sustainable protocols).

Both approaches consider the topic of sustainability to be a priority, as an effective response to environmental emergencies; it seems correct to assign to the term smart a meaning linked to intelligence in implementing strategies to combat climate change. Another common element, indirectly linked to sustainability, is energy: energy efficiency and the use of renewable energy sources are elements that characterize both approaches.

Looking at human beings in this context, although both approaches are concerned with ensuring a comfortable environment especially outside, the smart approach is a little more anthropocentric than the green approach that promotes a more ecological action, therefore more linked to the natural environment. It is important that flora and fauna are present but that is not enough, there is an underlying more constructive relationship between man and the environment which must be enhanced through a series of greening interventions in urban spaces which the green approach regularly lists, whereas the smart approach neglects them.

So far as the aspects related to economy are concerned, the smart approach conceives innovation that could refer to the circular economy or to the green economy but which can also refer to other areas.

The "business" opportunities of the green approach arise primarily from the choices linked to the opportunities that may emerge from the new approach (green economy). An element common to these two approaches is that of mobility: both promote sustainable mobility in a fairly shared view of how many to be respectful of the environmental ecosystem.

From here on in, the differences that emerge are omissions from the green approach which does not address the issue of quality of life unless strictly related to the environment: smart living and smart people contain elements that the green approach does not consider: it even avoids the issue of smart governance.

Table 15.2 shows a substantial overlap between the rating systems or environmental certification protocols (such as LEEDS®, BREEAM®, etc.) and the green city approach: so the smart city approach is well integrated with both, providing a synergistic completion.

Table 15.2 Comparison between green city strategy, smart city strategy and sustainable protocols strategy

Item	Green city	Smart city	Sustainable protocols
Adaptation climate change measures	•	•	•
Air quality	•	•	•
Circular economy	•	•	•
Energy efficiency and renewable energy	•	•	•
Environmental monitoring	•	•	•
Green buildings	•	•	•
Green growth	•		•
Greening urban architecture	•		•
Health protection	•	•	
Information and Communication Technologies (ICT)	•	•	•
Optimization of land use	•		•
Re-naturing city	•		•
Risk assessment	•		•
Security	•	•	•
Skills of people		•	
Smart economy		•	
Smart government		•	
Smart living		•	
Smart mobility	•	•	•
Social cohesion	•	•	
Sustainable development	•	•	•
Universal design	•	•	•
Waste management	•	•	•
Water management	•	•	•

Benevolo and Dameri (2013), in their paper concerning the relationship between smart city and green city, argue that "the key element that characterizes the smart city is technology, be it ICT or of another engineering-type, that apply to the typical infrastructures of the urban fabric, on the contrary the key element of the green city is the environment and its defence in all possible advances.

It is by intertwining these key elements and objectives that the overlap between smart city and green city is identified. Smart cities and green cities coincide where technology is used to reduce the impact of the city on the use of natural resources, energy consumption and pollution of the land, air and water.

In this overlap, it is possible to identify common positions to go towards a city that is both smart and green".

The need to re-naturalize the smart city approach emerges clearly from the thought of Grant (2017) when it states that: "*it is important that smart cities are as much about nature, health, and wellbeing as traffic flows, crime detection, and evermore efficient provision of utilities. Making the city more permeable to both wildlife and people is a process that could be informed by bringing sensors that monitor the movement of wildlife. Climate change adaptation using natural interventions is already on the agenda of many cities. However, the efficacy of green infrastructure types will be significantly improved through both detailed and wide-scale real-time measurement of temperature, humidity, evapotranspiration rate and flows. Combining this data with maps of hardship and deficiency will help cities to become smarter in the way they prioritize greening efforts*".

The green city approach has developed over time, experiences have been consolidated, and the smart city approach is more recent and is proposed, sometimes incorrectly, as a container for new technologies or as a synthesis of the real city with the digital city. Many cities follow *the same* path and so smart city and green city networks are born. Having analysed the two approaches and having verified the importance of a synergy between the two approaches, it is natural to raise a question? Why is it necessary to choose? If the goal is to combat climate change through a new balance between man and nature, it is better to exploit synergies and the greening of smartness can be the answer.

Regarding the synergies between the green city and the green city, many authors have gone so far as to define an integrated concept called sustainable smart city. Hoier and Wangel (2015) in their paper investigating the concept of sustainable cities argued that for the concept to have any useful meaning it should be more strictly defined than it has previously been. The authors suggest such a definition and bring up some of the concept's crucial challenges.

The difference between sustainable and smart cities is the topic covered in the paper by Ahvenniemi et al. (2017). Starting from the consideration that city assessment tools can be used as support for decision-making in urban development, since they provide assessment methodologies for cities to show the progress towards defined targets, the authors argue that a general goal of smart cities is to improve sustainability with the help of technologies and recommend the use of a more accurate term "smart sustainable cities" instead of smart cities. Taking into account the current large gap between smart city and sustainable city frameworks, the authors suggest that there is a need to develop further smart city frameworks or to re-define the concept of the smart city. The recommendation is that "the assessment of smart city performance should not only use output indicators that measure the efficiency of deployment of smart solutions but also impact indicators that measure the contribution towards the ultimate goals such as environmental, economic or social sustainability".

15.5 Conclusions

Speaking about smart cities today means referring to a model of the city in which, first of all, relations between citizens and institutions, between citizens and the world of economy are modified, and obviously among the citizens themselves. The social dimension of change, therefore, must be placed at the centre of attention for those who really want to integrate the dynamics in place and perhaps in this way be active in this process.

In this chapter, with a provocative title "greening the smartness of cities and communities", we have analysed the relationship between the green city approach and the smart city approach to urban planning, highlighting how the two approaches are absolutely compatible and synergistic, and indeed strategic if the goal is to demonstrate consistency with the United Nations SDG 11: make cities and human settlements inclusive, safe, resilient and sustainable.

We also reflected on the concepts of green and smart related to cities. Using these labels as an absolute definition may not be correct: cities cannot be green and intelligent but may tend to constantly improve the relationship between the built urban space (grey) and the natural urban space (green), as they may tend to an ever smarter way to imagine the city. We are talking about processes that evolve, as well as ways of living, citizens' perception of the importance of the environmental issue, starting with climate change) and technologies, of course. The green city approach and the smart city approach must inevitably merge, as both propose shared positive considerations. Whether the smart city approach should become greener or the green city approach should become smarter is not a problem: after all the end result is the same.

Nor do we believe it is useful to propose a new neologism, for example green smart cities or smart green cities: there are far too many neologisms to define the characteristics of cities. The important thing is to consolidate the objectives and find the way to reach them in the interest of man and nature. Technological innovation, generated by man, is welcome if its availability and its use are able to accelerate the process. The good news is that in many cases, this is already happening.

References

Ahvenniemi H, Huovila A, Pinto-Seppä I, Airaksinen M (2017) What are the differences between sustainable and smart cities? Cities 60(2017):234–245

Albino V, Berardi U, Dangelico RM (2015) Smart cities: definitions, dimensions, performance, and initiatives. J Urban Technol 22(1):3–21. https://doi.org/10.1080/10630732.2014.942092

Bakıcı T, Almirall E, Wareham J (2012) A smart city initiative: the case of Barcelona. J Knowl Econ 2(1):1–14

Benevolo C, Dameri RP (2013) La smart city come strumento di green development: il caso di Genova, Impresa e Progetto. Electron J 3:2013

Bibri SE, Krogstie J (2017) Smart sustainable cities of the future: an extensive interdisciplinary literature review. Sustain Cities Soc 31(2017):183–212

Brilhante O, Klaas J (2018) Green city concept and a method to measure green city performance over time applied to fifty cities globally: influence of GDP, population size and energy efficiency. Sustainability 10(6):2031. https://doi.org/10.3390/su10062031

Business Dictionary (2014) http://www.businessdictionary.com/definition/smart-city.html

Caragliu A, Del Bo C, Nijkamp P (2009) Smart cities in Europe, series research memoranda 0048. VU University Amsterdam, Faculty of Economics, Business Administration and Econometrics

Carli R, Dotoli M, Pellegrino R, Luigi Ranieri L (2013) Measuring and managing the smartness of cities: a framework for classifying performance indicators. In: 2013 IEEE international conference on systems, man, and cybernetics, pp 1288–1293

Dall'O' G (2014) Smart city, la rivoluzione intelligente delle città, Il Mulino. ISBN 978-888-15-25283-8

Dall'O' G, Bruni E, Panza A, Sarto L, Khayatian F (2017) Evaluation of citie's smartness by means of indicators for small and medium cities and communities: a methodology for Northern Italy. Sustain Cities Soc 34(2017):193–202

Deakin M (2013) From intelligent to smart cities. In: M. Deakin (ed) Smart cities: governing, modelling and analysing the transition, Taylor and Francis, p 15. ISBN 978-1135124144

Department for Business, Innovation and Skills, UK (2013) Smart Cities, Background paper BIS/13//1209

EC (2013) European innovation partnership on smart cities and communities—Strategic Implementation Plan, European Commission (EC), Communications, Networks, Content and Technology Publications, Brussels, Belgium

EP (2014) Mapping smart cities in the EU. European Union (EU), European Parliament's Committee on Industry, Research and Energy, Policy Department A: Economic and Scientific Policy, Brussels, Belgium

Giffinger R, Fertner C, Kramar H, Kalasek R, Pichler-Milanovic N, Meijers E (2007) Smart cities—ranking of European medium-sized cities. Smart cities. Centre of Regional Science, Vienna. Available on http://www.smart-cities.eu/download/smart_cities_final_report.pdf. Accessed 15 Sept 2019

Grant G (2017) Can smart cities be smart green cities? We'll see. Available on: https://www.thenatureofcities.com/2017/12/18/can-smart-cities-smart-green-cities-well-see/. Accessed 16 Sept 2019

Hall P (2000) Creative cities and economic development. Urban Stud 37(4):633–649

Hoier M, Wangel J (2015) Smart sustainable cities: definitions and challenges. In: ICT innovation for sustainability, Advanced in intelligent systems and computing, vol 310. https://doi.org/10.1007/978-3-319-09228-7_20

ISO/IEC (2015) Smart cities—preliminary report (2014). International Organization for Standardization and International Electrotechnical Commission (ISO/IEC). Genève, Switzerland

Ibrahim M, Adams C, El-Zaart A (2015) Paving the way to smart sustainable cities: transformation model and challenges. JISTEM J Inf Syst Technol Manag 12(3):559–576. ISSN online: 1807-1775. https://doi.org/10.4301/s1807-17752015000300004

Javidroozi V, Shah H, Cole A, Amini A (2015) Towards a city's systems integration model for smart city development: a conceptualization. Int Conf Comput Sci Comput Intell 2015:312–317

Komninos N (2011) Intelligent cities: variable geometries of spatial intelligence. Intell Build Int 3(3):172–188

Setis-Eu (2012) https://setis.ec.europa.eu/archive/technology-roadmaps. Accessed 15 Sept 2019

Techopedia (2019) https://www.techopedia.com/definition/31494/smart-city. Accessed 15 Sept 2019

Thite M (2011) Smart cities: implications of urban planning for human resource development. Human Resour Develop Int 14(5):623–631

Tratz-Ryan B, Finnerty B (2019) Hyper cycle for smart city technologies and solutions. Gartner, Inc. Stamford, USA. Available on https://www.gartner.com/en/documents/3955915/hype-cycle-for-smart-city-technologies-and-solutions-201. Accessed 15 Sept 2019

Chapter 16
Indicators and Rating Systems for Sustainable Smart Cities

Giuliano Dall'O'

Abstract The sustainability and the smartness of cities must be objectively measured starting from real observations. Only in this way is it really possible to measure progress or identify critical elements of the development policies adopted. This chapter addresses the issue of indicators and rating systems that can be constructed from them. After a general introduction on the indicators, international standards designed to promote sustainability in cities and to support actions related to SDGs will be analysed. Subsequently, the most widespread ranking systems will be considered, and some case studies of their application will be proposed.

16.1 Sustainability Indicators for Planning and Monitoring Actions

"*When you can measure what you are talking about, and express it in numbers, you can claim to know something; but if you can't measure it, if you can't express it with numbers, your knowledge will be poor and unsatisfactory*". This famous phrase by Lord Willian Thompson Kelvin expresses a concept that underlies any scientific approach: an approach that cannot be developed in a concrete way without the evidence of the data. The statement of Lord Kelvin helps us to introduce an important topic for the development of sustainability: that of indicators.

The indicators are fundamental to understand, in an objective and comparable way, the environmental, economic and social dynamics that occur within our cities or communities. Through these indicators, we can make an analysis of a phenomenon, monitor it and confirm whether our forecasts were correct and to understand what other actions we need to consider in order to reach our sustainable goals. Indicators are used in various disciplinary fields and with different expectations and habits.

There are hundreds of indicator definitions. Each traces an aspect, enhances a characteristic, but above all derives from a response to a request that comes from

G. Dall'O' (✉)
Architecture, Built Environment and Construction Engineering, ABC Department, Politecnico di Milano, Via G. Ponzio, 31, 20133 Milan, Italy
e-mail: giuliano.dallo@polimi.it

G. Dall'O' (ed.), *Green Planning for Cities and Communities*,
Research for Development, https://doi.org/10.1007/978-3-030-41072-8_16

different needs: sometimes general, sometimes particular and sometimes local. Those that interest us are part of the development of urban and territorial policies, linked to the principle of sustainability, which inspires every designer, be they architect or consulting engineer and urban planner.

The indicators can supplement the regular reports on the state of the environment and thereby contribute to the process of monitoring the progress of environmental policies in the various sectors. Such an approach makes it possible to regularly review progress achieved in achieving policy objectives and to communicate the results to stakeholders and to public opinion. Indicators are particularly important tools for the purposes of accountability and transparency.

This presupposes that they are few, relevant, reactive, simple and specific to the various policies.

The great interest in the subject of indicators has stimulated researchers to study this topic in a scientific way. The topic is addressed by Heink and Kowarik (2010), who analyse and discuss the use of indicators in ecology and environmental planning.

Huovilaa et al. (2019) compare seven recently published indicator standards for smart sustainable cities. A taxonomy was developed to evaluate each of their 413 indicators against five conceptual urban focuses (types of urban sustainability and smartness), ten sectoral application domains (energy, transport, ICT, economy, etc.) and five indicator types (input, process, output, outcome, impact) (Fig. 16.1).

The results clearly discriminate between indicator standards suited for evaluating the implementation of predominantly smart city approaches versus standards more focused on sustainability assessment. The authors provide guidance for city managers and policy makers to select those indicators and standards that best correspond to their assessment needs and goals and align with their stage in the implementation of a smart sustainable city.

To be useful for sustainable city and community planning purposes, an indicator must possess certain qualities or characteristics such as:

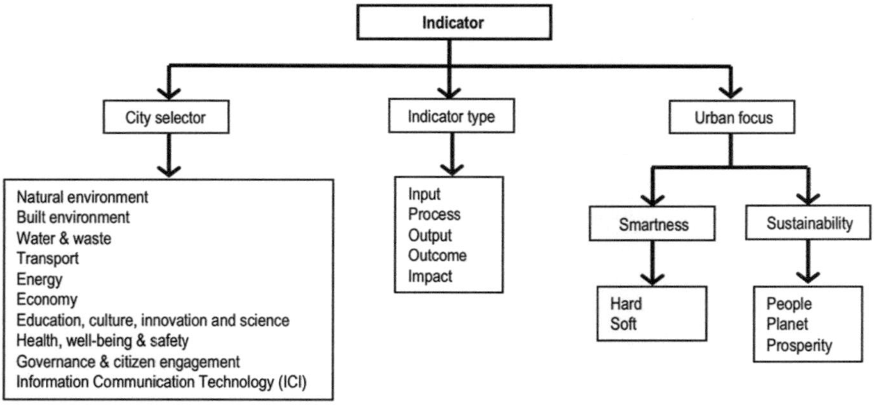

Fig. 16.1 Taxonomy for indicator analysis (Huovilaa et al. 2019)

- *pertinence* (must be able to meet the demands of sustainable goal setting);
- *relevance* (it must be able to effectively suggest and facilitate the construction of the decision-making process);
- *simplicity* (must be understandable and use clear and simple units of measurement);
- *comparability* (it must allow the comparison between different cases and along the time scale);
- *shareability* and *acceptability* (the choice of measures must go through a participatory evaluation process);
- *feasibility* (data must be available or otherwise easy to find);
- *transparency* (the process of constructing the indicators must be clear and the information underlying them available for verification);
- *replicability* (different subjects must be able to construct the extensive indicators obtaining the same values).

In Table 16.1, Huovilaa et al. (2019) provide a summary of indicator standards on sustainable cities. From the table, it is possible to observe how, at international level, one of the most important reference standards for sustainability issues in cities is currently ISO 37120: 2018 (ISO 2018a).

The state of progress, or regression, of SDGs in the European Union can be found in (EU 2019). In order to facilitate the implementation of the Sustainable Development Goals (SDGs) in the European Union, Eurostat annually provides an update of the indicators at national level regarding the Member States. The state of progress, or regression, of SDGs in the European Union can be found in (EU 2019).

16.2 The Indicators Proposed by ISO 37120

The ISO 37120 Standard "Sustainable cities and communities—Indicators for city services and quality of life" defined the first series of standardized urban indicators intended to provide a uniform approach to what had to be measured and how (ISO 2018a).

For the first time, cities were able to dialogue with each other using standardized and comparable data on global level, exchange information and learn from each other.

The document is focused on city services and the quality of life as a contribution to the sustainability of the city.

A set of standardized indicators provides an integrated approach to sustainable development and a uniform approach to what is measured, and how that measurement is to be undertaken.

The indicators can be used to track and monitor progress on city performances. The whole city system needs to be taken into consideration in order to achieve sustainable development. Planning for the future needs should take into consideration current use and efficiency of resources in order to better plan for tomorrow. The indicators considered in ISO 37120:2018 Standard have been developed in order to help cities:

Table 16.1 Summary of indicator standards on smart sustainable cities (Huovilaa et al. 2019)

Name, type and abbreviation	Main categories	Indicators (number)
ISO 37120:2018 Sustainable development of communities—indicators for city services and quality of life (ISO 2018a). (International non-mandatory standard)	Economy, education, energy, environment and climate change, finance, governance, health, housing, population and social conditions, recreation, safety, solid waste, sport and culture, telecommunication, transportation, urban/local agriculture and food security, urban planning, wastewater, water	104
ISO/DIS 37122:2018 Sustainable development in communities—indicators for smart cities (ISO 2018b). (Draft international standard)	Economy, education, energy, environment and climate change, finance, governance, health, housing, population and social conditions, recreation, safety, solid waste, sport and culture, telecommunication, transportation, urban/local agriculture and food security, urban planning, wastewater, water	85
ETSI TS 103 463 Key performance indicators for sustainable digital multiservice cities (ETSI 2017a)	People, planet, prosperity, governance	76
ITU-T Y.4901/L.1601 Key performance indicators related to the use of information and communication technology in smart sustainable cities (ITU 2016a). (Recommendation)	ICT, environmental sustainability, productivity, quality of life, equity and social inclusion, physical infrastructure	48
ITU-T Y.4902/L.1602 Key performance indicators related to the sustainability impacts of information and communication technology in smart sustainable cities (ITU 2016b). (Recommendation)	Environmental sustainability, productivity, quality of life, equity and social inclusion, physical infrastructure	30
ITU-T Y.4903/L.1603 Key performance indicators for smart sustainable cities to assess the achievement of sustainable development goals (ITU 2016c). (Recommendation)	Economy, environment, society and culture	52
Sustainable development goal 11+ monitoring framework (UN-Habitat et al. 2016)	UN SDG targets 11.1, 11.2, 11.3, 11.4, 11.5, 11.6, 11.7, 11a, 11b, 11c, 1.4, 6.3	18

- measure performance management of city services and quality of life over time;
- learn from one another by comparing across a wide range of performance measures; and,
- support policy development and priority setting.

"Maintaining, enhancing and accelerating progress towards improved city services and quality of life is also fundamental to the definitions of both smart cities and resilient cities. The document is intended to be implemented in conjunction with ISO 37122 and ISO 37123 to measure progress towards smart cities and resilient cities, respectively" (ISO 2018a).

The indicators are structured around themes. Recognizing the differences in resources and capabilities of cities worldwide, the overall set of indicators for city performance has been divided into "core" indicators (which are fundamental and strongly recommended), "supporting" indicators (less important) and "profile indicators" that are recommended to provide basic statistics and background information to help cities conduct peer comparisons.

The categories of indicators provided by the standard are the following: economy, education, energy, environmental and climate change, finance, governance, health, housing, population and social conditions, recreation, safety, solid waste, sport and culture, telecommunication, transportation, urban planning, wastewater and water.

16.3 Smart City Rankings

As has been seen, smart cities are a concentration of new technologies used at their best but at the same time places pervaded by widespread energy and environmental sustainability, from the correct and functional management of energy to that of waste or water. Then there is the social dimension that places man, his health, information, culture but also moments of leisure at the centre of interests. All this positivity, however, conceals a doubt shared by many: that smart cities express an abstract concept, a desire rather than an evolving reality.

The intelligence of a smart city to assume its credibility, to become something concrete, must be able to be analysed with objective assessment tools that have a scientific basis: anything that cannot be measured can at the most arouse emotions, but this may not be enough.

If the cities that already exist contain at least some of the elements that contribute to creating smartness, the goal is to have available tools, objective and necessarily shared, through which it can be measured.

The function of the indicators, analysed in the previous paragraphs, is consequently very important to giving substance to concepts that, if not objectively measured, can be perceived as abstract.

However, the indicators, if considered individually, risk not helping to understand the complexity of the city.

To objectively measure the sustainability or smartness of a city, it is necessary to proceed in steps:

- select, among all available indicators, those considered most strategic to monitor the city;
- group the "homogeneous" indicators by evaluation area (e.g. smart economy, smart environment, smart energy, smart people, smart mobility, etc.);
- establish an evaluation or ranking system which allows a score to be assigned to each evaluation area.

There is an objective difficulty in having to add indicators that refer to very different aspects, the use of these tools, however, is necessary in contributing to giving substance to the project of a smart city, to assess its current state, to be able to compare with other cities, to identify its strengths but also the areas of weakness, to be able to define goals in the medium and long term, to measure progress or, on the contrary, to verify the regressions.

There is no a single and shared classification system, but several projects with approaches that may be different: a common element is that they allow the definition of a single classification, but at the same time, they allow partial evaluations to be made within the individual evaluation areas. A city can be considered the model of environmental sustainability but at the same time be lacking in the social and economic fields. The evaluation of individual smartness can make it clear where it can be improved, and this is very important.

In defining a ranking system, it is appropriate to make some considerations regarding the cities to be considered.

Towns and cities must have a similar demographic class (e.g. up to 20,000 inhabitants, from 20,000 inhabitants to 100,000 inhabitants and so on). An evaluation ranking, in fact, cannot be used to compare, for example, a small town of 50,000 inhabitants with a large metropolis such as London or New York because some indicators might work for a small town but not work for a large net city and vice versa.

Aspects relating to the environment intended as a climatic zone should also be considered: energy indicators, in fact, are very much influenced by the climatic conditions in winter and summer.

Finally, homogeneity should also concern economic, social and cultural aspects.

In the set of cities that are chosen to build the ranking (obviously starting from the choice of indicators), it would therefore be useful to compare cities that are truly comparable, for example, within a well-defined territorial area.

It is then necessary to define the evaluation areas. In the diagram of Fig. 16.2, the six traditional evaluation areas have been considered, the graphic representation, called "radar" allows to compare, for each of them, the rating reached by the city in question, or reference city, with the rating level average of all the cities examined.

With this approach, it is possible to bring out potential or critical aspects of a specific city, even partially.

Each evaluation area is built on the basis of indicators that do not necessarily have to be numerically the same. There are more crowded areas and less crowded areas,

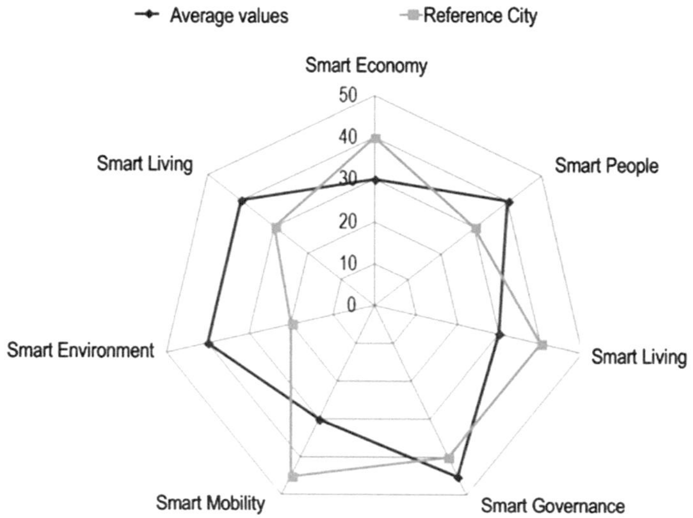

Fig. 16.2 Comparison of the area indicators between a city (reference city) and the city average of the examined set

and the important thing is that the indicators are all consistent with the evaluation area to which they refer.

Defining a ranking for a single area, with a necessarily limited number of indicators, inevitably introduces simplifications, but there are no other ways to measure sometimes complex phenomena.

Some indicators find their correct placement within a precise evaluation area, others do not. One can take, for example, those that define the energy efficiency of a city. The energy aspect in a city is important and even more so in a city that wants to be defined as "intelligent". The "smart energy" area does not exist in the six traditional assessment areas: energy efficiency indicators could belong to the "smart environment" area considering the close dependence between energy use from fossil sources and environmental impact, but they could also belong to the "smart governance" area if an improvement in urban energy efficiency is associated with a city's energy planning policies. Some rankings define a "smart energy" area, definitely enhancing this aspect.

Within the same evaluation area, the indicators can also be different from each other, and however, it is necessary to return a unitary frame of reference that defines the overall ranking achieved by the area itself. In other words, the individual indicators must be able to be added together in order to define a unique representative indicator of the area.

There are several methods of data aggregation: one of these consists in reshaping the value of the indicator in question on a scale from 0 to 1. Once the values of all indicators of that type (one per city) have been measured, a value of 0 is assigned to the worst one and a value of 1 to the best one. The indicators of all cities, for that same type, will necessarily be placed within these two values that define the extremes of the evaluation interval. To give more or less importance to an indicator within the same area, it is also possible to introduce correction with weighting coefficients.

With this criterion, each area obtains an overall score as the sum of the normalized, and hence, summable, scores of the possibly corrected individual indicators.

The overall ranking derives from the sum of the points of the individual areas. In the case where the objective is to evaluate the weight of the individual areas while maintaining a single scale, for example, from 0 to 50, as in the graph of Fig. 16.2, it is possible to proceed to a further normalization.

The need to measure the smartness of cities gave rise to various projects aimed at outlining a procedure for evaluating cities by means of indicators that support the rating of such smartness. The objective of these projects is comparing different cities based on the evaluation of both a partial rating (i.e. with reference to various evaluation areas like smart economy, smart mobility, smart people, smart environment, etc.) and an overall rating of urban smartness.

European Smart Cities (2017) is a European project dating back to 2007 that aimed at assessing the smartness of European medium-sized cities. It was financed by European Union funds and was conducted by the Polytechnic of Vienna in collaboration with the University of Ljubljana and the Technical University of Delft.

Currently, the project includes two schemes which take into account the demographic size of the 77 cities considered: European medium-sized cities (cities from 100,000 to 500,000 inhabitants) and larger European cities (cities from 300,000 to 1 million inhabitants).

An interactive assessment tool, available on an Internet platform [www.smart-cities.eu], allows developing the overall ranking but also partial rankings that take into account the urban smartness related to specific evaluation areas (economy, mobility, environment, people, governance). Each evaluation area is composed of indicators, which together sum up to 81. Each indicator has a specific weight. The interactive interface allows assessing the smartness of each city by basing on the rating of each area as well as of each individual indicator. As in all ranking modules, positive and negative aspects characterize the European Smart City Index.

Using the web platform, it is possible to compare more cities in order to see the differences. An example is given by Fig. 16.3 that compares the cities of Amsterdam, Malmö and Nice.

Via the web platform, it is also possible to obtain, for a city, the normalized values of the individual indicators that are proposed in dynamic graphs such as, for example, that of Fig. 16.4 which refers to the smart living area of the city of Malmö. It can be seen that some normalized values within the same area may be negative. A complete explanation of how the indicators were constructed, in particular, with regard to the normalization procedure, is contained in the project website (smart-cities.eu).

On the downside, the project is affected by the subjectivity of the criteria and definition of rules and by the lack of feedback from the municipalities assessed. The principal positive element of this project is the comparison of medium-sized cities with similar characteristics (Dall'O' et al. 2017).

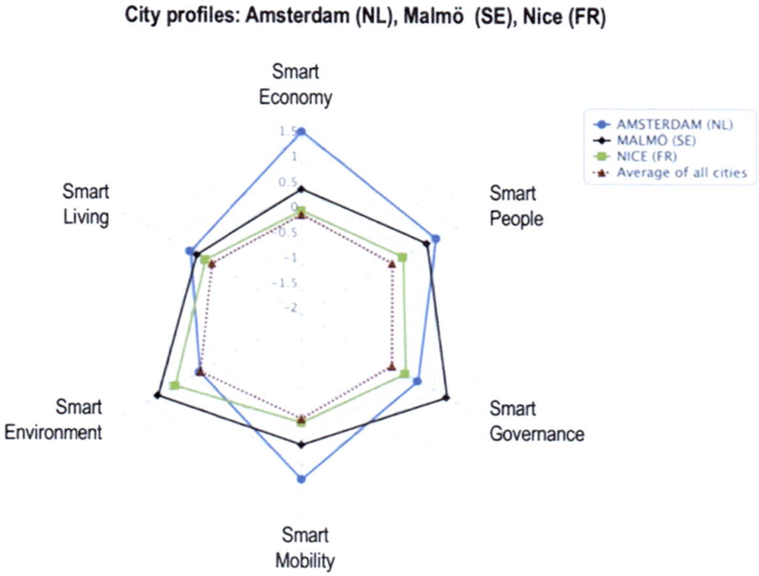

Fig. 16.3 Comparison between the smartness evaluation of three European cities divided by the six evaluation areas (www.smart-city.eu)

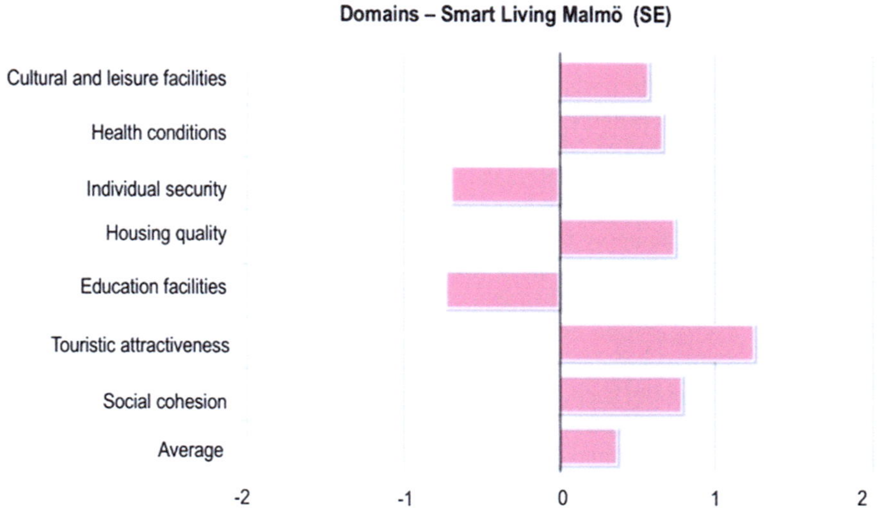

Fig. 16.4 Detailed overview of the normalized indicators of the city of Malmö for the smart living domain (www.smart-city.eu)

16.4 Rating Systems for Ranking Small and Medium Municipalities, with a Bottom-up Approach in Italy

When dealing with the topic of smart cities, it is relevant to consider also the evolution of cities towards models that place value on the sustainable development by promoting sustainable governance. The Covenant of Mayors project (http://www.covenantofmayors.eu/) which started in 2008 under the auspices of the European Union is an interesting initiative in this sense. The project represents a first step towards the implementation of planning measures to cut climatic emissions through the monitoring of their energy consumption and the drafting of energy budgets. In this way, the project paves the road for a debate on citizens' quality of life. A thorough environmental sustainability as conceived in the Covenant of Mayors project is an important matter to consider when talking of smart cities, other aspects are relevant too: the smart city approach offers a comprehensive view of the factors that are pertinent to citizens' quality of life.

These considerations are the basis of a methodological approach on the topic of the smart city developed by researchers of the Department of Architecture, Built environment and Construction Engineering (ABC) of the Polytechnic of Milan, aimed in promoting a smart city protocol for small and medium-sized municipalities (Dall'O' et al. 2017).

The following reasons, most of which are relevant also for other national contexts in the European Union, persuaded the authors to develop a methodology for the Italian situation:

- Most projects regarding smart cities concern large cities (e.g. Milan, Turin, Rome, Venice, Naples, etc.), and however, most of the Italian population is concentrated in small and medium-sized cities: 23% of the Italian population lives in cities with more than 100,000 inhabitants, 47.5% lives in towns with less than 20,000 inhabitants, while 31.3% lives in towns with less than 10,000 inhabitants;
- It is not possible to apply the same smartness indicators used for evaluating big cities to small and medium-sized towns. The assessment of smartness in these cases must be adapted and/or modified in order to meet the real needs;
- Small and medium-sized cities have the opportunity to support the development of groups of citizens moved by critical thinking, whose role is necessary to promote environmentally friendly projects like strategies for sustainable mobility (i.e. routes equipped for electric cars, cycle paths, bike sharing, car sharing, etc.);
- Information and communication technologies (ICT) have a significant contribution to make in virtually connecting companies, citizens and other entities, as well as providing services that enable them to limit travels to large urban centres;
- Several small and medium-sized towns and cities have long been committed to promoting the Sustainable Energy Action Plan (SEAP) within the framework of the Covenant of Mayors. The implementation of smart city models can constitute an important extension.

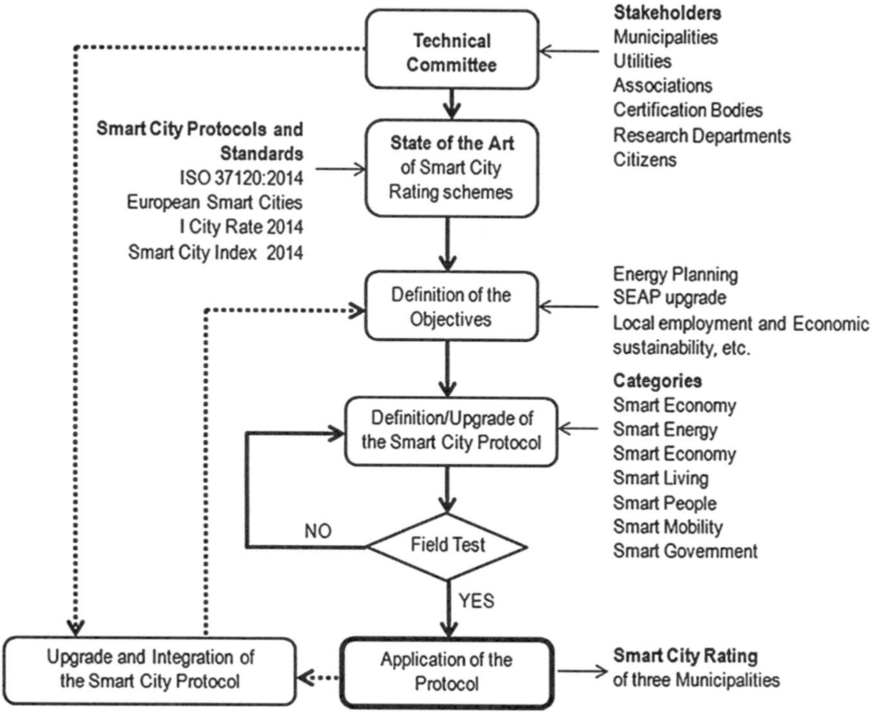

Fig. 16.5 Flow chart representing the methodology for the development of the smart city protocol (Dall'O' et al. 2017)

The purposed methodology should be considered as a way to expand the role of environmental and critical thinking in governance strategies already adopted by small and medium-sized cities around Europe.

Compared to existing protocols that use indicators based on national or regional databases, this methodology applies a specific type of analysis called "smartness audit", which assesses concrete results of policies promoted by municipalities in different evaluation areas.

Most smartness indicators are constructed starting from information that is locally available and already held or can be easily retrieved by the technical departments of the municipalities/local authorities involved in the audit. The systematic collection of this urban, social, economic and environmental information makes the protocol an effective monitoring tool for government policies.

According to what happens in the promotion and management of the SEAP, the methodological approach we adopted involved a large number of stakeholders. Sharing choices in governance policies and including different actors are key elements in the development of smart city models.

Figure 16.5 shows the steps adopted in the project. Initially, a Technical Committee was set up with the voluntary participation of different stakeholders among which municipalities, utility companies, associations, certification bodies, research departments, universities and citizens.

The Technical Committee had a strategic role in the development of the protocol: in addition to defining its technical directions in the initial steps, it was engaged in verifying the application of the protocol in the experimental stage and dealt with its periodic updates.

In the definition of the evaluations areas (or categories) and indicators contained in the protocol, the Technical Committee had the role of considering the needs expressed by stakeholders and the economic, societal and infrastructural features that characterize the small and medium-sized municipalities (approximately less than 50,000 inhabitants).

An important step of the method used to develop the protocol was the field test. The application of the protocol's indicators, many of which are proposed by the political representatives of the Technical Committee, require the retrieval of basic information. The purpose of the field test is to check whether this information is available and sufficiently reliable, as well as whether it is necessary to employ human and instrumental resources to build up administrative procedures aimed at finding such information.

In order to define the indicators the authors analysed two Italian smart city protocols, ICity Rate[1] and Smart City Index[2] and Standard ISO 37120:2014. Moreover, the action plans for sustainable energy processed by the municipalities that joined the European Covenant of Mayors project were examined in depth.

The above-mentioned tools were analysed to identify the following elements:

- the main features;
- the scope of application;
- the thematic areas within which the indicators were developed;
- the points of strength;
- the critical elements;
- the different types of indicators used across evaluation areas;
- the data source(s) useful for the calculation of the indicators;
- the complexity involved in finding the information to calculate the indicators;
- the units of measurement of the indicators;
- the relative weights assigned to the indicators;
- the standardization process for the indicators;
- the results of the analysis carried out;
- the degree of comparability of results over time or across cities.

[1] ICity Rate is a smart city protocol promoted by ICity Lab of Forum PA together with National Association of Italian Municipalities (ANCI) (https://www.forumpa.it/citta-territori/icity-rate-2018-la-classifica-delle-citta-intelligenti-italiane-settima-edizione/) (Accessed 23 September 2019).

[2] Smart City Index is the ranking of the 116 Italian chief towns created by the company between (https://www.csroggi.org/smart-city-index-2018-presentato-il-rapporto-biennale-elaborato-da-ey/) (Accessed 23 September 2019).

A consistent protocol suitable for the assessment of the smartness indicators was developed with the support of the Technical Committee. Initially, the Technical Committee conducted a recurrence analysis of the indicators used by ISO 37120, ICity Rate and Smart City Index and evaluated them according to specific criteria. In order to meet the needs of the project, the newly defined indicators had to be:

- mirroring the features of Italian cities;
- consistent with the size of small and medium towns and cities;
- normalized;
- characterized by easy calculation;
- accessible to the validated source of data for the calculation;
- suitable for and effective in representing specific features of small and medium cities;
- comparable over time or across cities.

The Technical Committee defined a set of indicators after discussing their usage over time, modified and revised them during several meetings which led to a finalized version. A total of 70 indicators were identified.

In a first phase, the Technical Committee analysed the indicators' recurrence within the protocols used as reference (ICity Rate and Smart City Index) and the ISO 37120 and started evaluating the selected indicators according to the above-mentioned criteria. In a second phase, the Technical Committee selected indicators that were already present in one of the protocols analysed or in the ISO 37120, that were relevant to evaluate the aspects characterizing the categories and that were consistent with the criteria outlined above.

In the last step, the list of indicators was enriched with newly introduced indicators that did not belong to those included in the reference projects and protocols. These indicators satisfied needs that were specific for the smart city protocol such as:

- indicators that responded to demands coming from territory;
- indicators that valorized new innovative solutions;
- indicators that valorized actions that are consolidated at local level;
- indicators and data already present in the Sustainable Energy Action Plan (SEAP) or Sustainable Energy and Climate Action Plan (SECAP);
- indicators that evaluate aspects related to the use of energy sources other than electrical sources;
- indicators that respond to the new integrated climate change approach of the SEAP or SECAP;
- indicators that enhance electrical mobility also in relation to national legislation.

The indicators are structured around seven evaluation areas (smart economy, smart energy, smart environment, smart governance, smart living, smart people and smart mobility) that allow the analysis of results related to a city considering a specific theme. In this way, a city can become able to identify and address priorities for future actions in specific contexts.

The authors applied the methodology to three medium-sized municipalities located in the Lombardy region near Milan: Carugate, Melzo and Pioltello. The indicators identified are aggregated in seven categories which are shown in Table 16.2 that also reports the rating attributed to the three case study municipalities in smartness auditing (the numbers higlighted in the table show the best ratings for each evaluation area).

The overall values attributed to the performance of the three municipalities are very close to each other. Albeit minimal, the differences between scores show interesting aspects related to the analysis of the three municipalities. The study of these aspects is further detailed during the activities to support the planning and the monitoring of the SEAP.

Carugate is the municipality with the highest overall score (35.12), as well as the highest scores for four categories out of seven. Instead, that same municipality scored rather low with respect to two of the categories, smart people and smart mobility. The presence of shopping centres that attract the population of neighbouring municipalities generates traffic problems which need to be solved by the town council.

Pioltello was attributed an overall score which is in between those of the other two municipalities. Interestingly, it scored quite highly in the categories in which Carugate was lagging behind. Instead, with respect to the smart economy category, the municipality needs to implement some improvements. This data spotlights a difference between Pioltello and Melzo since the second municipality scored the highest in this category.

Melzo had an overall score of 32.68 which is the lowest of the overall scores attributed to the three municipalities. The town had an excellent performance in the category of smart economy, especially due to the industrial vocation of the municipality. However, with respect to the other categories the municipality scored lower than Carugate and Pioltello. In particular, for the smart energy category it scored almost half the value of Carugate.

To summarize, the fact that Municipalities were attributed similar overall scores does not preclude the possibility of observing important differences by looking at partial data.

Table 16.2 Smart rating (performance) for the three case studies (Dall'O' et al. 2017)

Evaluation area	Carugate	Melzo	Pioltello
Smart economy	5.24	**5.98**	3.76
Smart energy	**5.55**	3.68	5.33
Smart environment	**6.64**	5.27	4.41
Smart governance	**6.75**	5.54	6.50
Smart living	**5.15**	4.03	4.71
Smart people	2.70	3.98	**4.57**
Smart mobility	3.08	4.18	**4.86**
Total score	35.12	32.68	34.13

In this methodology, the authors define the "audit smartness" as the set of activities that involve the collection of data and information needed to develop the indicators. One of the features of the adopted approach is to develop indicators based on real data derived from the local territorial context and provided by the municipalities involved, rather than statistical data.

One should encourage a good cooperation with the technicians responsible for the various relevant municipal departments (Territorial Department, Department of Urban Planning, Transportation Department, Department of Administration and Budget, Department of Culture, Department of Health, etc.) and foster the involvement and cooperation between different operators (administrative staff and technical staff) which normally act independently.

Based on this experience, the authors found that some information useful for calculating indicators is already available in organized databases, while others require a more complex analysis of documents (reports, drawings, etc.).

If the implementation of an audit system aimed at building smartness indicators encourages collaboration between the operators and personnel within the administrative structure (the municipality) and allows for an objective overall picture of the situation, this is already an important objective for justifying this type of procedure.

16.5 Conclusions

This chapter addresses the issue of indicators and rating systems that can be constructed from them. After a general introduction on the indicators, international standards designed to promote sustainability in cities and to support actions related to SDGs will be analysed. The chapter concludes proposing rating systems for ranking small and medium municipalities, with a bottom-up approach, analysing and discussing a practical experience.

In this methodology, the authors define the "audit smartness" as the set of activities that involve the collection of data and information needed to develop the indicators. One of the features of the adopted approach is to develop indicators based on real data derived from the local territorial context and provided by the municipalities involved, rather than statistical data.

One should encourage a good cooperation with the technicians responsible for the various relevant municipal departments (Territorial Department, Department of Urban Planning, Transportation Department, Department of Administration and Budget, Department of Culture, Department of Health, etc.) and foster the involvement and cooperation between different operators (administrative staff and technical staff) which normally act independently.

Based on this experience, the authors found that some information useful for calculating indicators is already available in organized databases, while others require a more complex analysis of documents (reports, drawings, etc.).

If the implementation of an audit system aimed at building smartness indicators encourages collaboration between the operators and personnel within the administrative structure (the municipality) and allows for an objective overall picture of the situation, this is already an important objective for justifying this type of procedure.

References

Dall'O' G, Bruni E, Panza A, Sarto L, Fazel K (2017) Evaluation of cities' smartness by means of indicators for small and medium cities and communities: a methodology for Northern Italy. Sustain Cities Soc 34:193–202

EU (2019) Sustainable development in the European Union: Monitoring report on progress towards the SDGs in an EU context 2019 edition. ISBN 978-92-76-00777-7, https://doi.org/10.2785/44964. Available on https://ec.europa.eu/eurostat/documents/3217494/9940483/KS-02-19-165-EN-N.pdf/1965d8f5-4532-49f9-98ca-5334b0652820. Accessed 22 Sept 2019

European Smart Cities (2017) http://www.smart-cities.eu/. Accessed 22 Sept 2019

European Telecommunications Standards Institute (2017a) ETSI TS 103 463 Key performance indicators for sustainable digital multiservice cities. Technical specification V1.1.1 (2017-07). http://www.etsi.org/deliver/etsi_ts/103400_103499/103463/01.01.01_60/ts_103463v010101p.pdf. Accessed on 15 June 2018

Heink U, Kowarik I (2010) What are indicators? On the definition of indicators in ecology and environmental planning. Ecol Indic 10(2010):584–593

Huovilaa A, Boschb P, Airaksinenc M (2019) Comparative analysis of standardized indicators for Smart sustainable cities: what indicators and standards to use and when? Cities 89(2019):141–153

International Standardization Organization (2018a) ISO 37120:2018 Sustainable cities and communities—indicators for city services and quality of life, 2nd edn (2018–07)

International Standardization Organization (2018b) ISO/DIS 37122 Sustainable cities and communities—indicators for smart cities. (Published 2018–06–06)

International Telecommunication Union (2016a) Key performance indicators related to the use of information and communication technology in smart sustainable cities

International Telecommunication Union (2016b) Recommendation ITU-T Y.4902/L.1602 Key performance indicators related to the sustainability impacts of information and communication technology in smart sustainable cities

International Telecommunication Union (2016c) Recommendation ITU-T Y.4903/L.1603 Key performance indicators for smart sustainable cities to assess the achievement of sustainable development goals

UN-Habitat, UNESCO, World Health Organization, UNISDR, UN Women, UNEP, et al. (2016) SDG goal 11 monitoring framework. http://unhabitat.org/sdg-goal-11-monitoring-framework/. Accessed on 19 Sept 2019

Chapter 17
Green BIM and CIM: Sustainable Planning Using Building Information Modelling

Giuliano Dall'O', Alessandro Zichi and Marco Torri

Abstract The design, construction and management of sustainable and intelligent buildings, neighbourhoods and cities require tools that can manage complexity in a completely different manner to the practices adopted to date. This chapter introduces the most advanced current approach, Building Information Modelling (BIM), focusing in particular on its use to support green and smart planning of buildings, neighbourhoods and cities. The initial part of the chapter introduces the general concepts underlying the BIM. The second part explores its application, the Green BIM, intended as the use of the tool to support the sustainable design of buildings. The chapter concludes with the introduction of a more recent concept, the City Information Modelling (CIM), intended as a common platform to design, build and manage the buildings, infrastructures and services of neighbourhoods and cities that evolve towards sustainable and smart schemes.

17.1 Introduction to Building Information Modelling

In recent years, Building Information Modelling or BIM has been frequently spoken of: it is viewed not only with great enthusiasm but also with some concern.

The enthusiasm stems from the fact that it is a new tool that takes full advantage of technological innovation, in particular Information Communication Technology (ICT) and the Internet and, thanks to this, it enables many professional designers to operate on common and shared projects, even when they are physically very distant.

G. Dall'O' (✉) · M. Torri
Architecture, Built Environment and Construction Engineering, ABC Department, Politecnico di Milano, Via G. Ponzio, 31, 20133 Milan, Italy
e-mail: giuliano.dallo@polimi.it

M. Torri
e-mail: marco.torri@polimi.it

A. Zichi
Consorzio CIS-E Scuola F.Lli Pesenti, P. zza Leonardo da Vinci, 32, 20132 Milan, Italy
e-mail: alessandro.zichi@polimi.it

G. Dall'O' (ed.), *Green Planning for Cities and Communities*,
Research for Development, https://doi.org/10.1007/978-3-030-41072-8_17

The concern, typical of designers accustomed to autonomous management of design, construction and management processes, is that the use of a tool as powerful as BIM can be too difficult.

A further concern is that interoperability, a key element of BIM, can generate complications in the exchange of data between design software and project management. Those who are very skeptical, but above all those who have always been used to working in a traditional way, consider BIM as a useless complication.

For designers who, even if they use software tools such as CAD or energy and structural simulation software for buildings and infrastructures, are in the habit of operating in the traditional way, the transition to BIM requires considerable effort, particularly in the initial phase. On the other hand, those designers who have a consolidated knowledge base, once they have overcome the main barrier, which is that of sharing work on a common platform, are able to obtain the best performance in using the BIM.

The use of BIM becomes natural for young designers who do not have previous professional experience and who at school and university have learned right from the start to design in BIM and above all to work in teams. The danger for young designers is that BIM is seen for what it is not: a machine capable of solving all the problems of design: a kind of black box in which it is sufficient to enter data to obtain results.

BIM is not a design methodology or an aggregate set of design methodologies but a powerful tool for synergistic integration of skills. Thus, designers also when operating in BIM must have all the necessary skills: the BIM contributes, if anything, to enhancing their potential in designing and managing projects, significantly reducing time and improving the overall quality of the final result.

Technological innovation is always unstoppable and the approach provided by the BIM is the most important technological innovation to date in the design and management of buildings and infrastructures. Once the initial barriers, partly technical but above all cultural and social, have been overcome, this new approach will spread rapidly, expressing all its potential. A common element of BIM and sustainable and smart planning is that the approach to these two modes cannot be considered a choice but the only possible option.

17.1.1 What is Building Information Modelling (BIM)

Between the end of 1970 and the beginning of 1980, the computer-aided design (CAD) systems increased their development potential so that the aerospace and manufacturing industries decided to implement this system through an active collaboration with the software house.

These software systems created for the aerospace and manufacturing industries were then adopted in the field of building construction for architectural design (e.g. AutoCAD®). Since that time, 2D drawing software and 3D modelling for the building industry have appeared on the market.

In 1986, Graphisoft presents the first Virtual Building Solution known as ArchiCAD®. Since 1986, unbounded progress has been made and since 2003, the expression BIM has become very popular especially following the debate organised by Jerry Laiserin (first promulgator of the term BIM) in the USA between the two software companies with the most widespread product introduction in the field of Building at that time, Autodesk® and Bentley.

The definition of BIM is not standardised and the demonstration thereof is the large number of traceable definitions in works published or circulating on the Internet.

Before analysing the most widely used definitions, it is useful to identify a concept that is at the base of the BIM and which is presented schematically in Fig. 17.1. In the traditional way of working, all the drawing works are involved in the process of planning, execution and maintenance of a construction interface among themselves, but not in a coordinated manner.

The technical or economic information within the process, made up of drawings, reports, economic tables, etc., undergoes continual modifications without leaving a trace of an overall and organic management.

With this type of organisation, which in any case is still applied today in most of the construction design processes, coordination is lacking but, above all, there is no single point for gathering information that is largely dispersed, or at least not updated during the process.

Working according to a BIM scheme means putting together, in a single digital platform, all the information relating to the project but also means obliging all the designers and operators involved to work synergistically as illustrated in the diagram in Fig. 17.1 on the right. There are three prerequisites for a design team to operate with a BIM platform:

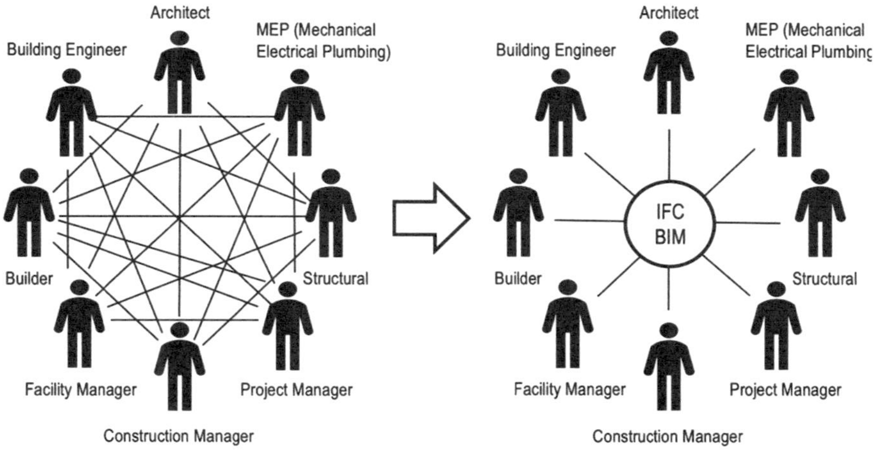

Fig. 17.1 Comparison between the traditional work method and the work method introduced by BIM for a construction project. Adapted from buildingSMART documentation. https://www.buildingsmart.org

- the availability of an information exchange network (Internet);
- the use of software able of exchanging data through shared exchange protocols (interoperability);
- the definition of shared level of details (LODs) of the project.

The integrated design supported by BIM is a collaborative method for the design of buildings or infrastructures and is considered the best process to pursue the objective of environmental sustainability: it allows one to have a complete view on the construction itself which, according to this scheme, is conceived as an interdependent system and not simply as the sum of several separate components.

The integrated design approach can, therefore, be defined as a process that considers the building throughout the entire life cycle, from the start-up of the construction site to the maintenance phases and at the end of its operating life (Fig. 17.2). The National BIM Standard-United States® (NBIMS-US™)[1] defines BIM as a "*digital representation of physical and functional characteristics of a facility. A BIM is a shared knowledge resource for information about a facility forming a reliable basis for decisions during its life-cycle; defined as existing from earliest conception to demolition*".

Building SMART, the community committed to creating and developing open digital ways of working for the built asset industry, considers BIM as "*A shared digital representation of physical and functional characteristics of any built object including buildings, bridges, roads, process plants* etc. *forming a reliable basis for decisions*", while ISO Standard ISO (2018b) considers BIM as "*a process involving the generation and management of digital representations of physical and functional characteristics of a facility. The resulting building information models become shared*

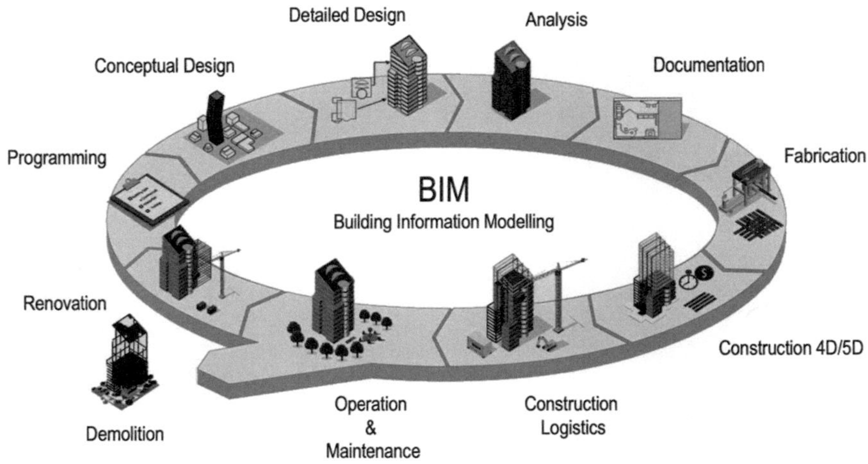

Fig. 17.2 Integration of all the life cycle phases of a construction project in the BIM approach

[1] https://www.nationalbimstandard.org.

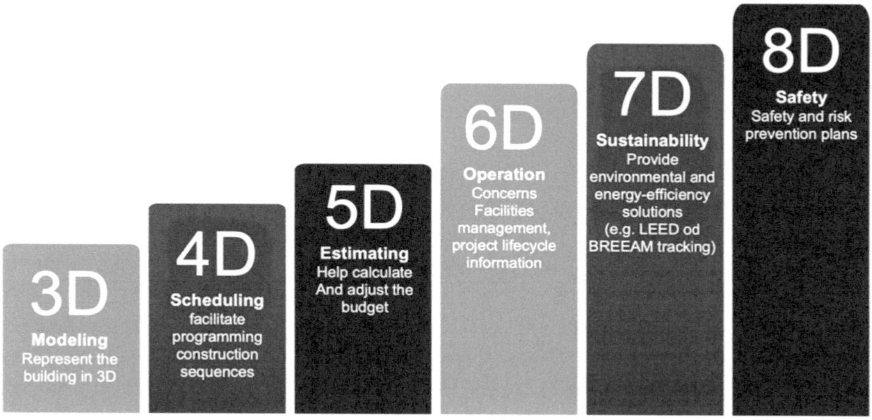

Fig. 17.3 Eight dimensions of BIM

knowledge resources to support decision-making about a facility from earliest conceptual stages, through design and construction, through its operational life and eventual demolition".

In a BIM design process, to the "objects" that make up the whole project are assigned items of information, additional to the geometric items (2D or 3D geometric), which can be used in different areas. BIM dimensions depend on BIM maturity levels and refer to the particular way in which they are linked to an information model. In Fig. 17.3, the current 8 dimensions of the BIM are illustrated schematically (others may be added in future):

- **BIM 3D, the Modelling dimension**, is the most familiar form of BIM and is the process of collecting graphic and non-graphic information to build 3D models and share them in a Common Data Environment (CDE). The technicians who collaborate in a project must provide precise inputs so that the potential critical issues can be identified and removed before the construction phase begins. BIM enables multidisciplinary teams to work together more effectively from a single source of correct information.
- **BIM 4D, the Scheduling dimension**, includes information on timing, e.g. installation time, duration of operation, care of materials, etc. The information contained allows the members of the project team to view the progress of a project in different phases. Working in this way means making a significant contribution to meeting deadlines.
- **BIM 5D, the Estimating dimension**, relates to cost information. With this information, the user is able to extract precise data from the model and to make the necessary modifications, for example, by choosing a component that, still satisfying the performance, has a lower cost or proposing an alternative design solution. Having cost information in the CDE helps with budget monitoring and cost analysis of the entire project.

- **BIM 6D, the Operations dimension**, is focused on information regarding the life cycle of the project and, therefore, considers aspects related to facility management. The data may include information provided by the manufacturer, including maintenance schedules, component configurations for optimal performance, expected component life, etc. This BIM dimension permits one to take better decisions, for example, on goods that last longer and make good economic sense. With the information contained in the 5D BIM, the facility personnel or facility managers can also pre-plan maintenance activities well in advance (preventive maintenance), thereby avoiding emergency situations which can arise from the sudden and unexpected breakage of components.
- **BIM 7D, Sustainability dimension**, contributes to improving energy efficiency and building sustainability. The information contained herein enables energy simulations to be performed, then analysing consumption with the aim of optimising resources and reducing environmental impacts from fossil fuels. Energy analyses can be performed throughout the design process in accordance with the logic of integrated design, interactively sharing data with energy simulation software. BIM 7D also facilitates the application of environmental certification protocols such as LEED® or BREEAM® and can therefore be considered to be the "green" dimension of BIM.
- **BIM 8D, Safety dimension**, it is the most recent dimension of BIM and deals with aspects related to the active and passive safety of the building. The information contained in this document contributes to risk analysis and to the elaboration of support documentation related to safety (e.g. safety plans) both during the construction phase and during the usage and maintenance phases.

BIM design adopts an "object oriented" approach. The different phases of a design process of a construction require the definition of the design details in different ways for the "objects" in the project (e.g. a wall, a window or a boiler). Within the framework of the methodological and conceptual approach that is characteristic of the BIM, particular emphasis is, therefore, placed on the "**Levels Of Detail**" or LODs.

The LODs should define precisely the level of in-depth analysis of the information that is contained within the digital model of the building complex. In this perspective, in particular, the Level Of Detail represents a reference point allowing all the people involved in the design process to specify the construction of a BIM model with a high degree of clarity about its contents and with the maximum reliability of information during all the different phases, from conception to the execution of the building (to which are added those of the eventual demolition at the end of the operating life).

At the international level, the definition of the characteristics of each individual Level Of Detail in relation to the type of element considered is a theme addressed with different approaches.

The British standardization body British Standard Institution (BSI) has produced a series of organically designed standards of great importance. The standards of the 1192 series (BS 1192-PAS 1192-2//3//4//5) and the others related to them, constitute

Table 17.1 Definition of LODs in a BIM process consistent with the AIA proposal

LOD 100	LOD 200	LOD 300	LOD 400	LOD 500

Model containing the performance requirements and site specifications. Conceptual model useful for the study of the whole building including base area and volume, initial cost estimate	Model containing generalised systems with approximate quantities, sizes, shapes, positions and orientations	Model of production or pre-construction, and for the purpose of planning. Accurate and coordinated method, useful for a more accurate estimate of costs	Accurate model, with construction requirements and specific construction elements	"As built" model of the building that shows the project as it was built

an important reference for the application of the BIM methodology, covering all the life phases of a building, from the planning to the construction and management phases.

The PAS Standards merge into the ISO 19650 Standards (ISO 2018b, c).

In the USA, the American Institute of Architects (AIA) has published a Level Of Detail framework for the AIA Protocol G202-2013 Building Information Modelling (AIA 2013), in which the term LOD refers to the level of development needed in relation to the contents of the elements of the model. The choice of using the definition "level of development" instead of "level of detail" is motivated by the fact that an element, although it may appear visually detailed, could actually be generic.

According to this document, the degree of development that can be achieved through the drafting of a building model defined via a BIM approach is divided into five levels (see Table 17.1):

- **LOD 100**: the element is represented in a generic manner or by means of a symbol but does not meet the requirements for the LOD 200; further information on the model element can be derived from other elements;
- **LOD 200**: the element is represented with generic quantity, size, shape, position and orientation; the element can also be linked to non-graphic information;
- **LOD 300/350**: the element is represented with dimensions, position and correct orientation/the element is represented with dimensions, position and correct orientation and interfaces correctly with the other elements of the model;
- **LOD 400**: the element is represented with correct dimensions, position and orientation, interfaces correctly with the other elements of the model, and contains details relating to fabrication, assembly and installation;
- **LOD 500**: the element faithfully reflects reality, and is a representation verified on site in terms of dimensions, shape, position, quantity and orientation.

At international level, standardization refers to the ISO 19650 standard, "Organization and digitisation of information about buildings and civil engineering works, including building information modelling (BIM)", whose first two parts have been published (ISO 2018b, c): they address, respectively, the general aspects of the BIM process and the information flow of the development phase of a real estate project.

From the point of view more properly of the life cycle of the real estate, then, the whole information flow is proposed in a synthetic and illustrative figure (see Fig. 17.4) with the highlighting of the various intermediate moments of evaluation, verification and approval, in which the client is also called to express his opinion about the satisfaction of the initially specified design requirements.

As regard Italy, the UNI 11337-4: 2017 Standard (UNI 2017) incorporates the fundamental terminological distinction between Level Of Detail (LOD), Level Of Geometry (LOG) and Level Of Information (LOI); in this perspective the LOD, the level of development of digital objects, is composed of LOGs, level of development of objects—geometric attributes, and Level Of Detail, level of development of objects—information attributes (Pavan et al. 2017). Figure 17.5 shows data flow in the digital process through the objects LODs.

The definition of the individual LODs, the Italian standard adopts the general scale that follows:

- **LOD A**: the entities are represented graphically through a symbolic geometric system or a representation of kind, taken as a reference without the constraint of geometry. The quantitative and qualitative characteristics are indicative;

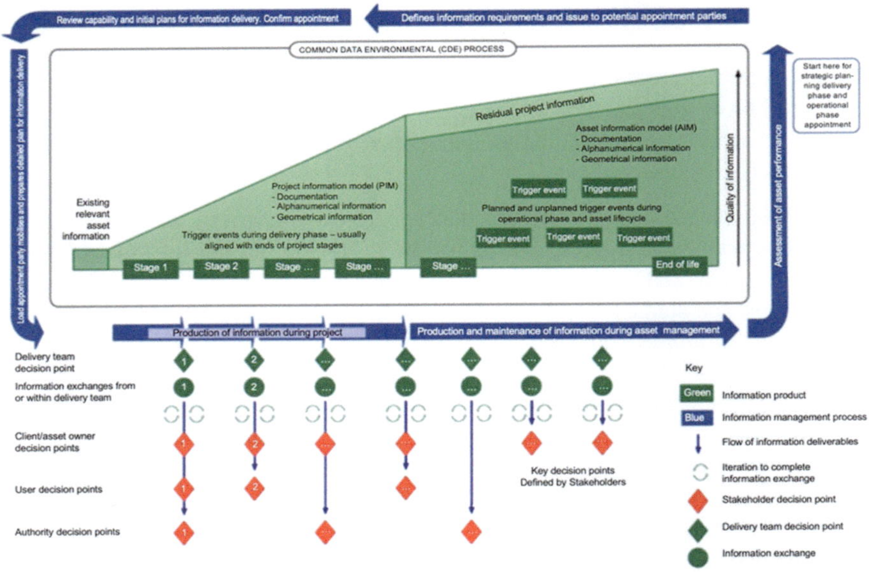

Fig. 17.4 Overview and illustration of the information management process (ISO 2018b)

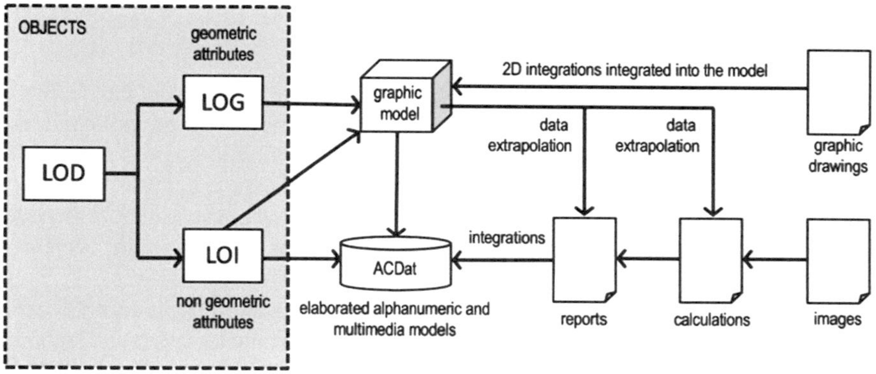

Fig. 17.5 Data flow in the digital process through the objects LODs (Pavan et al. 2017)

- **LOD B**: the entities are graphically virtualised as a generic geometric system or a space geometry. The qualitative and quantitative characteristics are approximate;
- **LOD C**: Entities are graphically virtualised as a defined geometric system. The qualitative and quantitative characteristics are defined in a generic manner in compliance with the limits of current legislation and the technical reference standards and are referable to a plurality of similar entities;
- **LOD D**: the entities are graphically virtualised as a detailed geometric system. The qualitative and quantitative characteristics are specific to a defined plurality of similar products. The interface with other specific construction systems is defined, including the approximate manoeuvring and maintenance dimensions;
- **LOD E**: the entities are graphically virtualised as a specific geometric system. The quantitative and qualitative characteristics are specific to a single production system, linked to the defined product. The level of detail relating to manufacture, assembly and installation is defined, including the specific manoeuvring and maintenance dimensions;
- **LOD F**: the objects express the virtualisation verified on the site of the specific production system executed/built. The quantitative and qualitative characteristics are those specific to the individual production system of the product laid down and installed. The management, maintenance and/or repair and replacement operations to be carried out throughout the life cycle of the building complex are defined for each individual product;
- **LOD G**: the objects express the updated virtualisation of the actual state of an entity in a defined time. The management, maintenance and/or repair and replacement operations to be performed throughout the life cycle of the building complex are defined for each product.

17.1.2 The Concept of Interoperability

If the information required is available when it is actually needed, and the quality of that information is appropriate, then the construction process can be improved. For this to happen, there must be a common understanding of building processes and of the information that is needed for them and results from their execution.

The Industry Foundation Classes (IFC), introduced in the next paragraph, provides a comprehensive reference for the totality of information within the life cycle of a constructed facility.

A fundamental role on promoting BIM worldwide is played by buildingSMART Technical Community[2] that is buying a wide variety of technical experts and software vendors. This includes the Industry Foundation Classes (IFC) data model, Model View Definitions (MVDs), the BIM Collaboration Format (BCF) and the building SMART Data Dictionary (bSDD) for the exchange of Building Information Models (BIMs) among different software products and platforms. The buildingSMART standards are often expressed as technologies which can be leveraged in existing industry tools and platforms, or serve as the basis for new ones.

The Industry Foundation Classes, or IFC, are an open international standard for BIM data that are exchanged and shared among software applications used by the various participants in the construction or facility management industry sector.

The standard ISO 16739-1:2018 (ISO 2018a), includes definitions that cover data required for buildings over their life cycle.

The last release, and upcoming releases, extends the scope to include data definitions for infrastructure assets over their life cycle as well. It is an open, international standard meant to be vendor-neutral, or agnostic, and usable across a wide range of hardware devices, software platforms, and interfaces for many different use cases.

More specifically, the IFC schema is a standardised data model that codifies, in a logical way:

- the identity and semantics (name, machine-readable unique identifier, object type or function);
- the characteristics or attributes (such as material, colour, and thermal properties);
- relationships (including locations, connections, and ownership);
- objects (like columns or slabs);
- abstract concepts (performance, costing);
- processes (installation, operations);
- people (owners, designers, contractors, suppliers, etc.).

The specification schema can describe how it is constructed, how it is installed and how it is managed. IFC is able to define building components, manufactured products, Heating ventilation Air Conditioning (HVAC)/electrical systems, as well as structural analysis models, energy analysis models, cost sharing, work programmes and much more.[3]

[2]https://technical.buildingsmart.org, Accessed 20 Sept 2019.

[3]Source of definitions https://technical.buildingsmart.org, Accessed 20 Sept 2019.

17.2 Managing Sustainability with Green BIM

The design of buildings or cities according to environmentally sustainable design (ESD) and design using Building Information Modelling (BIM) have several elements in common.

First of all, they are relatively "new" and in any case not completely consolidated approaches in current planning practice: even if strongly desired and often requested both by the real estate market and by the binding laws of some countries (e.g. in Italy) and in strong expansion they still encounter difficulties.

ESD and BIM share an integrated design vision that requires the collaboration of specialists with different skills: one should not forget that ESD introduces design principles and objectives, related to the environmental sustainability of the project, whilst BIM must be a tool and its purpose is to enhance the interoperability (not only of data but also of practices) of the project team, enhancing synergies. With this premise, the "marriage" between these two worlds was inevitable and took the term Green BIM almost naturally.

The application of building information modelling (BIM) and associated cooperation technologies (e.g. XML schemas, iCloud, etc.) represents a paradigm shift. The term "Green BIM" has, thus, been coined to describe the convergence of two key trends, green buildings and BIM (Gandhi and Jupp 2013).

17.2.1 A Definition of Green BIM

But what exactly is Green BIM? Given the great interest in both sustainable design (Green) and Building Information Modelling (BIM). A variety of definitions of Green BIM exists in the literature. The simplest definition is probably the one proposed by (Smart Market Report 2010): "*The use of BIM tools to help achieve sustainability and/or improving building performance objectives on a project*".

One of the first editorial contributions on the subject of Green BIM is that of Eddy Krygiel and Bradley Nies (Krygiel and Nies 2008) which, based on their direct experience in the architectural firm BNIM (Berkebile Nelson Immenschuh McDowell, Inc.) propose an integrated vision of the topic that combines the theme of sustainability with that of BIM. The authors demonstrate how the flexibility of BIM allows to enhance sustainable design, identifying a possible and convenient synergy between the BIM and the environmental certification protocols of buildings, in particular with the LEED® protocol.

Through the use of BIM, the authors show how the traditional process can change into work that integrates the entire design team and emphasise green solutions from the start.

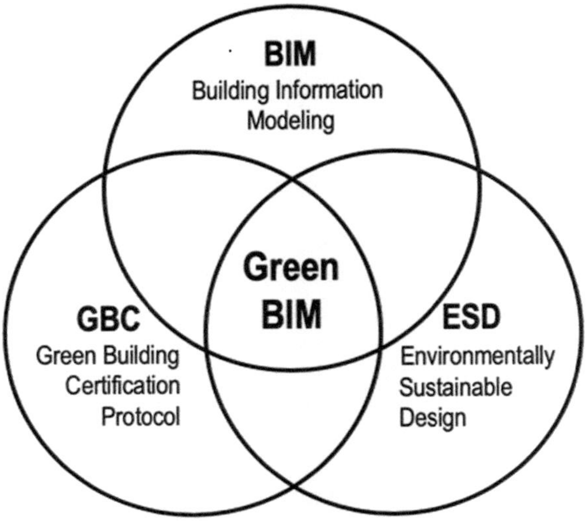

Fig. 17.6 Three pillars of Green BIM. Adapted from Shalini and Julie 2013

Gandhi and Jupp (2013), also based on an analysis of technical literature, emphasise the function of Green BIM as a tool to support the use of environmental certification protocols for buildings, infrastructures or neighbourhoods such as LEED® or BREEAM®. They identify three fundamental pillars that characterise the Green BIM (Fig. 17.6):

- environmentally sustainable design (ESD) principles;
- optimisation of the process of obtaining prerequisites and credits in order to obtain green building certification (GBC);
- integrated building systems and design processes supported by object-based modelling and analysis tools supported by BIM platform.

With this interpretation, the Green BIM is considered as an integrated process that allows one to automate and consequently to speed up the verification of the prerequisites and credits of a GBC process.

The objective of ESD is to eliminate negative environmental impact completely through skilful, sensitive design and should translate into more comprehensive, assessable sustainable design solutions for the built environment.

The link and the synergy between a BIM platform and the scheme of a certification protocol such as LEED® or BREEAM® is strategic: this is because many of the calculation and evaluation operations requested by the GBC, from the simple, but time-consuming, computational ones (e.g. the evaluation of the surfaces or the calculation of the number of devices whose installation is foreseen) to the more sophisticated ones (e.g. the modelling of summer and winter energy consumption or the calculation of shadows or natural lighting) can be largely automated.

Krygiel and Nies, authors of one of the first books on this topic (Krygiel and Nies 2008) indicated that in projects pursuing LEED® certification BIM can aid in the following aspects of sustainable design:

- building orientation (to select the best building orientation that results in minimum energy costs) Building massing (to analyse building form and optimise the building envelope);
- daylighting analysis;
- water harvesting (to reduce water needs in a building);
- energy modelling (to reduce energy needs and analyse renewable energy options such as solar energy);
- sustainable materials (to reduce material needs and to use recycled materials).

Other authors have gone into more detail on the theme of Green BIM: the results of a study of Azhar et al. (2011) indicate that documentation supporting LEED® credits may be directly or indirectly prepared using the results of BIM-based sustainability analyses software. The authors state that the BIM process could streamline the LEED® certification process and save substantial time and resources which would otherwise be required using traditional methods.

The Green Building Certification Protocols (GBC) are effective tools to facilitate the achievement of environmental sustainable design goals for buildings, infrastructures or urban areas. However, we can also consider Green BIM synergies that are not linked to the use of GBCs but that nevertheless enhance the environmental choices related to both the building or the infrastructure, and to the territory in which the building is located.

The energy efficiency of buildings, for example, is one of the elements that characterise sustainability in their life cycle. Maltese et al. (2017) confirm that the use of BIM to provide data for energy performance evaluation and sustainability assessment (Green BIM) could enable integrated design, construction and maintenance guiding towards Net Zero Energy buildings. Green BIM includes Building Energy Modelling dealing with energy performance to identify options for energy efficiency during the life cycle. The aim of paper is to investigate the opportunity to include the "green dimension" in BIM considering the more diffused rating systems. In paragraph 17.1.1, the size 7D of the BIM was introduced in order to consider the environmental aspects.

One reflection that can be made is that in the near future, just as it will no longer make sense to make a distinction between buildings and sustainable buildings (all buildings must be sustainable), it will no longer make sense to try to distinguish between BIM and Green BIM since the "sustainability dimension" will characterise all the components, all the systems and all the processes in the design, construction and management of buildings, infrastructures and urban areas. The Green BIM, therefore, could be seen as a transition phase of the BIM.

17.2.2 Green BIM for Improving Sustainability

The use of Green BIM for improving the sustainability of buildings, infrastructures and urban area projects is confirmed by the particularly extensive technical literature on this topic.

Building Information Modelling (BIM) is a digitalised technology under a collaborative working platform. The paper of Chong et al. (2017) provides a mixed review to determine the current state-of-the-art BIM development for sustainability. The result reveals that although a significant amount of research and development has been done on the use of BIM during various project phases, little work has been conducted on how it could be applied in refurbishment and demolition.

Certain significant insights and implications have been uncovered, namely: (a) new BIM tools are required for assessing sustainability criteria; (b) the need for improved interoperability among BIM software and energy simulation tools; (c) BIM uses in various aspects of refurbishment and demolition have to be streamlined and (d) an innovative procurement system is needed to adapt social sustainability into the project.

A critical review of BIM for green buildings is given by Lu et al. (2017). After an in-depth review of hundreds of journal articles published from 1999 to 2016 and 12 widely used types of BIM software, the study provides a holistic understanding and critical reflection on the nexus between BIM and green buildings, which is systematically illustrated by a "Green BIM Triangle" taxonomy (Fig. 17.7) that indicates that the nexus between BIM and green buildings needs to be understood based on three dimensions, namely project phases, green attributes and BIM attributes. According to the taxonomy, the paper systematically illustrates:

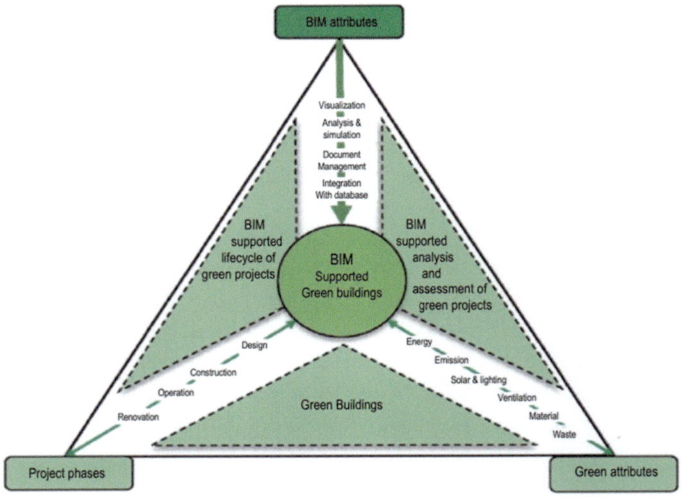

Fig. 17.7 Green BIM triangle taxonomy (Lu et al. 2017)

- the applications of BIM in supporting the design, construction, operation and retro-fitting processes of green buildings;
- the various functions of BIM for green building analyses such as energy, emissions and ventilation analysis;
- the applications of BIM in supporting green building assessments (GBA);
- research gaps and future direction of research in this area.

Through critical review and synthesis of BIM and green buildings based on evidence from both academic research and industrial practices, the authors provide important guidance for building researchers and practitioners to better align BIM development with green building development in future.

Shadram et al. (2016) present a framework that supports design decisions and enables assessment of the embodied energy associated with building materials supply chain based on suppliers of Environmental Product Declarations (EPDs).

The applicability of the framework is tested by developing a prototype and using it in a case study, which shows that a building's energy use and carbon footprint can be significantly reduced during the design phase by accounting for the impact of individual material in the supply chain.

According to the authors, several aspects of the framework made it possible to create a semi-automated process for assessing the embodied energy during the design development phase of a building's life cycle, including:

- The predefined recipes, which are stored in a generic database and linked to the inventory of the material EPDs.
- An integrated BIM process which incorporates the composed embodied recipes into the BIM tool by classifying the BIM object's attributes, facilitating the extraction of quantities from the BIM tool and their attribution to the associated composed, embodied recipes in order to enable the final assessment.
- An automated ETL process that serves as a core for processing and integration of BIM and EPD data throughout the evaluation, automating the tracking of distances between suppliers' facilities and the construction site as well as the quantity take-off from the BIM tool.

The use of ETL technology to ensure BIM-LCA interoperability is, therefore, one of the major contributions of this paper since its use tends to be overlooked to a large extent in the context of BIM and also among the prior BIM-LCA approaches which aimed to ensure BIM-LCA interoperability.

Ahmad et al. (2017) propose a study to develop a conceptual framework of a futuristic BIM-based Design Iteration (BIM-DIT) tool for selecting processes, systems, techniques and materials (PSTM) combinations during design.

Such development can support decision-making process during the design stage of residential buildings by assisting the design team in the generation of design alternatives. The discussion shows that the PSTM combination selection approach used for the BIM-DIT tool will help decision makers with precise knowledge of available options for achieving truly sustainable building projects.

17.2.3 Green BIM as a Synergy Tool Between BIM and GBCs

The growth of Green BIM reinforced the correlation between the growth of the green and sustainable certified building and the use of BIM tools. As the number of certified buildings is increasing, the need for the use of instruments which can accelerate the certification process is also increasing. The synergy between the potential offered by the BIM and the Green Building Certification Protocols (GBC) allows the rapid management of the entire design process, from conception to building operation management.

The most widespread experiences in this regard concern in particular the BIM coupling (Green BIM) and the LEED® protocols.

But how widespread is the use of Green BIM as a tool to quickly and reliably calculate the prerequisites and credits of a protocol such as LEED? The answer to this question is not simple because one is looking at a phenomenon characterised by an intense evolution. Some interesting data have been published by a research promoted by McGraw Hill Construction (Smart Market Report 2010). According to this study on users of Green BIM models who could be calculating LEED® credits, 42% of professionals believe that the technology has medium to high usefulness, 38% believe its usefulness is at a low level and 20% do not set any value on it. From the analysis a curious fact emerged: among firms that already use Green BIM procedures, and therefore already have the tools, and hence the ability and the skills to do these calculations, not all use Green BIM tools to calculate LEED credits.

About 42% of firms not currently practicing Green BIM are not using BIM for LEED® Credit calculation because they find exist tools easier to use and another 36% do not believe they have a need.

Nevertheless, Green BIM remains an emerging trend to the majority of the industry, and its full potential is yet to be explored according to recent market research reports. The hypothesis of the research carried on by Wu and Issa (2015) is that for companies to succeed in practicing green BIM, a holistic approach that strategically plans for BIM execution in green building projects is needed.

The BIM project execution planning guidelines (PEPG) is widely adopted today to offer general guidance and standardised workflow of strategic BIM implementation. Considering the prevalence of LEED® green building rating systems in global green building industry, the research is further refined to focus on LEED® projects as a case of exemplary usage. The goal of the research is to create a prototype that facilitates holistic and systematic green BIM practices based upon clearly defined business processes and execution planning.

Although the analysis mentioned above photographs a situation of about a decade ago, and in recent years, there has been a certain evolution, a reflection must be made on the difficulties that companies and professionals can have in adapting to the new design tools.

Research in recent years has not ceased, but rather has increased the interest in this topic.

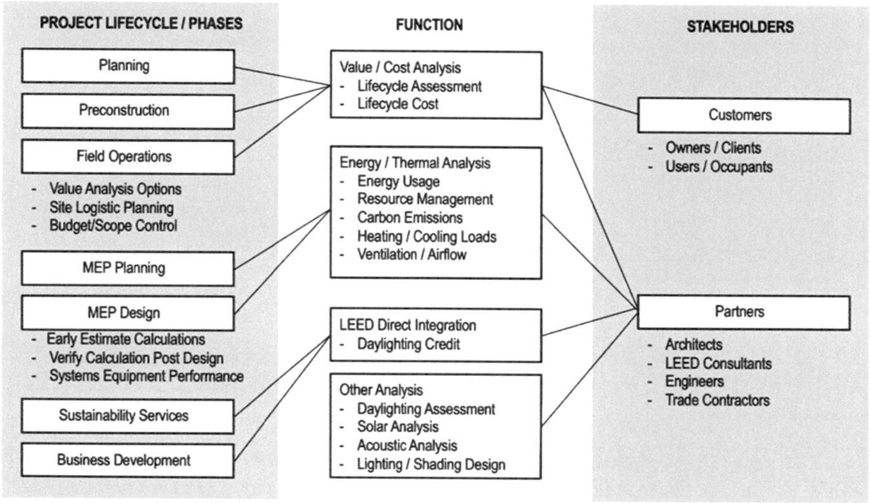

Fig. 17.8 Conceptual framework for BIM-based sustainability analysis (Azhar et al. 2009)

Azhar et al. (2009) in their exploratory research analysed three sets of building performance software namely Ecotect™, Green Building Studio™ (GBS) and Virtual Environment™: these are evaluated to gauge their suitability for BIM-based sustainability analysis.

An evaluation of the three above-mentioned software packages is performed with their pros and cons. An analytical weight-based scoring system is used for this purpose.

At the end, a conceptual framework is presented to illustrate how construction companies can use BIM for sustainability analysis and evaluate the LEED® rating of a building facility. A conceptual framework for BIM-based sustainability analysis is shown in Fig. 17.8.

This exploratory study indicates that BIM can facilitate the very complex processes of sustainable design such as daylighting and solar access, as well as automate the drudgery of activities like material take-offs, cost estimation and construction schedules, whilst capturing and coordinating information into a single integrated model.

Jalaei and Jrade (2015) describe a methodology that integrates BIM with the Canadian green building certification system (LEED®). The methodology depicts the implementation of a model that automatically calculates the compiled number of LEED® certification points and related registration costs for green and certified materials used in designing sustainable buildings, all within the concepts of BIM.

The authors state that using BIM in this methodology will help designers to invent and animate sustainable buildings in 3D mode easily and efficiently at the conceptual stage.

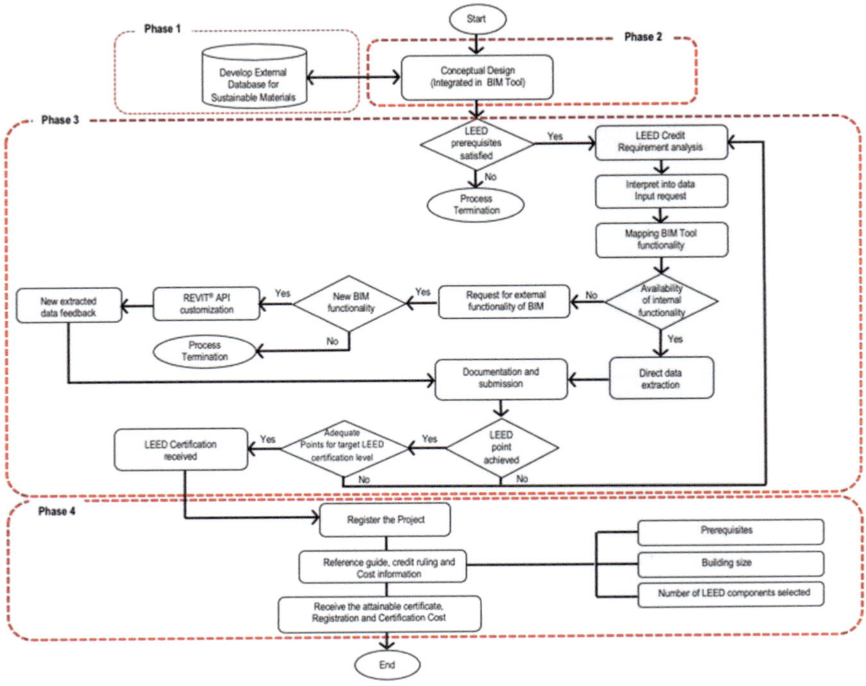

Fig. 17.9 Flowchart of the integration process (Jalaei and Jrade 2015)

The design information of the proposed sustainable building will be produced in a timely manner by using new plug-ins, which are developed for that purpose, and which will link the BIM model with an external database which stores sustainable materials and assembly groups. A real case project is presented to illustrate the usefulness and capabilities of the proposed model.

Figure 17.9 shows the flow chart of the integration process.

- Phase 1 consists of designing the model's relational database, needed when designing a sustainable building.
- Phase 2 focuses on customising BIM tool (i.e. Autodesk® Revit) to fit the modularity requirements of the model.
- Phase 3 includes the design and development of a green building certification module, which is linked to the BIM module.
- Phase 4 focuses on calculating the associated registration cost of the project with Canadian Green Building Council and the related certification costs.

The model helps in reducing one of the biggest barriers for going green by eliminating the documentation process. It is the first model that deals with LEED® categories, associated certification levels and soft cost estimation, by saving users' time and effort. The model concentrates on the soft cost. Furthermore it does not analyse the cost associated with the different materials or systems which can be used to attain each credit or point.

This study looked only at the integration of BIM and sustainability from the LEED® perspective, and only one LEED® rating system (LEED®-NC) was analysed. The results were, thus, limited to the LEED® instead of the general framework of sustainability.

17.3 Managing Green Planning with City Information Modelling (CIM)

17.3.1 From BIM to CIM

In the previous sections, the potentials of BIM have been analysed, highlighting the support that this tool can give in the active management of environmental sustainability goals: from the synergy between BIM, environmental sustainable design principles (ESD) and green building certification protocols (GBC) the Green BIM is born, a tool which can hence be considered as a natural evolution of BIM.

The BIM, with its features, and with its potential, has definitely opened up new avenues in project management in all its phases. After the arrival of BIM in design nothing will be the same as before.

The use of BIM, or of Green BIM, for the management of building or infrastructure projects is, albeit with some difficulty, an integral part of current design practice. The conception of a project with objects containing not only graphic but also performance, maintenance, economic information and much more (referred to the dimensions of the BIM), the interoperability between specialised software tools that, through shared protocols such as the IFC, can exchange data are achievements that cannot be relinquished.

Above all, one cannot forgo the new way of working introduced by BIM, which sees the synergy of the single IT platform converge with the efforts of designers, maintainers, managers and end-users.

In recent years, cities have evolved, albeit in different ways, towards more sustainable and smart models. This is a revolution not only from a technological or environmental point of view but also from a cultural and social standpoint. In cities, the way of living, of relating to other people, of moving about is changing. Just as a building or an infrastructure must be conceived, designed, built and managed in an integrated and synergic way, and the BIM does facilitate this approach, even the city needs to be considered in its complexity with a coordinated and synergistic approach, which of course takes into account all the aspects that characterise it.

In the same way that BIM has evolved to become Green BIM, is it possible to imagine its evolution on an urban scale? In other words, is it possible to imagine an extension of the BIM to a city or a community contest? If so, BIM could then become an effective tool to support green planning on an urban scale.

The use of BIM in an urban context is an objective that obviously arouses much interest and for the purpose, a new acronym has been created: City Integration Modelling (CIM). Before analysing the state of the art of CIM, it is necessary to make the premise that it is a piece of "work in progress": to achieve this important goal, different paths are being followed with different approaches. In addition, there are currently some pilot applications and a dissemination of the CIM will require further study and verification.

So CIM can be considered to be the latest advance from BIM. The potential of this tool for architects and planners in urban planning is considerable. With CIM, users are able to create and display 3D city model environments where data is available, acting as an analysis tool for various city components, including open data, sensors, Internet of Things (IoT) and even social media; the technicians and managers of the planning process are able to collaborate effectively and work on a set of 3D models. By using the CIM, City Councils are able to make decisions (e.g. the modification or approval of planning strategies) based on the knowledge of the objective information, for example, regarding traffic or noise issues or environmental impact of buildings and infrastructures.

The same goes for demolishing or modernising old buildings in the city. The data extracted from CIM could contribute to enabling cities to cope with population growth and rapid urbanisation.

The information contained in the CIM can be used not only to plan energy networks (e.g. district heating or cooling) or smart grids but also to manage all monitoring, control and maintenance functions.

There are essentially two possible ways to implement a CIM. The first way is to open a dialogue between two worlds which are far apart: that of the geographic information system (GIS), a computerised information system that provides acquisition, recording, management, consultation, return, sharing and presentation of information deriving from geographical data, with that of the BIM.

The greatest difficulty with this solution is that of identifying an "interoperability" between two protocols, the form files for the GIS files and the IFC files for the BIM, being very different.

The second way is to rely on software houses that already offer CIM platforms. The leading BIM software vendors, Autodesk and Bentley, have expanded their technology offerings to CIM.

Autodesk® solutions, including Revit and InfraWorks, support modelling for smart city implementation, with case studies in Copenhagen, Barcelona and Korea's U-City. Similarly, Bentley's CIM solutions have been widely used in large-scale city modelling projects in Montreal, Helsinki and Singapore.

"With technological development in place, user adoption will be the key to success. Project owners will play a driving role in this by requiring models that can capture all the relevant urban planning and infrastructure aspects of a building. Architects, planners, engineers, and construction firms will then be under pressure to further innovate in order to be able to meet the broader and more complex requirement.

As smart cities need intelligent infrastructure, CIM can revolutionise not only the planning and construction of buildings, but it also holds the key in the entire process of urban planning".[4]

17.3.2 The Structure of a CIM (City Integration Modelling)

The city is a very complex environment, in fact it represents not only a static model but a model in which dynamic aspects must be taken into consideration, such as the population, transport and flows of energy and matter within the city itself and exchanged with what is outside the city. Understanding the social, economic and environmental dynamics of a city requires the management of a large amount of information.

As suggested by Xu et al. (2014), to facilitate the understanding of city dynamics we can hypothesise a city structure subdivided into sub-modules, within which to aggregate homogeneous functions. In the diagram of Fig. 17.10, for example, seven modules have been identified: Buildings, Mechanical, Electrical and Plumbing (MEP), Mobility, City Services, Grey Infrastructures, Green Infrastructures and Water Body.

Consistently with the scheme presented by Xu et al. (2014), the information contained in each of the sub-modules can be managed using BIM technology and, subsequently, be geo-referenced using GIS technology.

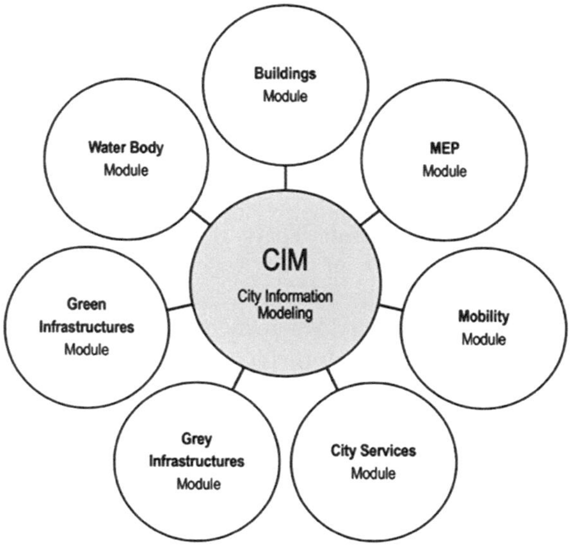

Fig. 17.10 Sub-modules of a city information modelling (CIM)

[4]Source: https://urbannext.net/zigurat/city-information-modelling-for-smarter-cities/ Accessed 20 Sept 2019.

The modules represented in Fig. 17.10 may contain the following information:

- **Buildings Module**. Contains information about buildings in the city; this information may relate to the geometric characteristics (volumes, surfaces), even the materials used and the performance characteristics (energy efficiency, water consumption, environmental impact, etc.). For recent buildings, one can have more information (e.g. energy performance certificate) or even a complete BIM, while for existing ones an audit is necessary, initially also containing essential data such as geometric characteristics or construction technology based on the year of construction.
- **MEP Module**. This module contains all the information available at the urban level about the plant infrastructures. (e.g. the sewage network, the drinking water addition network, the electricity grid, district heating or cooling, etc.)
- **Mobility Module**. This section contains information on the roads, the trend, the cost, etc. Based on the buildings' position, we can also find the city roads or traffic system in the city. In the city traffic information model, real-time road monitoring, traffic positioning and traffic management would be performed. This module, through the acquisition of data on urban traffic in an automatic way, permits the monitoring of the environmental sustainability of mobility and the supply of useful information to understand if some policies on sustainable mobility promoted by the municipal administration authority have been effective or not.
- **City Services Module**. This module contains the auxiliary structures of the city, such as public spaces, public services, and street furniture,. These auxiliary structures play an important role in people's lives, can provide entertainment and rest rooms for people and can also bring a lot of convenience to people (Xu et al. 2014).
- **Grey Infrastructures Module**. Contains the information regarding the infrastructures in the city: among these are to be found, for example, roads, elevated roads, bridges, tunnels, railway stations, etc.
- **Green Infrastructures Module**. The module contains information on the green in the city, for example, green ways, trees, vegetation, lawns, parks, urban forests, green roofs, etc. Information is important to ensure an active management of the green and to monitor its consistency.
- **Water Body module**. This part contains the water body of the city, such as rivers, lakes, etc. As we all know, water plays an important role in our lives, but water pollution is a serious problem, and the water that can be used by people is becoming ever less in quantity. Moreover, the water body is also important for the ecosystem. Collecting of this information makes it possible for everyone to make good use of water. There are some elements that have not been taken into consideration, such as the city refuse disposal system. Some of the existing information needed can be easily imported into the city information model (Xu et al. 2014).

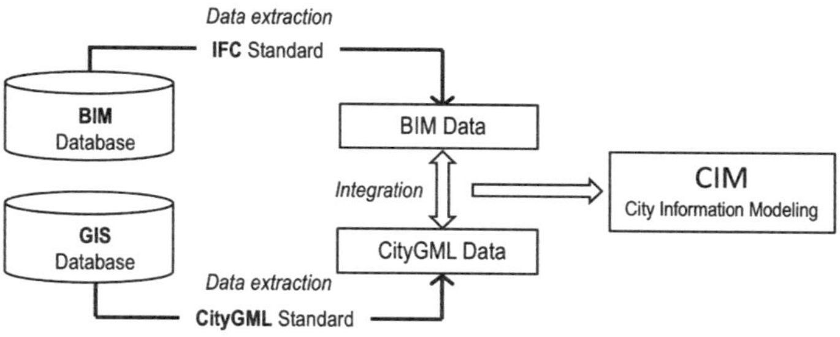

Fig. 17.11 Data collecting of BIM (Xu et al. 2014)

17.3.3 CIM as Integration of GIS and BIM

One of the major difficulties in the CIM is the modelling of information, which requires both external information (i.e. concerning the territory) and internal information (i.e. concerning the "objects" that make up the city as buildings. Infrastructures and services).

According to Xu et al. (2014), there are some approaches possible to solve this problem, one is to measure existing 3D objects, but this procedure requires a lot of work; the second approach is to use integrated CAD models and GIS models, but this method cannot provide internal information. A third, more feasible method consists in trying to automatically integrate BIM and GIS.

The concept is schematically illustrated in the diagram of Fig. 17.11 in which it is highlighted how the information contained in a CIM can derive from a mix of information, appropriately processed and integrated, coming from the BIM and GIS databases.

If the idea is very simple and understandable, its implementation is less straightforward since GIS and BIM are procedures that start from two very different needs and IT environments.

Geospatial applications are normally based on spatial objects as well as topological relationships. This information is usually classified in two sets which can be defined by two models: a geometric model that defines the objects and element types and a semantic model that defines entities, their non-spatial characteristics and their relationships.

Geometric models consist of different spatial objects (points, line strings, etc.) with the representation of their properties. Semantic models, however, consist of class definitions for the representation of spatial objects within the virtual 3D city models. They also specify spatial relationships with other objects and parts in applications like buildings, digital terrain models (DTMs), water bodies, transportation networks, vegetation and city furniture (El-Mekawy 2010).

The geometric information on buildings, infrastructures and the geospatial environment can be obtained from computer-aided design (CAD) drawings, or through geometric surveys with laser scanning methods or photogrammetric techniques.

Until a few years ago, most of the geometric information on buildings could be retrieved through 2D or 3D CAD modelling without having their features available. In these environments, the semantic representation was not considered. In recent years, this situation is changing also thanks to the spread of BIM, which is, however, only partially applied in more recent buildings.

Geospatial Information Systems (GIS) are used to model existing spatial objects in urban and regional areas.

The purpose of the GIS was to represent the urban scale and hence the models did not focus on the details. As a result of the fast technological development in the last decade, geospatial models have become increasingly important for modelling the world around us. Nowadays, the situation has changed: GIS systems have evolved, containing more and more information on the areas around the buildings, with 3D representations, that must represent the environment in more detail.

One of the greatest difficulties in modelling geospatial information is the collection of data for vast regional and urban areas: various approaches are adopted for the acquisition of data necessary for geospatial modelling:

- the first approach deals with measuring existing objects and constructing 3D models. For accurate measurement, 3D laser scanning technology and photogrammetry techniques are used;
- the second approach deals with the conversion of CAD 3D as well as 2D building models into geospatial models. The approach is the integration of CAD and GIS/Geospatial Information Systems in the target area;
- the third approach consists in obtaining a simplified geometric and semantic images of buildings and infrastructures models starting from the existing BIM models. The major challenge to this approach is the need for complete BIMs for the target areas, which are usually very time-consuming and expensive.

A critical aspect in operating by referring to the third approach, which in practice defines a path of integration between BIM and GIS, concerns the issue of interoperability between two environments which, as said, are very different from the point of view of data protocols.

Just as for the BIM, it was necessary to develop the open interchange protocol, the IFC, also for the GIS, following the need to have an open standard to be used for urban applications, the CityGML has been developed as a common model of semantic information, representing different urban and geographic 3D objects that can be shared among different applications.

CityGML was developed as an open data model with an XML-based format that can be used to store and exchange virtual 3D objects and city models between different applications.

CityGML also goes beyond the representation of only graphical appearance and specifies classes and relations for the most relevant topographical objects in cities and urban models. It specifies "City" to include not just its built structures, but also its elevation, vegetation, water bodies and more physical objects. Thus, CityGML allows sophisticated analysis, overlay tasks, decision support and thematic inquiries.

Moreover, it represents semantic, thematic, taxonomical and aggregation properties of the models (Kolbe and Bacharach 2006).

The integration between BIM and GIS, therefore, passes through an integration between the IFC protocol and the CityGML protocol, as shown in Fig. 17.11.

On the subject of CIM, in which the most critical issue arises from the integration of IFC and CityGML protocols, many authors, through research activities, have provided insights and interesting insights.

An interesting research work on this topic is contained in the thesis of El-Mekawy (2010) which focuses on the semantic integration of IFC and CityGML building models for bidirectional conversion. A unified model in study is defined as a superset model that is extended to contain all the features and objects from both IFC and CityGML building models.

The paper of Van Berlo and De Laat (2010) describes the development of a CityGML extension called GeoBIM to get semantic IFC data into a GIS context. The conversion of IFC to CityGML (including the GeoBIM extension) is implemented in the open source Building Information Model server.

In order to address the target problem of completing the semantic attribution of the 3D digital city model, a framework of integrating BIM technology into GIS is demonstrated in the paper of Xu et al. (2014) that proposes a new concept of City Information Modelling (CIM) with the goal of bringing great benefits to the urban construction and city management. In the paper, the data schema behind BIM and GIS (i.e. IFC and CityGML) are compared and mapped with each other. A case study of land planning of a campus is demonstrated to present the potential benefits of CIM.

Finally, Burry et al. (2015) in their paper discuss the challenges which designers face when modelling the expected behaviour of people: their movement and transactions around and within precinct scale development. Building Information Modelling (BIM) software philosophy contrasts with that of City Information Modelling (CIM)—the route by which we consider how precinct scale development, being somewhere between the two (BIM and CIM), requires a wholly different approach to information and behaviour modelling.

The authors offer evidence of the value of increasing the planners' analytical approach with the architects' synthesis from data, leading to meaningful speculations on otherwise unanticipated future scenarios for the precinct, far beyond expectations. Novel approaches to modelling behaviour at precinct scale suggest alternative readings of precincts, which require a wider set of approaches to Precinct Information Modelling (PIM) software development than simply an expansion of BIM.

17.4 Conclusions

The green planning of cities and communities requires first of all a strong determination to promote concrete actions to combat climate change. However, to obtain concrete results over time, there are two other requirements: on the one hand, a mix of multidisciplinary technical skills in the various areas involved (environmental, energy, economic and social), and on the other, tools capable of providing support in the management of complexity.

The use of Building Information Modelling (BIM) in the various scales, from that of the single building to that of urban space, should be considered in this context.

Green, therefore, sustainable and smart city planning require the management of a considerable amount of data. According to the BIM approach, the buildings or infrastructures in a city are not only geometric entities but are entities with which one must associate information that makes it possible to consider other aspects, e.g. their life cycle and, therefore, assess their impact on the environment.

The purpose of this chapter was to inform non-skilled operators and interested parties about the potential of BIM with an update on its most recent applications in the environmental sector, for example, Green BIM, and in the management of urban spaces through the most recent application of City Information Modelling (CIM).

The use of BIM, but above all the use of CIM, has long been in a mature phase from the point of view of protocols and hardware and software tools, but much remains to be done in order to be able to consider an application of these procedures as standard practice.

The software houses contribute greatly to this change by offering solutions with accessible and user-friendly interfaces but above all by constructively dialoguing with each other. About CIM, for example Autodesk®, one of the BIM technology leaders and ESRI®, the leader of the GIS applications declared to collaborate in contributing to solve the interoperability problems betweens protocol that belong to two worlds and this is positive.

A significant contribution in this transitional phase is that offered by international non-profit associations such as buildingSMART. Finally, many governments, starting from that of Italy, have promulgated laws that make the use of BIM compulsory, at least in certain sectors. The world, therefore, is going in this direction and the considerable efforts which will be necessary for an initial adaptation to this technology will be amply repaid over time.

References

Ahmada T, Aibinua A, Thaheem MJ (2017) BIM-based iterative tool for sustainable building design: a conceptual framework, international high-performance built environment conference—a sustainable built environment conference 2016 series (SBE16), iHBE 2016. Procedia Eng 180(2017):782–792

American Institute of Architects (2013) Document G202—2013 project building information modelling protocol form

Azhar S, Carlton WA, Olsen D, Ahmad I (2011) Building information modelling for sustainable design and LEED® rating analysis. Autom Constr 20(2):217–224

Azhar S, Brown J, Farooqui R (2009) BIM-based sustainability analysis: an evaluation of building performance analysis software. Available on https://www.researchgate.net/publication/237835268_BIM-based_Sustainability_Analysis_An_Evaluation_of_Building_Performance_Analysis_Software. Accessed 13 Oct 2019

Burry M, Karakiewicz JA, Holzer D, White M, Aschwanden GDPA, Kvac T (2015) BIM-PIM-CIM: the challenges of modelling urban design behaviours between building and city scales. Design modelling symposium copenhagen 2015. https://doi.org/10.1007/978-3-319-24208-8_34

Chong H-Y, Lee C-Y, Wang X (2017) A mixed review of the adoption of building information modelling (BIM) for sustainability. J Clean Prod 142:4114–4126

El-Mekawy M (2010) Integrating BIM and GIS for 3D city modelling: the case of IFC and CityGML. Licentiate thesis, Geoinformatics Division of the Department of Urban Planning and Environment Royal Institute of Technology (KTH) TRITA SoM 2010–11 ISSN: 1653-6126, ISRN KTH/SoM/10-11/SE, ISBN:978-91-7415-790-1

Ente Italiano di Normazione UNI (2017) UNI 11337–4:2017 - Edilizia e opere di ingegneria civile - Gestione digitale dei processi informativi delle costruzioni - Parte 4: Evoluzione e sviluppo informativo di modelli, elaborati e oggetti

Gandhi S, Jupp J (2013) Characteristics of green BIM: process and information management requirements. In: 10th product lifecycle management for society (PLM). Nantes, France, pp 596–605. https://doi.org/10.1007/978-3-642-41501-2_59.hal-01461909

International Organization for Standardization ISO (2018a) ISO 16739-1:2018—Industry foundation classes (IFC) for data sharing in the construction and facility management industries—Part 1: data scheme

International Organization for Standardization ISO (2018b) ISO 19650—organization and digitization of information about buildings and civil engineering works, including building information modelling (BIM)—information management using building information modelling—Part 1: concepts and principles

International Organization for Standardization ISO (2018c) ISO 19650—organization and digitization of information about buildings and civil engineering works, including building information modelling (BIM)—information management using building information modelling—Part 2: delivery phase of the assets

Jalaei F, Jrade A (2015) Integrating building information modelling (BIM) and LEED system at the conceptual design stage of sustainable buildings. Sustain Cities Soc J 18(November 2015):97–107

Krygiel E, Nies B (2008) Green BIM: successful sustainable design with building information modelling. Wiley Publishing, Inc, Hoboken

Kolbe T, Bacharach S (2006) CityGML: an open standard for 3D city models. http://www.directionsmag.com/articles/citygml-an-open-standard-for-3d-city-models/123103. Accessed 15 Sept 2019

Lu Y, Wu Z, Chang R, Li Y (2017) Building information modeling (BIM) for green buildings: a critical review and future directions. Autom Constr J 83:134–148

Maltese S, Tagliabue LC, Cecconi FR, Pasini D, Manfren M, Ciribini Angelo LC (2017) Sustainability assessment through green BIM for environmental, social and economic efficiency. In: International high-performance built environment conference—a sustainable built environment conference 2016 Series (SBE16), iHBE 2016

Pavan A, Mirarchi C, Giani M (2017) BIM: Metodi e strumenti, progettare, costruire e gestire nell'era digitale, Tecniche Nuove, ISBN: 978-88-481-3522-1

Smart Market Report (2010) GREEN BIM, how building information modelling is contributing to green design and construction. McGraw Hill Construction

Shadram SF, Johansson TD, Lu W, Schade J, Olofsson T (2016) An integrated BIM-based framework for minimizing embodied energy during building design. Energy Build 128(2016):592–604

Van Berlo L, De Laat R (2010) Integration of BIM and GIS: the development of the CityGML GeoBIM extension. In: Kolbe TH, König G, Nagel C (eds) 2011: advances in 3d Geo-information sciences. ISBN: 978-3-642-12669-7. In: Cartwright W, Gartner G, Meng L, Peterson MP (eds) ISSN: 1863-2246

Wu W, Issa RRA (2015) BIM execution planning in green building projects: LEED as a use case. J Manag Eng 31(1)

Xu X, Ding L, Luo H, Ma L (2014) From building information modelling to city information modelling. Journal of information technology in construction (ITcon), vol 19. Special issue BIM cloud-based technology in the AEC sector: present status and future trends, pp 292–307. http://www.itcon.org/2014/17